Unity Root Matrix Theory

Higher Dimensional Extensions

– RICHARD MILLER –

An environmentally friendly book printed and bound in England by
www.printondemand-worldwide.com

Mixed Sources
Product group from well-managed
forests, and other controlled sources
www.fsc.org Cert no. TT-COC-002641
FSC © 1996 Forest Stewardship Council

PEFC Certified
This product is
from sustainably
managed forests
and controlled
sources
www.pefc.org
PEFC
PEFC/16-33-415

This book is made entirely of chain-of-custody materials

www.fast-print.net/store.php

Unity Root Matrix Theory – Higher Dimensional Extensions
Copyright © Richard Miller 2012

ISBN 978-178035-296-1

First published 2012 by
FASTPRINT PUBLISHING
Peterborough, England.

Preface

This book extends the three-dimensional, Unity Root Matrix Theory (URMT), first published in Reference [1], to four and higher dimensions.

At its most basic, URMT is the study of the eigenvectors of a special type of square matrix, whose elements are all integers and isomorphic with the complex roots of unity. Just one of these eigenvectors is defined as invariant to special transformations applied to the matrix, whilst all the other eigenvectors freely evolve with these 'invariance transformations'. This rather dry definition hides a myriad of mathematical and physical structure in the form of the eigenvector solution to the matrix. This solution manifests itself as a discrete, infinite, n-dimensional lattice of points, i.e. that of a discrete, eigenvector space, with associated geometric properties and evolutionary behaviour.

URMT is cast as a discrete description because it is formulated entirely in integers, with its origins in number theory and integer, Diophantine equations. However, as the mathematical side of the theory developed, the similarities to mathematical physics, with a consistent physical interpretation, became so compelling that focus shifted to its study as a branch of discrete, theoretical physics.

Both number theoretic and physical issues were first published in the foundation book [1], subtitled 'Physics in Integers', so-named because of URMT's links between number theory and theoretical physics. However, this was all formulated in three dimensions, with a corresponding 3x3 unity root matrix and associated three-element eigenvectors. With the Special Theory of Relativity (STR) formulated in four dimensions, Kaluza-Klein theory in five, and String Theory currently favoured in eleven, extending URMT beyond three dimensions was inevitable, the challenge being to retain the existing mathematical and physical structure in the 3D formulation (URM3), whilst also expanding the links to modern, theoretical physics; the eventual goal being to provide a discrete formulation of fundamental physics.

When initially embarking upon the extension work it was thought that a 4D or higher theory could only work by abandoning much of the 3D theory, which was a prospect hardly worth considering. So intriguing is the ternary nature of URM3, with nature's own predominant 3D bias (three spatial dimensions, three families of particles etc.), failure to achieve the goal was considered to be a show-stopper. Thankfully it did not happen and the goals were achieved with a general, n-dimensional, formulation of URMT that manages to embrace URM3, whilst providing some new links to physics, with hints of yet more to come.

Most surprising and unexpected was the appearance of a very important feature in physics, namely compactification (or dimensional reduction) of higher dimensions. Using novel variational methods, initially developed in the first book, each extra dimension is seen to have an associated variational parameter that can be consistently associated with time. As this temporal parameter grows, i.e. as the eigenvector solution evolves, the higher dimensions are seen to shrink, and become too small to see directly. Such a property of shrunken (or compactified) dimensions is believed to be the reason why, for example, seven of the eleven dimensions of string theory are not observable. This is a major result for URMT's higher dimensional extensions and, as such, the last section of the book is devoted to its mathematical analysis.

That the physical application of compactification appears naturally in a solution to the higher dimensional formulation of URMT, whilst retaining all of URM3's geometric and physical properties, lends more support to the concept of URMT as a credible physical theory of a discrete, infinite, n-dimensional space, and lays the foundations for further physical development of the theory.

References

[1] Unity Root Matrix Theory, Physics in Integers, R. J. Miller, FastPrint Publishing, 2011, ISBN 978-184426-974-7,

http://www.fast-print.net/bookshop/823/unity-root-matrix-theory-physics-in-integers

This book is broken into six separate papers, each paper is given a specific reference #1 to #6 as follows:

[1],#1 Unity Root Matrix Theory Foundations

[1],#2 see [2], below

[1],#3 Geometric and Physical Aspects

[1],#4 Solving Unity Root Matrix Theory

[1],#5 Unifying Concepts

[1],#6 A Non-unity Eigenvalue

[2] Pythagorean Triples as Eigenvectors and Related Invariants, R. J. Miller, 2010, a free document available for download, www.urmt.org.

[3] Unity Root Matrix Theory, Overview. An overview of the six papers published in [1], a free document available for download, www.urmt.org.

[4] W.G. Bickley, R.S.H.G.Thompson, MATRICES Their Meaning and Manipulation, The English Universities Press Ltd. 1964

[5] Sadri Hassani, Foundations of Mathematical Physics, Prentice-Hall International Editions, 1991, ISBN 0-13-327503-5.

[6] I.Niven, S.Zuckerman, H.L.Montgomery, An Introduction to the Theory of Numbers, 5th Edition, John Wiley & Sons, Inc 1991. ISBN 0-471-54600-3.

[7] Unity Root Matrix Theory web site, **www.urmt.org**.

References have been kept extremely minimal for a book of this size, simply because the Web has largely superseded the necessity to reference general academic textbooks. As such, only one linear algebra text is referenced [4], one mathematical physics text [5], and one number theory text [6]; otherwise, the reader is advised to search the web. In particular, as regards work in the field of Unity Root Matrix Theory, since this is a completely new subject area (less than three years old), and to the author's best knowledge, the only currently available text is, indeed, the first book [1], plus some free material [2] and [3]. The reader may wish to visit the web-site [7] occasionally since new, free material is added every few months.

Abbreviations

gcd : greatest common divisor

STR : Special Theory of Relativity

URM : Unity Root Matrix

URMT : Unity Root Matrix Theory

URM3,4,5 : URMT 3x3, 4x4, 5x5 Matrix formulation

URM n : the $n \times n$ matrix formulation of URMT, $n \geq 2$

Acknowledgements

I would like to thank my girlfriend Deborah Armstrong for her continuing support; thanks to my Parents for everything, including some proof reading by my father; thanks also to Dr Harbinder Ghataure for advice and assistance; and finally, Micro SciTech Ltd. for providing the finance. This second book took considerably more work than the first and my girlfriend, in particular, had to endure many occasions where my mind was completely buried in the work. Thank you all.

Overview

This book is split into two parts, Part I starts with a straightforward extension from three to four dimensions, using 4x4 matrices and associated four-element eigenvectors, Part II then extends to five dimensions and beyond

The extension to four dimensions, termed 'URM4', doubles the number of dynamical variables, i.e. the integer elements of the unity root matrix, from three in URM3 to six in URM4. Just as in URM3, each dynamical variable also has a related conjugate, giving twelve in total.

In URMT, the non-singular determinant condition on the eigenvector equation leads to the characteristic equation, i.e. a polynomial in the eigenvalue. This polynomial is interpreted in URMT as an energy conservation equation, termed the Dynamical Conservation Equation (the DCE), and accordingly split into a kinetic and Potential term. It soon becomes clear that the URM4 Potential term is far more complicated than URM3's equivalent, whilst the kinetic term is more of a logical extension and not so mysterious. Nevertheless, the form of the URM4 Potential makes it clear that obtaining solutions in URM4 is not going to be so easy as URM3.

Despite the increased complexity that comes with a 4x4 matrix, some simplifications can be made by retaining several URM3 properties and 'embedding' the URM3 unity root matrix within the URM4 matrix, Section (2). In this form, with some additional URM3 features also retained, progress can be made that preserves the original eigenvalue C of URM3 and adds a new zero eigenvalue, thus giving two of four possible eigenvalues. Note that URM3, in its most general form, only mandates a single eigenvalue C, with a corresponding eigenvector $\mathbf{X}$ satisfying the eigenvector equation $\mathbf{AX} = C\mathbf{X}$. It is only when additional URM3 Pythagoras conditions are engaged is a complete set of three eigenvectors obtained, and much the same process is also employed in URM4.

The eigenvalue C is of paramount importance in URMT as its finite, non-zero, integer value keeps URMT from descending into singularity problems. Since it is a scalar quantity, with units of speed, its square represents energy (per unity mass) and it appears in numerous scalar invariants arising from inner product relations amongst the eigenvectors; such equations are classed as conservation equations in URMT, not least the DCE.

Most URMT eigenvectors and their reciprocals are obtained using the 'residual matrix method', Section (3). This method forms a matrix polynomial by factoring the Cayley-Hamilton polynomial. The resulting matrix polynomial has the special property that it is also equivalent to the outer product of an eigenvector with its reciprocal eigenvector - this outer product being related to projection operators in linear algebra. The residual method, combined with 'embedding' principles in Section (2), can be applied to all existing URM3, three-element eigenvector solutions, to obtain four-element, eigenvector solutions for URM4. In actuality, the 3-element URM3 solution $\mathbf{X}$, for eigenvalue C, remains as per URM3, whilst its reciprocal is a true, 4-element, non-trivial vector. Both the eigenvector $\mathbf{X}_0$ and its reciprocal $\mathbf{X}^0$ are also four-element eigenvectors.

Whilst the method of embedding URM3 in URM4 can give URM4 eigenvector solutions, practically all of URM3's physical interpretation is based upon additional simplifying 'Pythagoras conditions'. These conditions make all eigenvectors, for non-zero eigenvalues, Pythagorean triples (URM3) or quadruples (URM4). With STR in mind, it is a requirement that the URM4, four-element eigenvectors are Pythagorean quadruples with a Minkowski metric. In fact, it is really five-element vectors (Pythagorean quintuples), with a non-zero, relativistic interval, that are required, which is the realm of URMT's 5-dimensional formulation (URM5), studied in Part II.

The UMRT Pythagoras conditions on the dynamical variables relate to the standard dynamical variables with their conjugate forms. The relations are very similar to URM3 and can be seen to generalise to any number of dimensions. By applying these conditions, progress can be made on obtaining quadratic, Pythagorean eigenvector solutions. At this point URMT can split two ways: either pursuing the embedding method under URM3 Pythagoras conditions; or secondly, developing URMT purely under URM4 Pythagoras conditions, but not necessarily embedding URM3. Indeed, a couple of solutions, 'PS+RU' and '2a2p1', are explored in sections (5) and (6) following each of these two paths, but both have their limitations or issues.

Unsurprisingly, the best method to obtain a complete set of eigenvectors, without undesirable consequences, is to combine both methods by embedding URM4 under URM4 Pythagoras conditions, with the two additional constraints of a zero Potential energy (in the DCE) and a special form for the URM4 dynamical variables - these latter two conditions are added with further physical aspects in mind. Put all these methods and conditions together and the process is known as 'lifting' in URMT. The method of lifting, Section (7), enables a complete, n-dimensional eigenvector solution to be obtained, which inherits all URM3 Pythagorean properties plus more. Using lifting as a basis, URM4 is extended to URM5, followed by a general n-dimensional solution supplied in Part II. The solution is actually known as the zero Potential solution, rather than the 'lifted' solution, because of its physical importance. It is revisited again after an excursion into variational methods in URMT, Section (8).

Having obtained a satisfactory solution with a zero Potential energy, focus shifts to variational methods in URMT. Although obtaining one-off solutions is a good start, for complex behaviour any solution needs to evolve parametrically, e.g. with respect to time. This translates in URMT to seeing how a complete eigenvector solution varies as the unity root matrix varies. The standard URMT variational method is to keep the single eigenvector $\mathbf{X}$ completely invariant, such that $\mathbf{AX} = C\mathbf{X}$ always holds, whilst varying the matrix in every possible way to see how the other, non-invariant eigenvectors change.

In fact, variational methods in URM3 were used extensively in the first book to obtain complete eigenvector solutions from scratch, with the invariance of $\mathbf{X}$ enshrined in the 'Invariance Principle', which is a postulate of URMT. This principle is retained, but the complexity of the variational method in 4D and 5D becomes so much greater than URM3 that variations are applied after the initial solutions are found. Adopting this philosophy, variational methods are subsequently applied in Sections (9) to (11) to the three solutions in Sections (5) to (7), which were determined by analytic, not variational, methods. A comparison of their relative merits, when subject to variations, show that only the zero Potential solution is considered worthy of further development.

From a physical perspective, a zero Potential is highly desirable in URMT as the solution can be interpreted as that of a point moving at constant speed (that of the eigenvalue C) with no force acting upon it by virtue of a constant Potential. That the Potential is also zero, and not just constant, is an algebraic consequence. The most important point is that the separate kinetic and potential energy terms remain constant, i.e. invariant, with no kinetic/potential energy interchange. The entire, n-dimensional formulation in this book strives to achieve solutions that have this property because, basically, it is the simplest possible physical solution, i.e. the uniform motion of an object with no force acting. This is rather underselling URMT since the eigenvectors also have a consistent physical interpretation as acceleration, velocity and position vectors, with calculus properties relating them. Furthermore, the trajectory through the discrete eigenvector lattice is also curved and can be interpreted as free-fall motion along a curved null geodesic. That such physical aspects naturally emerge from pure mathematical origins is quite some feat, especially given URMT is essentially a discrete linear algebra problem rooted in number theory.

Having successfully applied variational methods to a 4D, zero Potential solution, such that the Potential remains invariant to certain, parametric variations, it becomes clear that the same form of solution can be extended to five and higher dimensions. However, before commencing this dimensional extension, the association of the variational parameters, as being temporal in nature, is discussed. In fact, throughout most of the earlier sections, this temporal association is made quite clear when comparing solutions with the standard physical association of URMT, i.e. that which ascribes eigenvectors as accelerations, velocities and positions. The unity root matrix itself comprises dynamical variables with units of velocity, and its characteristic eigenvalue equation is that of an energy conservation equation (per unit mass). Indeed, it is the temporal nature that allows URMT solution to be discussed as evolving, and such evolutionary aspects are discussed in the last section of Part I.

Part II builds upon the work in Part I by extending the invariant, zero Potential solution to five and higher dimensions. With a five dimensional eigenvector solution obtained, it is obvious that the solution generalises very easily. Each variational temporal parameter has its own dimension and a casual observation shows that as the solution evolves, i.e. the temporal parameters grow, the size of all excess dimensions (those greater than three) shrinks, becoming zero in the infinite time limit. In fact, it is an accelerating URM3 solution, relative to a constant velocity in all excess dimensions, that gives a novel twist to the physical phenomena known in mathematical physics as compactification or dimensional reduction, i.e. URMT's compactification is not so much shrinkage of excess dimensions, but rapid expansion of those of URM3, which actually converge to a 2D conical subspace of three dimensions, i.e. a discrete cone surface. The book thus concludes with an analysis of the compactification behaviour, Section (14), which verifies that, indeed, for long evolutionary timescales, all excess dimensions are effectively seen to disappear.

The analysis of Compactification completes this second book, and with it the mathematical foundations of URMT. However, there is more to come, much of which will appear in an anticipated third book that will shift the focus away from mathematical foundations toward expanding upon URMT's physical properties.

Table of Contents

Part I

Part I

(1) Foundations

(1-1) Determination of a 4x4 Unity Root Matrix

The first stage in the extension of URMT to higher dimensions is the natural step from three to four dimensions, algebraically expressed using 4x4 matrices instead of the 3x3 matrices of URM3, [1]. It is also formulated such that it can be reduced back to URM3 by zeroing various elements where necessary. This makes URM4 an evolution of URM3 rather than a new formulation starting from scratch. With this in mind, a 4x4 matrix $\mathbf{A}$ is defined as follows, where the top row and the first column are to be determined, and the 3x3 sub-matrix, starting at the second row and column, is simply the URM3, 3x3 unity root matrix, now denoted by $\mathbf{A}_3$ to differentiate it from the 4x4 matrix $\mathbf{A}$, i.e.

$$(1.1) \quad \mathbf{A} = \begin{pmatrix} - & - \\ - & \mathbf{A}_3 \end{pmatrix}$$

where $\mathbf{A}_3$ is reproduced below from Appendix (A)

$$(A1a) \quad \mathbf{A}_3 = \begin{pmatrix} 0 & R & \overline{Q} \\ \overline{R} & 0 & P \\ Q & \overline{P} & 0 \end{pmatrix}, \text{URM3}$$

$(A1b) \quad P,Q,R \in \mathbb{Z}, \ (P,Q,R) \neq (0,0,0)$

$(A1c) \quad \overline{P},\overline{Q},\overline{R} \in \mathbb{Z}, \ (\overline{P},\overline{Q},\overline{R}) \neq (0,0,0).$

The elements of $\mathbf{A}_3$ are the dynamical variables P,Q,R and their conjugates $\overline{P},\overline{Q},\overline{R}$; see Appendix (A) for their full definitions.

3

The lead diagonal of $\mathbf{A}$ (1.1) is defined to be all zero to retain the zero trace of $\mathbf{A}_3$, i.e.

(1.2) $Tr(\mathbf{A}) = Tr(\mathbf{A}_3) = 0$,

and therefore the 1st element of $\mathbf{A}$ (1.1) is zero. The primary reason for an all-zero lead diagonal and consequent zero trace is that, when later expanding the $\mathbf{A}$ matrix eigenvector equations, $\mathbf{AX} = C\mathbf{X}$ (1.7), each element of the four-element eigenvector $\mathbf{X}$ is expressed purely as a function of the other three elements, but not itself. A zero trace is also desirable because, under additional Pythagoras conditions considered later in Section (4), it leads to symmetric eigenvalues, e.g. $\pm C$, and repeated zero eigenvalues, which together form the basis of all current higher-dimensional solutions, as in Part II of this book.

The top row of $\mathbf{A}$ (1.1), columns 2 to 4, is populated with three new variables S, T and U, and the left column, rows 2 to 4, with their conjugates $\overline{S}$, $\overline{T}$ and $\overline{U}$, so that the URM4 $\mathbf{A}$ matrix becomes

(1.3)

(1.3a) $\mathbf{A} = \begin{pmatrix} 0 & S & T & U \\ \overline{S} & 0 & R & \overline{Q} \\ \overline{T} & \overline{R} & 0 & P \\ \overline{U} & Q & \overline{P} & 0 \end{pmatrix}$

(1.3b) $S, T, U, \overline{S}, \overline{T}, \overline{U} \in \mathbb{Z}$.

The justification for this choice of notation primarily stems from the neatness of the results it provides, where 'neatness' covers permutation symmetry in the dynamical variables in expressions such as the characteristic (eigenvalue) equation, see Section (1-4). It is noted that, unlike the rows and columns of $\mathbf{A}_3$ (A1a), the top row and left column of $\mathbf{A}$ (1.3a) are not mixed functions of standard variables S, T, U and their conjugates $\overline{S}, \overline{T}, \overline{U}$. Of course, when originally developing URM4, this was not the case and mixed variants were considered, e.g. mixing a conjugate variable $\overline{T}$ on the top row with standard variables S, U, as in a top row of $\begin{pmatrix} 0 & S & \overline{T} & U \end{pmatrix}$. However, as noted, the results are more neatly expressed with the unmixed definition, (1.3a).

Although the S,T,U variables and conjugates $\overline{S},\overline{T},\overline{U}$ have not yet been defined, the **A** matrix still retains its URM3 Hermitian nature, superficially at least, i.e. it is equal to its transpose conjugate when the condition is imposed that conjugation of all conjugate elements $\overline{S},\overline{T},\overline{U}$ returns the standard, unconjugated forms S,T,U. This is the same as saying that conjugation is its own inverse, i.e. $\overline{\overline{S}} = S$, $\overline{\overline{T}} = T$ and $\overline{\overline{U}} = U$, as is the case (to within a congruence) in URM3 for dynamical variables P,Q,R - see the conjugate relations (A2g) to (A2i).

Lastly, note that S,T,U can all be simultaneously zero, i.e. $(S,T,U) = (0,0,0)$ is allowed (even if trivial), and similar remarks also apply to the conjugates $\overline{S},\overline{T},\overline{U}$. This is unlike the conditions (A1b) and (A1c) on the URM3 dynamical variables: in the case of URM3, the dynamical variables satisfy their non-zero constraints (1.1c) and (1.1d) for genuine, algebraic reasons, and not just by definition, as discussed in [1],#3 under the topic of 'no singularities'.

(1-2) The URM4 Eigenvector **X**

An eigenvector **X** of matrix **A** (1.3a), for eigenvalue C, see (1.7) further below, is defined with URM3 compatible elements x,y,z (as per URM3 eigenvector $\mathbf{X}_3$) and an additional fourth coordinate w as the first coordinate, i.e.

$$(1.4) \; \mathbf{X} = \begin{pmatrix} w \\ \mathbf{X}_3 \end{pmatrix},$$

where $\mathbf{X}_3$ is the familiar URM3 eigenvector to matrix $\mathbf{A}_3$ (A1a), for the same for eigenvalue C,

$$(A3) \; \mathbf{X}_3 = \begin{pmatrix} x \\ y \\ z \end{pmatrix} \; \text{URM3.}$$

Expanded in full, **X** (1.4) is therefore

$$(1.5) \quad \mathbf{X} = \begin{pmatrix} w \\ x \\ y \\ z \end{pmatrix}.$$

Additionally, the elements w, x, y, z are generally co-prime, i.e. they have no greatest common divisor (gcd) other than unity, which makes the eigenvector $\mathbf{X}$ 'primitive' (I12), expressed as the following 'co-primality criterion'

$$(1.6) \quad \gcd(w, x, y, z) = 1.$$

(1-3) The Dynamical Equations

The 4x4 dynamical equations are defined for a general eigenvalue C as follows, as per URM3

$$(1.7) \quad \mathbf{AX} = C\mathbf{X}, \ C \in \mathbb{Z}, \ C \geq 1.$$

Using the definition of $\mathbf{A}$ (1.3a), the dynamical equations expand to

$$(1.7a) \quad Cw = Sx + Ty + Uz$$

$$(1.7b) \quad Cx = \bar{S}w + Ry + \bar{Q}z$$

$$(1.7c) \quad Cy = \bar{T}w + \bar{R}x + Pz$$

$$(1.7d) \quad Cz = \bar{U}w + Qx + \bar{P}y$$

Evidently, if $w = 0$, $(S, T, U) = (0,0,0)$ and $(\bar{S}, \bar{T}, \bar{U}) = (0,0,0)$, then $\mathbf{A}$ and $\mathbf{X}$ reduce to their URM3 equivalents, (A1a) and (A3) respectively, and the dynamical equations are those of URM3 (A5). In fact, only the two conditions $w = 0$ and $(S, T, U) = (0,0,0)$ are required to obtain the URM3 dynamical equations. Furthermore, if only $w = 0$, URMT enters the realm of 'embedding', Section (2), and the associated 'lifting' of solutions, Section (7). Both of these topics, embedding and lifting, are fundamental to obtaining higher dimensional solutions in URMT.

In the general treatment of URM3, only the unity eigenvalue $C = 1$ and associated eigenvectors need be considered, with a non-unity eigenvalue treated as a special case. Furthermore, only this single eigenvalue, unity or otherwise, is considered when not under

Pythagoras conditions. However, as probably anticipated, the majority of interest in URM3 currently lies in simplifying Pythagoras conditions, where all eigenvectors are important, and the same will also be seen to be true of URM4.

Once a non-zero eigenvalue C ($C \geq 1$ (1.7)) is imposed, the URM3 dynamical variables P, Q, R cannot all be trivially zero, i.e. $(P, Q, R) \neq (0,0,0)$ (A1b). By conjugate relations, (A2g) to (A2i), neither are $\overline{P}, \overline{Q}, \overline{R}$ all zero (A1c). One or two of the three P, Q, R can possibly be zero, but not all three simultaneously. This also constrains the elements w, x, y, z of vector $\mathbf{X}$ (1.5) in the same way. That no vector in URMT can comprise all zeros, and therefore have zero magnitude, is an algebraic consequence of mandating $C \geq 1$. Effectively it means URMT has no singularities and, naturally, this is a highly desirable physical attribute.

(1-4) The Characteristic Equation

To study all the eigenvalues λ, the characteristic equation is given by the usual non-singular condition on the eigenvalue equation $\mathbf{AX} = \lambda \mathbf{X}$, i.e.

(1.8) $\det(\mathbf{A} - \lambda \mathbf{I}) = 0$, the DCE - see (1.14) further below,

which expands to

(1.9) $\lambda^4 + a_2 \lambda^2 + a_1 \lambda + a_0 = 0$,

for constant coefficients a_2, a_1, a_0 defined by

(1.10)

(1.10a) $a_2 = -(P\overline{P} + Q\overline{Q} + R\overline{R} + S\overline{S} + T\overline{T} + U\overline{U})$

(1.10b)
$$a_1 = -(PQR + \overline{P}\overline{Q}\overline{R})$$
$$-(RS\overline{T} + \overline{R}ST)$$
$$-(Q\overline{S}U + \overline{Q}S\overline{U})$$
$$-(\overline{P}\overline{T}U + PT\overline{U})$$

$$a_0 = (P\overline{P}S\overline{S} + Q\overline{Q}T\overline{T} + R\overline{R}U\overline{U})$$

(1.10c)
$$- (PRS\overline{U} + \overline{P}R\overline{S}U)$$
$$- (QR\overline{T}U + \overline{Q}\overline{R}T\overline{U})$$
$$- (\overline{P}\overline{Q}S\overline{T} + PQ\overline{S}T)$$

Note that there is no cubic term in (1.9) by virtue of an all-zero lead diagonal in **A** (1.3a) and consequent zero trace. Furthermore, as will be seen shortly, if $a_1 = 0$ then (1.9) effectively reduces to a quadratic in λ^2. Taken together, these two points are good reason to keep the lead diagonal all zero, as mentioned earlier.

The complete characteristic equation is homogeneous of fourth degree, and hence the coefficients a_2, a_1, a_0 are of quadratic, cubic and quartic degree respectively in the dynamical variables.

Since $\lambda = C$ is the single, mandated eigenvalue under URM3, and also now URM4, the characteristic equation could be factored to give a cubic. Furthermore, by symmetry considerations of the eigenvalues (symmetric about zero), reminiscent of URM3 under Pythagoras conditions, the eigenvalue $\lambda = -C$ could also be imposed. This would then leave a quadratic after factorisation, and given all four eigenvalues sum to zero by virtue of a zero trace (1.2), the two remaining eigenvalues would also be symmetric, say $\lambda = \pm D$. Unlike C, which is always non-zero in URM3, this may or may not be the case for both C and D in URM4 until further investigation. Nevertheless, C is defined as positive, non-zero and, additionally, real not complex. The value of D is left undecided for now and could even be complex, something new to URMT.

In general, because the unity root matrix always has a zero trace, the sum of the eigenvalues is zero, and with $\lambda_1 = C$, the remaining three eigenvalues always therefore satisfy

(1.11) $\lambda_2 + \lambda_3 + \lambda_4 = -C$.

With symmetric eigenvalues in mind, it is desirable to eliminate the linear term, coefficient $a_1 = 0$, which would leave the characteristic equation as a quadratic form in λ^2. Examination of the components of a_1 (1.10b) hint that this is a URM4 equivalent of the URM3 zero Potential condition, $V_3 = 0$ (A24), since a_1 contains the URM3

8

Potential V_3 (A9). This then implies some type of URM4-equivalent Pythagorean conditions; see Section (4). Assuming a_1 is zero reduces the characteristic equation (1.9) to

(1.12) $\lambda^4 + a_2\lambda^2 + a_0 = 0$, $a_1 = 0$.

For general URMT, the assumption $a_1 = 0$ may be rather strong, but given a_1 (1.10b) is a function of six dynamical variables and their six conjugates, this gives quite a few parameters to work with in trying to meet the condition. Having said that, this extra constraint on the dynamical variables is not explicitly applied but, just as in URM3, the main focus is on simplifying Pythagoras conditions, Section (4-1), which implicitly make a_1 zero anyhow.

An alternative simplification to the above is to make coefficient $a_0 = 0$ instead of $a_1 = 0$; this would then automatically give a zero eigenvalue $\lambda = 0$ to go with the predefined eigenvalue $\lambda = C$. Factoring the characteristic equation for $\lambda = 0$ leaves a cubic

(1.13) $\lambda(\lambda^3 + a_2\lambda + a_1) = 0$,

which reduces to a quadratic when additionally factoring $\lambda = C$.

In fact, both these simplifications are key to URM4, and now discussed again in terms of the following Dynamical Conservation Equation.

(1-5) The Dynamical Conservation Equation

In URM3 the characteristic equation is known as the Dynamical Conservation Equation (DCE), which is analogous to an energy conservation equation, and is consequently separated into a kinetic energy term (K) and Potential energy (V) as in

(1.14) $C^2 = K + V$, the DCE, equivalent to (1.8).

The total, conserved energy E is thus

(1.14b) $E = C^2$.

The URM4 development and language is no different in this respect albeit, as will be seen, the form of the Potential is more complicated in URM4 than in URM3.

By substituting the single, mandated eigenvalue $\lambda = C$ into the characteristic equation (1.9), and rearranging and dividing throughout by C ($C > 0$ by definition), the most general form of the URM4 DCE is obtained

(1.15) $C^2 = -a_2 - \left(\dfrac{a_1}{C} + \dfrac{a_0}{C^2} \right)$.

Comparing this with the DCE (1.14), then the kinetic and Potential terms are defined as

(1.16) $K = -a_2$,

(1.17) $V = -\left(\dfrac{a_1}{C} + \dfrac{a_0}{C^2} \right)$.

For later comparison with URM4, the URM3 equivalent kinetic term K_3 and Potential term V_3 are given by

(A8) $K_3 = P\overline{P} + Q\overline{Q} + R\overline{R}$, URM3

(A9) $V_3 = \dfrac{(PQR + \overline{P}\,\overline{Q}\,\overline{R})}{C}$, URM3

Looking at coefficient a_2 (1.10a), the kinetic term K (1.16) is evidently of the desired quadratic form, as per URM3 (A8), but now includes the extra URM4 dynamical variables S, T, U and their conjugates $\overline{S}, \overline{T}, \overline{U}$

(1.18) $K = P\overline{P} + Q\overline{Q} + R\overline{R} + S\overline{S} + T\overline{T} + U\overline{U}$, URM4.

The Potential V (1.17) is, however, quite different to the URM3 Potential V_3 (A9). Nevertheless, both Potentials are of the correct physical units, which are the same as the kinetic term K and total energy term E, i.e. a quadratic in the dynamical variables; see the 'Standard Physical Interpretation', Appendix (J). This is because the eigenvalue C is of the same physical units as the dynamical variables, and equated with a velocity quantity such that the energy C^2 (1.14b) is of physical units velocity squared, i.e. energy per unit mass.

Looking at coefficient a_1 (1.10b), this coefficient is seen to split into the URM3 component V_3 (A9) and an extra URM4 term a_1', as follows:

(1.19a) $\dfrac{a_1}{C} = -V_3 - \dfrac{a_1'}{C}$,

where a_1' is defined as

(1.19b) $a_1' = -(RS\overline{T} + \overline{R}\overline{S}T) - (Q\overline{S}U + \overline{Q}S\overline{U}) - (\overline{P}\overline{T}U + PT\overline{U})$.

The a_0 coefficient is of quartic degree in the dynamical variables and completely new to URM4, i.e. it has no URM3 legacy components, unlike a_1, which contains the URM3 Potential V_3.

Whilst the compound URM4 Potential (1.17) appears rather more involved, and consequently unwieldy compared with its URM3 counterpart V_3 (itself considered rather abstract as a physical quantity), it is not currently used in its full form, where both a_0 and a_1 are non-zero. Instead, one or both of the two coefficients is invariably set to zero as a simplification. In fact, URM4 splits into the following three cases according to the settings of a_0, a_1 and a_1', where the method of 'URM3 Embedding' is fully described in Section (2), and URM4 Pythagoras conditions will be explained a bit later:

(1.20)

(1.20a) URM3 Embedding, Section (2)

$V = V_3 = -\dfrac{a_1}{C}$, $a_1 \neq 0$, $a_1' = 0$, $a_0 = 0$.

(1.20b) URM3 Embedding under URM3 Pythagoras conditions

$V = V_3 = 0$, $a_1 = 0$, $a_1' = 0$, $a_0 = 0$.

(1.20c) URM4 Pythagoras conditions, Section (4), which includes URM3 Pythagoras conditions as a subset

$V = -\dfrac{a_0}{C^2}$, $a_1 = 0$, $a_0 \neq 0$

The reason that the case $a_1 \neq 0$ and $a_0 \neq 0$ is not studied is that, from a physical perspective, the three cases (1.20) appear satisfactory to include both existing URM3 properties and add new URM4 properties. There is also another, more straightforward reason, which is that the full URM4 Potential (1.17) is just too complicated. Were it not for the fact that the cases (1.20) are fruitful in their results, then the excuse of 'too complicated' would be insufficient justification to avoid considering the full case where $a_1 \neq 0$ and $a_0 \neq 0$, which has not been entirely eliminated and could well provide some interesting and new results, but is beyond the scope of the current URMT development work.

Keeping with an integer-only formulation of URM4, the definition of the Potential V in (1.20c) assumes a_0 is divisible by C^2. As will be seen later, this is at least the case under Pythagoras conditions, where a_0 is also a perfect square.

Of course, the work so far still leaves K and V as functions of the coefficients a_0, a_1 and a_2, which are, as yet, undetermined, i.e. no URM4 analytic solution for the coefficients, in terms of the dynamical variables, has yet been given. Neither has any solution in the coordinates and eigenvectors been given, just a discussion on possible simplifications. Ultimately, solving URM4 for all dynamical variables and coordinates is difficult, if not impossible, in the most general case; fortunately, it doesn't seem necessary to get interesting results. All higher-dimensional URMT development work to date, as reported in this book, concentrates on simplifying assumptions, which seem sufficient to achieve the aims as stated at the beginning of the book. These simplifications, based upon cases (1.20), are considered throughout the remainder of this book.

Inserting $\lambda = 0$ into the characteristic equation (1.9) gives the determinant of $\mathbf{A}$ (1.3a), which is just coefficient a_0, and hence,

(1.21) $\det(\mathbf{A}) = a_0$.

In URM3, the determinant of $\mathbf{A}_3$ is related to the Potential V_3, i.e.

(A11) $\det(\mathbf{A}_3) = CV_3$,

but comparing (1.21) with (1.17), it is evident $\det(\mathbf{A})$ is only the same as the Potential V when $V = a_0$ and $a_1 = 0$, i.e. cases (1.20b) and (1.20c) where $a_1 = 0$ and $a_0 \neq 0$. Furthermore, as in [1],#2, the URM3 Potential V_3 is zero under URM3 Pythagoras conditions whereas now, in URM4, it is not necessarily zero, as in case (1.20c). In

12

fact, the URM4 Potential will be seen to always be greater than or equal to zero, but not negative, unlike in URM3 where the Potential can generally be positive, zero or negative. Lastly, a non-zero URM4 Potential, even when under URM4 Pythagoras conditions, makes **A** invertible, which under URM3 Pythagoras conditions (A18) is never possible since the Potential V_3 is always zero as a consequence of the conditions.

(1-6) Loss of Unity Root Nature?

In going from 3x3 to 4x4 matrices, the dynamical variables no longer have a natural definition as unity roots, unlike URM3, and if the unity root property is desirable in URM4, which it certainly is, then it has to be imposed. This loss of unity root property may seem almost catastrophic given the subject 'unity root' appears in the title! Nevertheless, although there is no obvious connection of URM4 with unity roots, the connection is imposed by retaining the original definition of the URM3 dynamical variables P,Q,R and $\overline{P},\overline{Q},\overline{R}$ as unity roots, with the new URM4 dynamical variables S,T,U and $\overline{S},\overline{T},\overline{U}$ left unconstrained. To understand this natural loss of the unity root nature of dynamical variables in URM4, it is necessary to review how the property arises in URM3.

In URM3 the dynamical equations (A5), reproduced below from Appendix (A), are a set of three equations in 3 unknowns, albeit only two are linearly independent by the definition of the eigenvalue problem

(A5) URM3 dynamical equations

(A5a) $Cx = Ry + \overline{Q}z$

(A5b) $Cy = \overline{R}x + Pz$

(A5c) $Cz = Qx + \overline{P}y$.

From these equations it is seen that each one of the three coordinates x,y,z is dependent on the other two, i.e. they are all interdependent on one another. Important to the interpretation of the dynamical variables as unity roots when $C = \pm 1$ (or, more generally, power residues when $|C| > 1$), is the fact that taking residues of each equation, modulo a coordinate, gives a congruence directly linking the other two

13

coordinates. For example, for unity eigenvalue $C = 1$, taking residues of (A5b) and (A5c), modulo x, gives the following two congruences linking y and z.

(1.22)

(1.22a) $y \equiv Pz \pmod{x}$, $C = 1$

(1.22b) $z \equiv \overline{P}y \pmod{x}$, $C = 1$

These congruences are then raised to the nth order, $n \geq 2$, as an extension to the theory (and not a pre-condition), i.e.

(1.23)

(1.23a) $y^n \equiv P^n z^n \pmod{x}$

(1.23b) $z^n \equiv \overline{P}^n y^n \pmod{x}$.

The congruences can always be satisfied if the dynamical variables are then defined as having the unity root property:

(A2a) $P^n \equiv +1 \pmod{x}$, $C = 1$

(A2d) $\overline{P}^n \equiv +1 \pmod{x}$, $C = 1$.

Returning to the linear congruences (1.22), then multiplying (1.22a) by $\overline{P}$ and equating with (1.22b) implies that, for $\gcd(x, z) = 1$ (A3),

(1.24) $1 - P\overline{P} \equiv 0 \pmod{x}$,

which is satisfied by the following conjugate relation (A2g), whilst retaining the unity the root definition (A2d)

(A2g) $\overline{P} \equiv P^{n-1} \pmod{x}$.

Returning to URM4, if the same process is applied to the 4x4 dynamical equations, reproduced below, there is no natural pairing of, say, y and z with x, as in congruences (1.22)

(1.7)

(1.7a) $Cw = Sx + Ty + Uz$

(1.7b) $Cx = \overline{S}w + Ry + \overline{Q}z$

(1.7c) $Cy = \bar{T}w + \bar{R}x + Pz$

(1.7d) $Cz = \bar{U}w + Qx + \bar{P}y$.

Taking residues of (1.7), modulo x, for those equations where x appears on the right in (1.7) gives

(1.25a) $Cw = Ty + Uz \,(\mathrm{mod}\,x)$

(1.25b) $Cy = \bar{T}w + Pz \,(\mathrm{mod}\,x)$

(1.25c) $Cz = \bar{U}w + \bar{P}y \,(\mathrm{mod}\,x)$.

Whilst the coordinates remain such that each depends on the other three, the congruence relations become messier. Instead of simply giving two equations relating two coordinates, e.g. y with z as in (1.22), they now relate y with z and w. As a consequence there is no neat, clear-cut path to defining any simple unity root or power residue definitions involving only the dynamical variables, as in (A2).

One such method, considered in the early development, was to set two of the unity roots to 0, e.g. Q and T. This gives the following dynamical equations:

(1.26a) $Cw = Sx + Uz$

(1.26b) $Cx = \bar{S}w + Ry$

(1.26c) $Cy = \bar{R}x + Pz$

(1.26d) $Cz = \bar{U}w + \bar{P}y$.

Although each of these equations now has a more desirable form, i.e. each single coordinate on the left is dependent upon only two others on the right, it still doesn't give the symmetry in (1.22). For example, taking residues modulo x of (1.26a) and (1.26c) gives

(1.27)

(1.27a) $Cw = Uz \,(\mathrm{mod}\,x)$

(1.27b) $Cy = Pz \,(\mathrm{mod}\,x)$

But, unlike (1.22), the left and right sides of (1.27) are not exclusive functions of y and z, but still involve an extra coordinate w. This means the two equations in (1.27) cannot be equated with each other to give an expression like (1.24), which involves only the dynamical variables (barring the coordinate modulus), and ultimately relates dynamical variables with unity roots.

Without labouring the point, a thoughtful consideration of the dynamical equations concludes that URM4, in its most general form, does not naturally accommodate a unity root definition of its dynamical variables.

In a sense, it points to a beauty in simplicity of URM3, and makes one wonder whether pursuing URM4 or higher, is going to give the desirable properties seen in URM3. Nevertheless, whilst this may seem to make URM4 lack any of the simplicity and succinctness of URM3, it is very simple to mandate (define) that the URM4 dynamical variables, P, Q, R and $\overline{P}, \overline{Q}, \overline{R}$, retain their URM3 definition as unity roots (or power residues), whilst allowing new dynamical variables S, T, U and $\overline{S}, \overline{T}, \overline{U}$ to take on a more arbitrary nature. Indeed, it is not even necessary that S, T, U and P, Q, R have the same physical units. This retention of URM3 within URM4, whilst gaining new URM4 properties (especially extra dimensionality and more local variational terms), is precisely what the embedding and lifting methods in Sections (2) and (7) aim to achieve, and is a very important, if not the most important facet of URMT beyond URM3.

(1-7) Summary

To summarise this first section, a 4x4, four-dimensional unity root matrix $\mathbf{A}$ (1.3a) has been defined in terms of the existing 3x3, URM3 matrix $\mathbf{A}_3$ (A1a), with the addition of three new dynamical variables S, T, U, and their conjugates $\overline{S}, \overline{T}, \overline{U}$ (1.3b), to complement the existing URM3 dynamical variables. A vector $\mathbf{X}$ (1.5), comprising the four elements w, x, y, z (termed coordinates), is defined as an eigenvector of matrix $\mathbf{A}$ for eigenvalue C, i.e. $\mathbf{AX} = C\mathbf{X}$ (1.7).

The URM4 characteristic equation, $\det(\mathbf{A} - \lambda\mathbf{I}) = 0$ (1.8), is expanded as a quartic polynomial (1.9) in the eigenvalue, and the coefficients a_2, a_1, a_0 (1.10) associated with a kinetic energy term K (1.16) and a Potential energy term V (1.17), in accordance with the standard physical interpretation of URMT given in Appendix (J). The characteristic equation (1.9) is then re-written using these energy terms to become the Dynamical Conservation Equation, 'the DCE', (1.14).

It is noted in the last section (1-6) that, unlike the URM3 dynamical variables, P, Q, R and $\overline{P}, \overline{Q}, \overline{R}$, the URM4-specific variables, S, T, U and $\overline{S}, \overline{T}, \overline{U}$, are not naturally associated with having unity root properties.

So far, only a single eigenvalue C and eigenvector $\mathbf{X}$ (1.5) have been defined for matrix $\mathbf{A}$, with no actual solution given for either the dynamical variables or the coordinates w, x, y, z of $\mathbf{X}$. The remainder of the book now focuses on obtaining all the eigenvalues, eigenvectors and solutions for the dynamical variables, such that they satisfy the DCE (1.14).

This completes the foundations of URM4. Of course, this is only an extension to four dimensions and 4x4 matrices. However, this small step in going from three to four dimensions is sufficient to provide insights into the way forward for extension of URMT to an arbitrary number of dimensions. As such, the remainder of Part I concentrates on studying the four-dimensional case, with Part II exploring URM5 and beyond to the general, n-dimensional case.

(2) Embedding URM3 in URM4

Given the rich mathematical and physical structure of URM3, any extension to URM4 would be hard-pressed not to encompass compatibility with URM3. As such, a simple method to proceed is to make URM4 look much like URM3, but with the extra geometric structure that comes with more dimensions.

(2-1) The Embedded Matrix and Eigenvector

As discussed prior in Section (1), a trivial solution to the URM4 dynamical equations (1.7) is just the URM3 solution when $w = 0$ and $(S,T,U) = (0,0,0)$, i.e.

$$(2.1) \quad \mathbf{AX} = \begin{pmatrix} 0 & 0 & 0 & 0 \\ \overline{S} & 0 & R & \overline{Q} \\ \overline{T} & \overline{R} & 0 & P \\ \overline{U} & Q & \overline{P} & 0 \end{pmatrix} \begin{pmatrix} 0 \\ x \\ y \\ z \end{pmatrix} = C \begin{pmatrix} 0 \\ x \\ y \\ z \end{pmatrix},$$

$w = 0$, $(S,T,U) = (0,0,0)$,

which simply reduces to the equivalent URM3 eigenvector equation (A4), using the definition of $\mathbf{X}_3$ (A3),

$$(2.2) \quad \mathbf{AX} \sim \mathbf{A}_3\mathbf{X}_3 = C\mathbf{X}_3.$$

Notice that $\overline{S}, \overline{T}$ and $\overline{U}$ here are unconstrained and do not have to be zero by virtue of a zero first element in $\mathbf{X}$.

Whilst $w = 0$ and $(S,T,U) = (0,0,0)$ reduces to URM3, this is almost too trivial, and something less trivial is required. The first compromise is to keep $w = 0$ in $\mathbf{X}$, but leave S, T and U unconstrained. Doing so retains $\mathbf{A}$ as per its original definition (1.3a) and modifies $\mathbf{X}$ (1.5) to become

$$(2.3)\ \mathbf{X} = \begin{pmatrix} 0 \\ x \\ y \\ z \end{pmatrix}, \quad w = 0.$$

Using the URM3 definition for $\mathbf{X}_3$ (A3), $\mathbf{X}$ is also written in block form as

$$(2.4)\ \mathbf{X} = \begin{pmatrix} 0 \\ \mathbf{X}_3 \end{pmatrix}.$$

Using $\mathbf{A}$ (1.3a) and $\mathbf{X}$ (2.3), the eigenvector equation $\mathbf{AX} = C\mathbf{X}$ (1.7) now evaluates to

$$(2.5)\ \mathbf{AX} = \begin{pmatrix} Sx + Ty + Uz \\ Cx \\ Cy \\ Cz \end{pmatrix}.$$

(2-2) The Orthogonality Condition

Looking at (2.5), if the following orthogonality condition is imposed

$$(2.6)\ Sx + Ty + Uz = 0,$$

then

$$(2.7)\ \mathbf{AX} = C \begin{pmatrix} 0 \\ x \\ y \\ z \end{pmatrix} = C\mathbf{X},$$

which is just the URM4 dynamical equations (1.7).

Thus, a URM3 compatible set of equations can be obtained with a non-zero top row for $\mathbf{A}$, by determining a conjugate, orthogonal vector $\overline{\mathbf{X}}$ defined as

$$(2.8)\ \overline{\mathbf{X}} = \begin{pmatrix} 0 & S & T & U \end{pmatrix},$$

and such that the following inner product, orthogonality condition holds, as per (2.6),

(2.9) $\overline{\mathbf{X}} \cdot \mathbf{X} = 0$.

Note that 'conjugate' vectors are row vectors, and are the transpose conjugate of their column vector equivalents, whereby the conjugation operation conjugates each element of the column vector, i.e. with $\mathbf{X}$ defined by (2.3) then the transpose conjugate $\overline{\mathbf{X}}$ is

(2.10) $\overline{\mathbf{X}} = \begin{pmatrix} 0 & \bar{x} & \bar{y} & \bar{z} \end{pmatrix} = \begin{pmatrix} 0 & S & T & U \end{pmatrix}$ by (2.8).

This definition has surreptitiously introduced three new conjugate coordinates $\bar{x}, \bar{y}, \bar{z}$ without formally defining what they actually are, unlike the conjugate dynamical variables $\overline{P}, \overline{Q}, \overline{R}$, which are related to their standard forms P, Q, R by the conjugate relations (A2g) to (A2i). Except under Pythagoras conditions (URM3 (A18) and URM4, Section (4)), no specific relation between $\bar{x}, \bar{y}, \bar{z}$ and x, y, z is given other than $\bar{x}, \bar{y}, \bar{z}$ is a triple such that the vector $\overline{\mathbf{X}}$ (2.10) satisfies the orthogonality condition (2.9). This manifests itself in (2.20) further below, and is the only definition currently required of $\bar{x}, \bar{y}, \bar{z}$.

Defining the URM3 conjugate vector $\overline{\mathbf{X}}_3$ as

(2.11) $\overline{\mathbf{X}}_3 = \begin{pmatrix} S & T & U \end{pmatrix} \Rightarrow \overline{\mathbf{X}}_3 = \begin{pmatrix} \bar{x} & \bar{y} & \bar{z} \end{pmatrix}$ by (2.10)

then $\overline{\mathbf{X}}$ (2.10) is also written in block matrix form as

(2.12) $\overline{\mathbf{X}} = \begin{pmatrix} 0 & \overline{\mathbf{X}}_3 \end{pmatrix}$

Note that the conjugate of $\overline{\mathbf{X}}$, i.e. $\overline{\overline{\mathbf{X}}}$, is simply $\mathbf{X}$ since the transpose conjugate of the row vector $\overline{\mathbf{X}}$ is a column vector identical to $\mathbf{X}$, when the conjugate of each of the elements of $\overline{\mathbf{X}}$ are also the elements of $\mathbf{X}$; see Appendix (E). Here the elements of $\overline{\mathbf{X}}$ are the conjugate coordinates $\bar{x}, \bar{y}, \bar{z}$, and so their conjugates are expressed as $\bar{\bar{x}}, \bar{\bar{y}}, \bar{\bar{z}}$, i.e. the conjugate of a conjugate. Whilst $\bar{x}, \bar{y}, \bar{z}$ have not been given a rigid definition (see above), in general URMT it is a definition (or convention) that the conjugate of a conjugate variable returns the original variable, i.e. conjugation is its own inverse,

(2.13) $\bar{\bar{x}} = x$, $\bar{\bar{y}} = y$, $\bar{\bar{z}} = z$, see also (A40).

21

Using $\mathbf{X}$ (2.4) and $\overline{\mathbf{X}}$ (2.12), then the URM4 orthogonality condition (2.9) reduces to the same condition between URM3 vectors $\overline{\mathbf{X}}_3$ (2.11) and $\mathbf{X}_3$ (A3), i.e.

(2.14) $\overline{\mathbf{X}}_3 \cdot \mathbf{X}_3 = 0$

which conforms with (F1), Appendix (A).

In block matrix form, $\mathbf{A}$ (1.3a) is now written as

$$(2.15) \quad \mathbf{A} = \begin{pmatrix} 0 & \overline{\mathbf{X}}_3 \\ \mathbf{X}_3 & \mathbf{A}_3 \end{pmatrix},$$

and using the URM3 orthogonality condition (2.9), the dynamical equations $\mathbf{AX} = C\mathbf{X}$ become

$$(2.16) \quad \mathbf{AX} = \begin{pmatrix} \overline{\mathbf{X}}_3 \cdot \mathbf{X}_3 \\ \mathbf{X}_3 \end{pmatrix} = C \begin{pmatrix} 0 \\ \mathbf{X}_3 \end{pmatrix}.$$

Of course, keeping with the URM3 Hermitian-like property (A41) for $\mathbf{A}$, the first column of $\mathbf{A}$ has to be the transpose conjugate of the first row, so that the first column is simply the column vector $\mathbf{X}$ (the transpose conjugate of $\overline{\mathbf{X}}$), i.e.

$$(2.17) \quad \mathbf{X} = \begin{pmatrix} 0 \\ \overline{S} \\ \overline{T} \\ \overline{U} \end{pmatrix}.$$

With these points in mind, and comparing $\mathbf{X}$ (2.3) with (2.17), a change in the notation of the dynamical variables S, T, U and $\overline{S}, \overline{T}, \overline{U}$, to those of the coordinates x, y, z and conjugates $\overline{x}, \overline{y}, \overline{z}$, is defined as follows, where the swap between standard variables S, T, U and conjugate coordinates $\overline{x}, \overline{y}, \overline{z}$ is intentional,

(2.18)

(2.18a) $\overline{S} = x$

(2.18b) $\overline{T} = y$

(2.18c) $\overline{U} = z$

Assuming self-conjugacy of S, T and U, i.e. $\overline{\overline{S}} = S$ etc, then in accordance with (2.11),

(2.18d) $S = \overline{x}$

(2.18e) $T = \overline{y}$

(2.18f) $U = \overline{z}$.

Using this notation, then matrix $\mathbf{A}$ (1.3a) is now written as

$$(2.19)\ \mathbf{A} = \begin{pmatrix} 0 & \overline{x} & \overline{y} & \overline{z} \\ x & 0 & R & \overline{Q} \\ y & \overline{R} & 0 & P \\ z & Q & \overline{P} & 0 \end{pmatrix},$$

and using eigenvectors $\mathbf{X}$ (2.4) and $\overline{\mathbf{X}}$ (2.10), the orthogonality condition (2.9) becomes

$$(2.20)\ \overline{\mathbf{X}} \cdot \mathbf{X} = \overline{\mathbf{X}}_3 \cdot \mathbf{X}_3 = x\overline{x} + y\overline{y} + z\overline{z} = 0.$$

With this orthogonality, the dynamical equations $\mathbf{AX} = C\mathbf{X}$ (1.7) are once again satisfied.

The orthogonality condition (2.20) is the nearest equivalent to the coordinate conservation equation in [1],#1, reproduced below

$$(2.21)\ 0 = x^n + y^n - z^n + xyz.k(x, y, z).$$

Most importantly though, (2.20) has no dependence on an integer exponent n. Under URM3 Pythagoras conditions, $\overline{x} = x$, $\overline{y} = y$ and $\overline{z} = -z$ (A39a), and equation (2.20) is simply a restatement of the Pythagoras equation, i.e. $0 = x^2 + y^2 - z^2$ (F1). However, nowhere is there a restriction to exponent $n = 2$ in the orthogonality relation (2.20). It is emphasized that x, y, z and conjugates $\overline{x}, \overline{y}, \overline{z}$ are not necessarily Pythagorean triples. In geometric terms, $\overline{\mathbf{X}}$ ($\sim \overline{\mathbf{X}}_3$) is simply any vector in the plane normal to $\mathbf{X}$ ($\mathbf{X}_3$). In URM3 this is a familiar 2D planar surface, in URM4 it is a 3D hyperplane.

Of particular note, if a set of conjugate coordinates $\bar{x}, \bar{y}, \bar{z}$ could be found related to the standard coordinates as follows, for integer exponent $n > 2$,

(2.22) $\bar{x} = x^{n-1}$, $\bar{y} = y^{n-1}$, $\bar{z} = -z^{n-1}$,

then (2.20) would be an FLT counter-example. Given no such counter-example exists, as famously proved by Wiles 1995, then no such conjugate coordinates $\bar{x}, \bar{y}, \bar{z}$, satisfying (2.22), exist.

In the case of Pythagoras, i.e. exponent $n = 2$, the relations (2.22) become

(2.23) $\bar{x} = x$, $\bar{y} = y$, $\bar{z} = -z$, for $n = 2$,

which are simply the URM3 conjugate relations (A39a) in the coordinates, under URM3 Pythagoras conditions, and the URM4 conjugate vector $\overline{\mathbf{X}}$ (2.12) becomes an embedding of the URM3 conjugate vector $\mathbf{X}^{3-}$ (A35c), since $\overline{\mathbf{X}}_3 = \mathbf{X}^{3-}$ (A37a), i.e.

(2.24) $\overline{\mathbf{X}} = \begin{pmatrix} 0 & \mathbf{X}^{3-} \end{pmatrix}$.

(2-3) Consideration of Physical Dimensionality

In URM3, the coordinates x, y, z are not particularly thought of as the same type of quantity (not of the same physical units) as the dynamical variables P, Q, R. Furthermore, the conjugate coordinates $\bar{x}, \bar{y}, \bar{z}$ may not necessarily be of the same units as either the coordinates x, y, z or the dynamical variables P, Q, R. As a consequence, the $\mathbf{A}$ matrix (A1a) in URM4 may contain a mix of different physical quantities. Whilst acceptable on a purely mathematical basis, this mix of quantities is not acceptable on physical grounds. The solution to obtain a consistent matrix $\mathbf{A}$ (2.19) of elements, all with the same physical units, can be achieved by multiplying the left column coordinates x, y, z, and their top row conjugates $\bar{x}, \bar{y}, \bar{z}$, by conversion factors such that they match the same physical quantity represented by the dynamical variables in the embedded matrix $\mathbf{A}_3$ (A1a). Two such conversion factors, integer parameters s and $\bar{s}$, are introduced into URM4 to give a new matrix $\mathbf{A}$ as follows:

$$(2.25) \quad \mathbf{A} = \begin{pmatrix} 0 & \overline{s}\overline{\mathbf{X}}_3 \\ s\mathbf{X}_3 & \mathbf{A}_3 \end{pmatrix}, \quad s, \overline{s} \in \mathbb{Z}.$$

The physical units of the elements are now such that

$$(2.26) \quad unit(s\mathbf{X}_3) = unit(\overline{s}\overline{\mathbf{X}}_3) =$$

$$units(sx, sy, sz) = units(\overline{s}\overline{x}, \overline{s}\overline{y}, \overline{s}\overline{z}) =$$

$$units(\mathbf{A}_3) = units(\mathbf{A}) = \text{velocity } LT^{-1}.$$

Indeed, further in this book, the URM3 dynamical variables P, Q, R (units of velocity) are assigned to the elements of $\overline{\mathbf{X}}$, whilst the elements of $\mathbf{X}$ remain the familiar coordinates x, y, z, with units of acceleration, in accordance with the standard physical interpretation of URMT, Appendix (J). This means that the units of s are time T, and $\overline{s}$ is dimensionless, such that the units of all three $\mathbf{X}$, $\overline{\mathbf{X}}$ and $\mathbf{A}_3$ remain those of the dynamical variables, i.e. velocity LT^{-1}.

(2-3a) Variational Parameters

The two new integer parameters s and $\overline{s}$, introduced above, not only serve as physical conversion factors, but are actually forms of variational parameter in URMT. Variational methods are very important in URMT and visited many times throughout this book, starting at Section (7) and continuing thereafter to the end. To explain a bit more, a short digression from the subject of embedding follows, offering some justification as to how two such parameters can be easily introduced without having any effect on the dynamical equations.

Matrix $\mathbf{A}$ (2.25) can be split into two component matrices as follows

$$(2.27) \quad \mathbf{A} = \begin{pmatrix} 0 & 0 \\ 0 & \mathbf{A}_3 \end{pmatrix} + \begin{pmatrix} 0 & \overline{s}\overline{\mathbf{X}}_3 \\ s\mathbf{X}_3 & 0 \end{pmatrix}.$$

The first matrix term on the right is the initial value of $\mathbf{A}$ for $s = \overline{s} = 0$, and denoted by $\mathbf{A}'$, i.e.

$$(2.28) \quad \mathbf{A}' = \begin{pmatrix} 0 & 0 \\ 0 & \mathbf{A}_3 \end{pmatrix}, \quad s, \overline{s} = 0.$$

25

The second matrix on the right of (2.27) is a form of variational matrix, denoted by Δ and defined here as

$$(2.29) \quad \Delta = \begin{pmatrix} 0 & \overline{s}\mathbf{X}_3 \\ s\mathbf{X}_3 & 0 \end{pmatrix}.$$

Most importantly, using orthogonality (2.14), the variational matrix Δ annihilates eigenvector $\mathbf{X}$ as in

$$(2.30) \quad \Delta\mathbf{X} = \begin{pmatrix} \overline{s}\mathbf{X}_3 \cdot \mathbf{X}_3 \\ 0 \end{pmatrix} = \begin{pmatrix} 0 \\ 0 \end{pmatrix}.$$

Using the two matrix definitions for $\mathbf{A}'$ (2.28) and Δ (2.29), then $\mathbf{A}$ (2.27) is written in the standard URMT variational form as

$$(2.31) \quad \mathbf{A} = \mathbf{A}' + \Delta.$$

By virtue of the annihilation property (2.30), the effect of Δ and $s, \overline{s}$ on the transform $\mathbf{A}$ (2.31) is therefore zero, i.e.

$$(2.32) \quad \mathbf{A}\mathbf{X} = \mathbf{A}'\mathbf{X} + \Delta\mathbf{X} = \mathbf{A}'\mathbf{X} = C\mathbf{X}.$$

In [1] there was a single, global 'delta' variational parameter, denoted by δ or m (where $\delta = -m$), and given as the mapping

$$(2.33a) \quad \mathbf{A} \rightarrow \mathbf{A} + \delta\Delta$$

or

$$(2.33b) \quad \mathbf{A} \rightarrow \mathbf{A} - m\Delta.$$

Being a single parameter, the variational parameter δ affects every element of the matrix Δ and is thus considered a 'global' change. Hence, (2.31) is fully termed a 'global, delta variational transformation'. Conversely, here in URM4, the two parameters s and $\overline{s}$ internally (locally) scale different elements of Δ (2.29), and are thus termed a 'local' transformation in URMT; noting that URM3 also has three local transformations, denoted by '$\eta, \delta, \varepsilon$' (8.4). In fact, in Section (8), it is shown that URM4 has eight possible local transformations, albeit they are not used in full.

With **A** in its single matrix form (2.25), then the URM3 Hermitian-like property of **A** ([1],#1]) still holds, assuming $\bar{\bar{s}} = s$ and given $\bar{\bar{X}}_3 = X_3$ (A37c) and $A_3 = \overline{A}^T{}_3$ (A41), i.e. the transpose conjugate of **A** evaluates as

$$(2.34) \quad \overline{A}^T = \begin{pmatrix} 0 & \bar{s}\overline{X}_3 \\ \overline{\bar{s}}\overline{X}_3 & \overline{A}_3^T \end{pmatrix} = \begin{pmatrix} 0 & \bar{s}\overline{X}_3 \\ sX_3 & A_3 \end{pmatrix} = A,$$

and so **A** is equal to its transpose conjugate, i.e.

$$(2.35) \quad \left(\overline{A}\right)^T = A.$$

The exact conjugate relation between s and $\bar{s}$ is left undefined for now, with both s and $\bar{s}$ assumed unrelated. Invariably, however, as in URM3 under Pythagoras conditions, the relationship becomes $\bar{s} = -s$ (7.4).

Given that s and $\bar{s}$ are arbitrary, integer parameters, which have no effect on the dynamical equations for eigenvector **X** (2.32), then they will be retained in the workings herein. Because they play a role as variational parameters they do, however, affect all eigenvectors not yet determined, other than **X**.

(2-3b) Conjugate Coordinates

Returning to the general subject of embedding, the most important facet of this URM3 extension work, so far, is the natural incorporation of conjugate coordinates $\bar{x}, \bar{y}, \bar{z}$ into URM3 without any pre-conditions on their form (barring they satisfy orthogonality (2.20)) and, equally importantly, this puts them on an equal basis with the standard coordinates.

The coordinate equation (2.21) is now written as a simple contraction of a vector **X** (2.4) and its conjugate $\overline{X}$ (2.12), as in the orthogonality condition (2.20), to give a conserved, scalar quantity of zero. Whereas before, in [1], the coordinate equation was expressed solely in terms of the standard coordinates, and it was only towards the end of [1], in paper 5, that conjugate coordinates were introduced, primarily as a neat way to write the Pythagoras equation.

In fact, it is considered in general URMT that the only correct way to write a conservation equation in the coordinates is, indeed, as the sum of quadratic terms comprising one standard and one conjugate coordinate, e.g. (2.20), and not to write it in terms of only one set x, y, z or the conjugate set $\bar{x}, \bar{y}, \bar{z}$ as in, say, the coordinate equation (2.21). In [1], this was only truly done from the outset in the URM3 DCE (A7), which was correctly written in terms of both standard dynamical variables P, Q, R and their conjugates $\bar{P}, \bar{Q}, \bar{R}$ as in

(A7) $\quad C^2 = P\bar{P} + Q\bar{Q} + R\bar{R} + (PQR + \bar{P}\bar{Q}\bar{R})/C$, URM3.

There is a beauty in the symmetry of both the coordinate equation (2.20) and the DCE (A7) in that they are symmetric upon interchange of the standard forms of variables with their conjugate forms. Furthermore, they are also symmetric to a permutation in any of two their three variables, either x, y, z in (2.20), or P, Q, R in (A7), i.e. they retain the same form when swapping, say, x with y or P with Q, which is known as a permutation symmetry [5]. This also applies to a permutation in the conjugate forms $\bar{x}, \bar{y}, \bar{z}$ in (2.20) or $\bar{P}, \bar{Q}, \bar{R}$ in (A7). Such symmetries in URMT's conservation equations are highly desirable from a physical perspective since it means there are no preferred coordinates amongst x, y, z or dynamical variables P, Q, R. It also means the formulation can be done in either the standard or conjugate forms.

All the work so far may seem a cheat in that both $\mathbf{X}$ (2.3) and $\overline{\mathbf{X}}$ (2.10) are only actually three non-zero elements x, y, z, with the fourth element w equal to zero, and therefore nothing fundamentally new has been achieved. However, it is shown in Section (3) that the complete eigenvector solution, with non-trivial, four-element vectors (I10), can be generated using the 'residual matrix method' (Section (3)), on the additional assumption of orthogonality (2.9) between $\mathbf{X}$ and $\overline{\mathbf{X}}$. Before this, however, the eigenvalues are examined, and it is shown that a new, zero eigenvalue naturally arises in URM4 as a consequence of the aforementioned orthogonality.

(2-4) Eigenvalues

Returning to the original characteristic equation (1.9), reproduced below,

(1.9) $\quad \lambda^4 + a_2\lambda^2 + a_1\lambda + a_0 = 0$, then the constant coefficients a_2, a_1, a_0 (1.10) are now redefined using the coordinates x, y, z in place of $\bar{S}, \bar{T}, \bar{U}$, and $\bar{x}, \bar{y}, \bar{z}$ in place of S, T, U, to give

(2.40a) $a_2 = -(P\bar{P} + Q\bar{Q} + R\bar{R}) + \bar{x}x + \bar{y}y + \bar{z}z$

$$a_1 = -(PQR + \bar{P}\bar{Q}\bar{R})$$

(2.40b)
$$\begin{aligned} &-(R\bar{x}y + \bar{R}x\bar{y})\\ &-(Qx\bar{z} + \bar{Q}\bar{x}z)\\ &-(\bar{P}y\bar{z} + P\bar{y}z) \end{aligned}$$

$$a_0 = (\bar{x}xP\bar{P} + \bar{y}yQ\bar{Q} + \bar{z}zR\bar{R})$$

(2.40c)
$$\begin{aligned} &-(PR\bar{x}z + \bar{P}\bar{R}x\bar{z})\\ &-(QR\bar{y}\bar{z} + \bar{Q}\bar{R}\bar{y}z)\\ &-(\bar{P}\bar{Q}\bar{x}y + PQx\bar{y}) \end{aligned}$$

Using the URM3 kinetic term K_3 (A8), and the orthogonality condition (2.20), then the a_2 coefficient (2.40a) simplifies nicely to the negative of the URM3 kinetic energy, i.e.

(1.16) $a_2 = -K = -K_3$.

The first term in coefficient a_1 is the URM3 Potential V_3 (A9) multiplied by the eigenvalue C so that

(2.41)
$$\begin{aligned} a_1 &= -V_3 C\\ &-(R\bar{x}y + \bar{R}x\bar{y})\\ &-(Qx\bar{z} + \bar{Q}\bar{x}z)\\ &-(\bar{P}y\bar{z} + P\bar{y}z) \end{aligned}$$

This expression for a_1 can be simplified using the URM3 dynamical equations (A5), and multiplying (A5a) by $\bar{x}$, (A5b) by $\bar{y}$, and (A5c) by $\bar{z}$ to give

(2.42)
$$\begin{aligned} &Cx\bar{x} + Cy\bar{y} + Cz\bar{z} =\\ &(R\bar{x}y + \bar{Q}\bar{x}z) + (\bar{R}x\bar{y} + P\bar{y}z) + (Qx\bar{z} + \bar{P}y\bar{z}) \end{aligned}.$$

Using the orthogonality relation (2.20) implies the left of (2.42) is zero so that

(2.43) $(R\bar{x}y + \bar{Q}\bar{x}z) + (\bar{R}x\bar{y} + P\bar{y}z) + (Qx\bar{z} + \bar{P}y\bar{z}) = 0$.

Substituting this result into (2.41) reduces a_1 to just the Potential term

(2.44) $a_1 = -V_3 C$.

Continuing with the a_0 coefficient (2.40c), it is now shown that this coefficient a_0 is actually zero if the original URM3 DCE (A7) is retained.

(2-5) A Zero Eigenvalue

Substituting for a_2 and a_1 in terms of the kinetic and Potential energies, K (1.16) and V_3 (2.44) respectively, into the characteristic equation gives

(2.50) $\lambda^4 - K\lambda^2 - V_3 C\lambda + a_0 = 0$,

and since $\lambda = C$ is an eigenvalue, then this becomes

(2.51) $C^4 - KC^2 - V_3 C^2 + a_0 = 0$.

Dividing throughout by C^2, and using the URM3 DCE $K + V_3 = C^2$ (A12), implies that

(2.52) $a_0 = 0$.

With $a_0 = 0$, the characteristic equation (2.50) now becomes

(2.53) $0 = \lambda^4 - \lambda^2 K - \lambda V_3 C$.

This immediately factors giving an additional eigenvalue $\lambda = 0$ to go with the pre-defined eigenvalue $\lambda = C$:

(2.54) $0 = \lambda(\lambda^3 - K\lambda - V_3 C)$.

In other words, by retaining URM3 compatible dynamical equations (A5) and the DCE (A12), coupled with a new orthogonality condition (2.20), this forces URM4 to have an additional, zero eigenvalue and subsequent eigenvector. Remember, in URM3, a zero eigenvector (I18) only materialises when under URM3 Pythagoras conditions. Under

30

general URM3, with no Pythagoras conditions, there is no eigenvalue $\lambda = 0$ and no subsequent eigenvector; the only eigenvector of note in general URM3 is **X** for the $\lambda = C$ factor.

Notes

The Pythagoras conditions referred to so far are those of URM3 (A18), and not URM4 Pythagoras conditions, which have not yet been encountered, see Section (4); URM3 conditions are actually a subset of URM4 conditions.

Unlike the general URM4 discussion in the earlier section (1-5), there is no constraint $a_1 = 0$ (1.20c). Instead, enforcing URM3 compatibility via the DCE (A12) and dynamical equations (A5), and adding an orthogonality condition (2.20), makes $a_0 = 0$ and gives a zero eigenvalue.

(2-6) The other two eigenvalues

With two eigenvalues, $\lambda = 0$ and $\lambda = C$, found using the URM3 embedding scheme, the other two of the four possible eigenvalues can be obtained by factoring (2.53) with $\lambda(\lambda - C)$, and substituting for K in terms of V_3 from (A12), i.e. $K = C^2 - V_3$, to give

(2.60) $0 = \lambda(\lambda - C)(\lambda^2 + C\lambda + V_3)$.

This could not have been better since the remaining quadratic factor is identical to that remaining in the URM3 characteristic equation (a cubic), when the linear factor $(\lambda - C)$ is removed from the URM3 cubic equation, i.e.

(2.61) $0 = (\lambda^2 + C\lambda + V_3)$.

The two eigenvalue solutions to (2.61) are thus already known from URM3 work, as detailed in [1],#6 in the section entitled 'What are the other two eigenvalues?'. Given that the quadratic (2.61) contains the Potential V_3, these two eigenvalues will vary accordingly with V_3, whereas the first two eigenvalues, $\lambda = C$ and $\lambda = 0$, are constant and independent of V_3.

The URM3 solution to (2.61) in [1],#6 is repeated here as follows: firstly, if the URM3 Potential V_3 is zero, as it is under URM3 Pythagoras conditions, then (2.61) simply factors as

(2.62) $0 = \lambda(\lambda + C)$, $V_3 = 0$,

and the two eigenvalue solutions are $\lambda = 0$ and $\lambda = -C$.

Thus, given the first two eigenvalues are $\lambda = 0$ and $\lambda = +C$, then all four eigenvalues are as follows, termed the 'URM4 Pythagorean eigenvalues' (I14)

(2.63) $\lambda = 0,0$ and $\lambda = \pm C$ when $V_3 = 0$.

They are given a special name because they occur repeatedly throughout the book and are the main case of interest, especially in Part II.

When not under URM3 Pythagorean conditions, with a non-zero URM3 Potential, i.e. $V_3 \neq 0$, then the quadratic factor (2.61) is solved using the standard quadratic solution to give the two roots

(2.64) $\lambda = \dfrac{-C \pm \sqrt{C^2 - 4V_3}}{2}$.

Because V_3 is dependent on C (in the denominator of (A9)), its value can always be made negative by appropriate choice for the sign of C. The Potential can also be varied by a variational transformation in the dynamical variables, see Section (8), whilst leaving the coordinate solution in x, y, z invariant. A chief reason to do this is to make the discriminant, $C^2 - 4V_3$ in (2.64), a perfect square. This can be done if the Potential V_3 is parameterised as follows, for arbitrary integer S.

(2.65) $V_3 = -S(S + C)$, $S \in \mathbb{Z}$.

Note that this integer S is not the same as the dynamical variable S (1.3b), but the symbol is retained here since it is used in [1] where there is no ambiguity.

Substituting for V_3 from (2.65) into (2.64) gives the other two eigenvalues, denoted by λ_3 and λ_4, and parameterised in terms of S as

(2.66a) $\lambda_3 = -(C + S)$

(2.66b) $\lambda_4 = S$.

Using these two eigenvalues λ_3, λ_4, plus the first two eigenvalues $\lambda_1 = C$ and $\lambda_2 = 0$, it is confirmed that the sum of all four eigenvalues, i.e. the trace (1.2) of matrix $\mathbf{A}$, is zero as in

(2.67) $\lambda_1 + \lambda_2 + \lambda_3 + \lambda_4 = Tr(A) = C - (C + S) + S = 0$.

So, for a non-zero Potential, $V_3 \neq 0$, the four URM4 eigenvalues are, under URM3 embedding conditions (1.20a),

(2.68) $\lambda_1 = C , \lambda_2 = 0, \lambda_3 = -(C + S)$ and $\lambda_4 = +S$ when $V_3 \neq 0$.

Unlike URM3, URM4 also considers complex, integer eigenvalues. Suppose the eigenvalue C is even, i.e.

(2.69) $C = 2p$, for some $p \in \mathbb{Z}$,

then the discriminant $C^2 - 4V_3$ becomes

(2.70) $C^2 - 4V_3 = 4(p^2 - V_3)$,

and the quadratic solution now reads

(2.71) $\lambda = -p \pm \sqrt{p^2 - V_3}$.

By defining the potential V_3 as the sum of two squares in p and a new integer q , as follows:

(2.72) $V_3 = p^2 + q^2$, $q \in \mathbb{Z}$,

then the discriminant (2.71) is a perfect, negative square q^2 , and two complex eigenvalue solutions are then obtained

(2.73) $\lambda_{3,4} = -p \pm iq$.

33

Unsurprisingly, if the two roots are complex, then they are the conjugate of each other.

Given that the trace of **A** is zero (1.2), then the sum of the eigenvalues is zero and so, as per (2.67),

(2.74a) $0 + C + \lambda_3 + \lambda_4 = 0$

i.e.

(2.74b) $\lambda_3 + \lambda_4 = -C$.

In which case, by (2.73), the eigenvalues are of the form,

(2.75) $\lambda_{3,4} = -\dfrac{C}{2} \pm iq$, where $p = -\dfrac{C}{2}$ by (2.69).

If $C = -1$ then the eigenvalues are

(2.76) $\lambda_{3,4} = \dfrac{1}{2} \pm iq$, $p = \dfrac{1}{2}$,

where, using (2.72), the imaginary part q is given by

(2.77) $q = \sqrt{V_3 - p^2}$

Given that the real part of the eigenvalues $\lambda_{3,4}$ (2.76) is always $\frac{1}{2}$ when $C = -1$, then this eigenvalue solution brings URMT into the realm of the Riemann Hypothesis. Of course, constant C is defined as $+1$ (1.7), not -1. However, a value of $-C$ is legitimate, e.g. under URM3 Pythagoras conditions (A18), so it is plausible.

Given that the matrix **A** has Hermitian-like properties (2.35), and is thus considered a Hermitian operator, the reader may be interested to know that there is a conjecture, namely the Hilbert-Polya conjecture, which links the Riemann Hypothesis with Hermitian Operators. Unfortunately, this is well beyond the scope of this book, but a search of the web is recommended for further study.

Notes

Because the characteristic equation $0 = \lambda^4 - \lambda^2 K - \lambda V_3 C$ (2.53) has no constant term, i.e. it is zero, the product of the eigenvalues is also zero, which is confirmed since one of the eigenvalues, λ_2, is zero, i.e.

(2.78) $\lambda_1 \lambda_2 \lambda_3 \lambda_4 = 0$.

The product of just the two other eigenvalues λ_3 and λ_4 is

(2.79) $\lambda_3 \lambda_4 = S(C + S) = -V_3$,

and, as the Potential V_3 in (2.79) is thus a quadratic function of parameter S, there are always two values of S for a specific value of V_3, which could be real or complex.

There is, however, a cautionary note to all the above analysis using parameter S, as first noted in [1],#6: "*Given the discriminant can always be made a perfect square for arbitrary, integer parameter S, and hence the Potential and eigenvalues are also always integer, then it may seem that it is very easy to always construct three integer eigenvalues. This is, however, largely illusory because the Potential is constructed from six dynamical variables, all of which must satisfy their unity root definitions, and it is not easy to work backward from a known value for the Potential V_3 to get integer unity roots.*"

(2-7) A Summary of Embedding

With the goal of retaining as many of the URM3 properties as possible, the URM3, three-element eigenvector $\mathbf{X}_3$ (A3) is embedded into the URM4, four-element eigenvector $\mathbf{X}$ (2.3), with a first element w of zero. This vector $\mathbf{X}$ is defined to be an eigenvector of matrix $\mathbf{A}$ (2.19), which satisfies both the URM3 dynamical equations $\mathbf{A}_3 \mathbf{X}_3 = C \mathbf{X}_3$ (A4), for eigenvalue $\lambda = C$, and the equivalent URM4 equations $\mathbf{A}\mathbf{X} = C\mathbf{X}$ (1.7).

By enforcing the URM3 dynamical equations (A5) and the DCE (A7), an orthogonal vector $\overline{\mathbf{X}}$ (2.10) is defined that satisfies the orthogonality condition $\overline{\mathbf{X}} \cdot \mathbf{X} = 0$ (2.9). Given the definition of $\mathbf{X}$ (2.3), then $\overline{\mathbf{X}}$ is defined to be an embedding of any URM3 conjugate vector $\overline{\mathbf{X}}_3$, orthogonal to $\mathbf{X}_3$ (A3), such that $\overline{\mathbf{X}}_3 \cdot \mathbf{X}_3 = 0$ (2.20). To satisfy all these conditions and definitions, the top row $(0, S, T, U)$ of matrix $\mathbf{A}$ (2.19) becomes the conjugate vector $\overline{\mathbf{X}}$ (2.10), and forces the new URM4 dynamical variables S, T, U to be defined in terms of the URM3 conjugate coordinates $\bar{x}, \bar{y}, \bar{z}$ as vector $\overline{\mathbf{X}}_3$ (2.11). In turn, to satisfy Hermitian property (2.35) for matrix $\mathbf{A}$, its first column is defined to be the conjugate (which implicitly involves a transposition) of the top row $\overline{\mathbf{X}}$, i.e. $\overline{\overline{\mathbf{X}}}$, which is just $\mathbf{X}$ (2.3) given $\overline{\overline{\mathbf{X}}} = \mathbf{X}$, in keeping with URM3's rules that the conjugate of a conjugated variable returns the original, unconjugated variable, e.g. (A37c).

All these considerations lead to definition (2.15) for matrix $\mathbf{A}$. However, on physical grounds, this is not dimensionally satisfactory and two scalar, conversion factors s and $\bar{s}$ are added to $\mathbf{A}$ (2.25), defined to make all elements of $\mathbf{A}$, i.e. the dynamical variables, have the same, common, physical units (2.26), i.e. those of velocity under the standard physical interpretation, Appendix (J). These two new scalars, s and $\bar{s}$, have no effect on the dynamical equations (2.32), i.e. the equations are invariant to them. Consequently, s and $\bar{s}$ act as variational parameters in URMT, and serve as arbitrary parameters in the complete eigenvector solution, as will be developed in subsequent sections.

Adhering to all the above, which amounts to retaining the URM3 dynamical equations (A5), the DCE (A7), plus orthogonality (2.9), shows that there must exist a second eigenvalue λ_0, in addition to $\lambda = C$, which is identically zero, with an associated eigenvector $\mathbf{X}_0$, where $\mathbf{A}\mathbf{X}_0 = 0$ by definition.

Having ascertained that there are now two eigenvalues, $\lambda = C$ and $\lambda = 0$, with eigenvectors $\mathbf{X}$ and $\mathbf{X}_0$ respectively, their associated reciprocal row eigenvectors, $\mathbf{X}^+$ and $\mathbf{X}^0$ (Appendix (E)), are now determined in the next section using the 'Residual Matrix Method'.

(3) The Residual Matrix Method

The Residual Matrix Method, as first outlined in [1], is a favoured method in URMT to determine eigenvectors, as opposed to the more usual direct method of algebraic manipulation. Its key advantage is that it gives both an eigenvector and its reciprocal, avoiding any rational expressions in the process, and thereby keeping all URMT results in integers without having to rescale.

Note that the term 'residual' is unique to URMT, and coined only for want of a name for the process, given no specific name can be found in the literature. The method itself is not unique to URMT and the actual theory is mentioned in a few texts, e.g. [4], and is related to 'purification', i.e. removal of factors from the Cayley-Hamilton polynomial (see below). It is found discussed in the context of numerical evaluation of eigenvalues, where the numerical process is termed 'Richardson's Purification Process' - references: 1) Numerical Analysis, D.R. Hartree, 2nd Edition, Oxford at the Clarendon Press; 2) Phil. Trans. Royal Society 242 (1950), 439. However, the residual method used herein is not a numerical method, and only mentioned because of the seeming scarcity of references to this method in the literature.

(3-1) Solving for the Eigenvectors

The residual method derives from the Cayley-Hamilton theorem (or 'Hamilton-Cayley' in [5]), which basically says that a matrix satisfies its own characteristic equation. Given the characteristic equation for a URM3 embedded solution factors as in (2.60), reproduced below for eigenvalues $\lambda = C$ and $\lambda = 0$

(2.60) $0 = \lambda(\lambda - C)(\lambda^2 + C\lambda + V_3)$, then by the Cayley-Hamilton theorem, the characteristic equation (2.6) can be re-written in terms of matrix $\mathbf{A}$ (2.19) as follows, where $\mathbf{I}$ is the 4x4 identity matrix

(3.1) $0 = \mathbf{A}(\mathbf{A} - C\mathbf{I})(\mathbf{A}^2 + C\mathbf{A} + V_3\mathbf{I})$.

Note that the URM3 Potential V_3 (A9) is used here as part of the embedding solution; see cases (1.20a) and (1.20b) in Section (1) and, more generally, Section (2).

According to the residual matrix method, the residual matrices $\mathbf{E}$ ($\lambda = C$) and $\mathbf{E}_0$ ($\lambda = 0$) are given by the following matrix polynomials, a short explanation follows:

(3.2a) $\mathbf{E} = \mathbf{A}(\mathbf{A}^2 + C\mathbf{A} + V_3\mathbf{I})$

(3.2b) $\mathbf{E}_0 = (\mathbf{A} - C\mathbf{I})(\mathbf{A}^2 + C\mathbf{A} + V_3\mathbf{I})$.

(3.5) As a reminder of their derivation in [1],#2, the residual matrix for eigenvalue λ is formed from the matrix polynomial that is left after factorisation of the Cayley-Hamilton polynomial (3.1) by the factor $\mathbf{A} - \lambda\mathbf{I}$. Hence $\mathbf{E}$, for eigenvalue $\lambda = C$, is the matrix polynomial factored by $(\mathbf{A} - C\mathbf{I})$, giving (3.2a). Likewise $\mathbf{E}_0$, for eigenvalue $\lambda = 0$, is the same matrix polynomial factored by $\mathbf{A}$, giving (3.2b).

The residual matrix $\mathbf{E}$ is also equivalent to the outer (or dyadic (I3)) product of an column eigenvector with a row eigenvector, i.e. for the eigenvalue $\lambda = C$ it is the outer product of the eigenvector $\mathbf{X}$ (1.5) and its reciprocal eigenvector $\mathbf{X}^+$, which is defined by $\mathbf{X}^+\mathbf{A} = C\mathbf{X}$, cf. the definition of $\mathbf{X}$ by $\mathbf{A}\mathbf{X} = C\mathbf{X}$ (1.7), i.e.

(3.3) $\mathbf{E} = \mathbf{X}\mathbf{X}^+$.

Although both $\mathbf{X}$ and $\mathbf{X}^+$ are usually unknowns, in URMT the vector $\mathbf{X}$ is, invariably, already known, either by algebraic evaluation or, in the case of embedding, predefined as the URM3 equivalent $\mathbf{X}_3$ vector (2.4). Furthermore, under URM3 Pythagoras conditions, it can be any embedded, arbitrary Pythagorean triple. On the other hand, the reciprocal row eigenvector $\mathbf{X}^+$ is not generally known. However, since $\mathbf{E}$ can be calculated by (3.2a), and the vector $\mathbf{X}$ is pre-defined, e.g. (2.4), then $\mathbf{X}^+$ can be found from $\mathbf{E}$ and $\mathbf{X}$, using (3.3).

As an aside note, the residual matrix $\mathbf{E}$, as defined by the outer-product (3.3), is actually a form of projection matrix or projection operator, which is usually described in the linear algebra literature under 'Spectral Theory' or 'Spectral Decomposition', see [5], for example.

Firstly, to determine $\mathbf{E}$ using $\mathbf{A}$ (2.25), then the CA term in (3.2a) is simply

$$(3.4)\ CA = \begin{pmatrix} 0 & C\overline{s}\overline{\mathbf{X}}_3 \\ Cs\mathbf{X}_3 & CA_3 \end{pmatrix},$$

The 4x4 identity matrix $\mathbf{I}$, scaled by the URM3 Potential V_3, is written in block form as

$$(3.5)\ V_3\mathbf{I} = \begin{pmatrix} V_3 & 0 \\ 0 & V_3\mathbf{I}_3 \end{pmatrix},$$

and $\mathbf{A}^2$ evaluates to

$$(3.6)\ \mathbf{A}^2 = \begin{pmatrix} \overline{s}\overline{\mathbf{X}}_3\mathbf{X}_3 & \overline{s}\overline{\mathbf{X}}_3\mathbf{A}_3 \\ s\mathbf{A}_3\mathbf{X}_3 & s\overline{s}\mathbf{X}_3\overline{\mathbf{X}}_3 + \mathbf{A}_3^2 \end{pmatrix}.$$

Using the embedding assumption $\overline{\mathbf{X}}_3\mathbf{X}_3 = 0$ (2.14), and the URM3 dynamical equations $\mathbf{A}_3\mathbf{X}_3 = C\mathbf{X}_3$ (A4), then $\mathbf{A}^2$ simplifies to

$$(3.7)\ \mathbf{A}^2 = \begin{pmatrix} 0 & \overline{s}\overline{\mathbf{X}}_3\mathbf{A}_3 \\ sC\mathbf{X}_3 & s\overline{s}\mathbf{X}_3\overline{\mathbf{X}}_3 + \mathbf{A}_3^2 \end{pmatrix}.$$

Thus, the quadratic, matrix polynomial $(\mathbf{A}^2 + CA + V_3\mathbf{I})$ in $\mathbf{E}$ (3.2a) evaluates to

$$(3.8)\ \mathbf{A}^2 + CA + V_3\mathbf{I} = \begin{pmatrix} V_3 & \overline{s}\overline{\mathbf{X}}_3\mathbf{A}_3 + \overline{s}C\overline{\mathbf{X}}_3 \\ 2sC\mathbf{X}_3 & s\overline{s}\mathbf{X}_3\overline{\mathbf{X}}_3 + \mathbf{X}_3\mathbf{X}^{3+} \end{pmatrix},$$

where the term $\mathbf{X}_3\mathbf{X}^{3+}$ in the bottom right is the URM3 equivalent residual matrix $\mathbf{E}_3 = \mathbf{X}_3\mathbf{X}^{3+}$, i.e.

$$(3.9)\ \mathbf{E}_3 = \mathbf{X}_3\mathbf{X}^{3+} = \mathbf{A}_3^2 + CA_3 + V_3\mathbf{I}_3, \text{ see [1],\#2 for } C = 1,$$

and the URM3 row eigenvector $\mathbf{X}^{3+}$ is the vector of scale-factors α_3, β_3, and γ_3, as in $\mathbf{X}^{3+} = \begin{pmatrix} \alpha_3 & \beta_3 & \gamma_3 \end{pmatrix}$, (A13), where $\mathbf{X}^{3+} \sim \mathbf{X}^3$.

(3-2) The Quadratic Residual Polynomial

Before proceeding further, it is worth noting in advance that, under certain conditions discussed further below, the quadratic expression in $\mathbf{A}$ (3.8) is sufficient to derive the URM4 eigenvectors rather than having to calculate the full cubic expression (3.2a). The details are as follows:

If the URM3 Potential V_3 is zero, (URM3 Pythagoras conditions (A24)), then the characteristic equation (2.60) factors with eigenvalues $\lambda = \pm C, 0, 0$

(3.20) $0 = \lambda^2 (\lambda - C)(\lambda + C)$.

In this case, there is a URM3 row eigenvector $\mathbf{X}^{3-}$ (A35c), for eigenvalue $\lambda = -C$, such that $\mathbf{X}^{3-}\mathbf{A}_3 = -C\mathbf{X}^{3-}$ by definition (A36c). Most importantly, by the rules of matrix algebra (F25), this eigenvector $\mathbf{X}^{3-}$ is orthogonal to $\mathbf{X}_3$, and thus serves as a vector $\overline{\mathbf{X}}_3$ in (2.12) to satisfy orthogonality (2.9), i.e.

(3.21)

(3.21a) $\overline{\mathbf{X}}_3 = \mathbf{X}^{3-}$

and

(3.21b) $\mathbf{X}^{3-}\mathbf{X}_3 = 0$, (F1) and (F25).

hence

(3.21c) $\overline{\mathbf{X}}_3\mathbf{X}_3 = 0$.

Note that it is not enough to just have a zero URM3 Potential $V_3 = 0$, but that the conjugate vector $\overline{\mathbf{X}}_3$ must also be the URM3 eigenvector $\mathbf{X}^{3-}$ to matrix $\mathbf{A}_3$, and satisfying $\mathbf{X}^{3-}\mathbf{A}_3 = -C\mathbf{X}^{3-}$ (A36c).

Substituting $\mathbf{X}^{3-}$ for $\overline{\mathbf{X}}_3$ in the top right term $\overline{s}\mathbf{X}_3\mathbf{A}_3 + \overline{s}C\mathbf{X}_3$ of (3.8), and using the eigenvector definition for $\mathbf{X}^{3-}$ (A36c), makes this top-right term completely zero. Furthermore, with V_3 also zero by assumption, the entire quadratic term (3.8), reduces to the following matrix, with a top row of all zeros,

$$(3.22) \quad \mathbf{A}^2 + CA = \begin{pmatrix} 0 & 0 \\ 2sC\mathbf{X}_3 & s\bar{s}\mathbf{X}_3\mathbf{X}^{3-} + \mathbf{X}_3\mathbf{X}^{3+} \end{pmatrix}, \quad V_3 = 0.$$

This is significant because it is already in a form which can be expressed as the outer product of two vectors, $\mathbf{X}$ and $\mathbf{X}^+$, i.e.

$$(3.23) \quad \mathbf{X}\mathbf{X}^+ = \mathbf{A}^2 + CA, \quad V_3 = 0.$$

The first vector $\mathbf{X}$ has, of course, already been defined by (2.4). By comparing (3.20) and (3.23), the vector $\mathbf{X}^+$ is thus evaluated as the following 1x4 row vector, where $\mathbf{X}^{3-}$ (A35c) and $\mathbf{X}^{3+}$ (A13) are both 1x3 row vectors, with the first element of $\mathbf{X}^+$ being the scalar $2sC$

$$(3.24) \quad \mathbf{X}^+ = \begin{pmatrix} 2sC & s\bar{s}\mathbf{X}^{3-} + \mathbf{X}^{3+} \end{pmatrix}.$$

In outer product form, $\mathbf{X}\mathbf{X}^+$ is thus

$$(3.25) \quad \mathbf{X}\mathbf{X}^+ = \begin{pmatrix} 0 \\ \mathbf{X}_3 \end{pmatrix} \begin{pmatrix} 2sC & s\bar{s}\mathbf{X}^{3-} + \mathbf{X}^{3+} \end{pmatrix}.$$

(3.26a) Note that under this quadratic scheme, using the standard physical interpretation, Appendix (J), the product $\mathbf{X}\mathbf{X}^+$ has physical units of velocity squared, i.e. kinetic energy per unit mass. Using the cubic polynomial (discussed shortly), it would be a cubic function of the velocity, and no longer, therefore, physically represent a quadratic energy term. This is important in the physical development of URMT, as will be seen more and more as the book proceeds. The physical units of the terms are as follows:

(J3) $units(\mathbf{X}, \mathbf{X}_3, \mathbf{X}^{3-}) = $ acceleration $= LT^{-2}$

(J4) $units(C) = $ velocity $= LT^{-1}$

(J5) $units(\overline{\mathbf{X}}, \bar{s}\overline{\mathbf{X}}_3) = $ velocity $= LT^{-1}$

(J6) $units(\mathbf{A}_3) = $ velocity $= LT^{-1}$

(J10a) $units(s, \bar{s}) = $ time $= T$ when $\overline{\mathbf{X}}_3 = \mathbf{X}^{3-}$ (3.21a)

(J11) $units(\mathbf{X}^+, \mathbf{X}^{3+}) = length = L$

(J12) $units(\mathbf{XX}^+) = velocity\ squared = L^2 T^{-2}$, for (3.23)

(3.26b) $units(2sC) = units(s\bar{s}\mathbf{X}^{3-}) = units(\mathbf{X}^{3+}) = length = L$.

By (3.23), multiplying the outer product form $\mathbf{XX}^+$ by $\mathbf{A}$ is equivalent to multiplying the quadratic polynomial (3.22) by $\mathbf{A}$ to give the cubic polynomial for $\mathbf{E}$ (3.2a). Doing so, and using definition $\mathbf{AX} = C\mathbf{X}$, gives

(3.27) $\mathbf{A}(\mathbf{XX}^+) = (\mathbf{AX})\mathbf{X}^+ = (C\mathbf{X})\mathbf{X}^+ = C(\mathbf{XX}^+)$.

It is seen that the product $\mathbf{A}(\mathbf{XX}^+)$ is simply the outer product $\mathbf{XX}^+$ multiplied by the eigenvalue C, that is the matrix $\mathbf{E} = \mathbf{XX}^+$ is an 'eigenmatrix' of $\mathbf{A}$. This should not be surprising since the matrix $\mathbf{E}$ comprises four, 4x1 eigenvectors $\mathbf{X}$, by the definition of the outer product $\mathbf{XX}^+$. However, the main point is that this multiplication by $\mathbf{A}$ is the same as evaluating $\mathbf{E}$ using the cubic polynomial expression in (3.2a), except it produces no new eigenvector that wasn't already present in the quadratic polynomial (3.22), and the extra multiplication by $\mathbf{A}$ simply scales the eigenvector $\mathbf{X}$ by eigenvalue C. Hence the quadratic expression (3.22) is sufficient, under a zero, URM3 Potential $V_3 = 0$, and for eigenvector $\mathbf{X}^{3-}$ (A35c), to calculate the URM4 eigenvectors under the embedding scheme $\overline{\mathbf{X}}_3 = \mathbf{X}^{3-}$ (3.21a).

Notes

That vector $\mathbf{X}^+$ (3.24) is a row eigenvector satisfying its defining equation

(3.28) $\mathbf{X}^+\mathbf{A} = C\mathbf{X}^+$,

can be verified as follows, by substituting $\mathbf{X}^{3-}$ for $\overline{\mathbf{X}}_3$ in $\mathbf{A}$ (2.25) and multiplying out $\mathbf{X}^+\mathbf{A}$ to give

(3.29) $\mathbf{X}^+\mathbf{A} = \left((s\bar{s}\mathbf{X}^{3-} + \mathbf{X}^{3+})s\mathbf{X}_3 \quad 2s\bar{s}C\mathbf{X}^{3-} + (s\bar{s}\mathbf{X}^{3-} + \mathbf{X}^{3+})\mathbf{A}_3\right)$.

Using the following URM3 relations:

(3.30a) $\mathbf{X}^{3-}\mathbf{X}_3 = 0$ by orthogonality (3.21b)

(3.30b) $\mathbf{X}^{3+}\mathbf{X}_3 = 2C^2$ by the URM3 Potential Equation (F4)

(3.30c) $\mathbf{X}^{3-}\mathbf{A}_3 = -C\mathbf{X}^{3-}$ by definition (A36c)

(3.30d) $\mathbf{X}^{3+}\mathbf{A}_3 = C\mathbf{X}^{3+}$ by definition (A14), ($\mathbf{X}^{3+} \sim \mathbf{X}^3$),

then the product $\mathbf{X}^+\mathbf{A}$ (3.29) simplifies to

(3.31) $\mathbf{X}^+\mathbf{A} = C\big((2sC \quad s\bar{s}\mathbf{X}^{3-} + \mathbf{X}^{3+}\big)$,

which, comparing with $\mathbf{X}^+$ (3.24), proves $\mathbf{X}^+\mathbf{A} = C\mathbf{X}^+$ (3.28).

Of course, the Potential V_3 is always zero under URM3 Pythagoras conditions, but it is stressed these URM3 conditions are not the same as URM4 Pythagoras conditions (4.1), merely a subset. The point of this remark being that, whilst in a URM3 world there will be the two eigenvectors $\mathbf{X}_3$ ($\sim \mathbf{X}_+$) and $\mathbf{X}^{3-}$, both of which are Pythagorean triples, this will not necessarily be the case in URM4, and the eigenvectors $\mathbf{X}$ and $\mathbf{X}^+$ are not Pythagorean quadruples unless the full URM4 Pythagoras conditions (4.1) are also satisfied, in which case the URM3 conditions will also be satisfied by default since they are a subset of the URM4 conditions. This point will be revisited again in Section (7) when discussing the subject of 'lifting' solutions under URM4 Pythagoras conditions. However, for now, attention returns to the full, cubic polynomial computation of $\mathbf{E}$ (3.2a).

(3-3) The Cubic Residual Polynomial

Multiplying the quadratic polynomial (3.8) once again by $\mathbf{A}$, and using orthogonality (2.9) and the URM3 dynamical equation (A4), then the full cubic expression for the URM4 residual matrix $\mathbf{E}$ (3.2a) becomes

(3.40) $\mathbf{E} = \begin{pmatrix} 0 & 0 \\ s(2C^2 + V_3)\mathbf{X}_3 & \mathbf{X}_3(s\bar{s}\overline{\mathbf{X}}_3\mathbf{A}_3 + 2s\bar{s}C\overline{\mathbf{X}}_3 + C\mathbf{X}^{3+}) \end{pmatrix}$.

Defining the 3-element, row eigenvector $\overline{\mathbf{Y}}$ as

(3.41) $\overline{\mathbf{Y}} = s\bar{s}\overline{\mathbf{X}}_3\mathbf{A}_3 + 2s\bar{s}C\overline{\mathbf{X}}_3 + C\mathbf{X}^{3+}$,

then the bottom right term is the outer product $\mathbf{X}_3\overline{\mathbf{Y}}$, and the residual matrix is more concisely written

(3.42) $\mathbf{E} = \begin{pmatrix} 0 & 0 \\ s(2C^2 + V_3)\mathbf{X}_3 & \mathbf{X}_3\overline{\mathbf{Y}} \end{pmatrix}$.

In this form, using $\overline{\mathbf{Y}}$ (3.41), it is evident that every column of the residual matrix $\mathbf{E}$ is simply a multiple of the eigenvector $\mathbf{X}$ (2.4), i.e. the outer product:

$$(3.43) \quad \mathbf{E} = \begin{pmatrix} 0 \\ \mathbf{X}_3 \end{pmatrix} \left(s(2C^2 + V_3) \quad \overline{\mathbf{Y}} \right).$$

The evaluation of each of the terms in $\overline{\mathbf{Y}}$, i.e. $\overline{\mathbf{X}}_3 \mathbf{A}_3$, $2C\overline{\mathbf{X}}_3$ and $C\mathbf{X}^{3+}$, is as follows:

The product $\overline{\mathbf{X}}_3 \mathbf{A}_3$ is new to URM4, and not in URM3. One of the advantages of the conjugate vector $\overline{\mathbf{X}}_3$ is that, by orthogonality (3.21c), two of its three conjugate elements are completely free parameters. The only constraint that all three elements have to satisfy, as a triple $\overline{\mathbf{X}}_3 = (\overline{x}, \overline{y}, \overline{z})$, is the orthogonality condition (2.20). One such choice is

$$(3.44) \quad \overline{x} = -C, \quad \overline{y} = R, \quad \overline{z} = \overline{Q},$$

which defines the vector $\overline{\mathbf{X}}_3$ as

$$(3.45) \quad \overline{\mathbf{X}}_3 = \begin{pmatrix} -C & R & \overline{Q} \end{pmatrix}.$$

With this choice of $\overline{\mathbf{X}}_3$, the inner product $\overline{\mathbf{X}}_3 \mathbf{X}_3$ is given by

$$(3.46) \quad \overline{\mathbf{X}}_3 \mathbf{X}_3 = -Cx + Ry - Qz = 0,$$

and evaluates to zero since the term on the right is the rearranged, URM3 dynamical equation $Cx = Ry + \overline{Q}z$ (A5a) for the standard coordinate x.

Thus, $\overline{\mathbf{X}}_3$ (3.45) satisfies the orthogonality condition (2.20) as required. Two other similar forms for $\overline{\mathbf{X}}_3$ are:

$$(3.47) \quad \overline{\mathbf{X}}_3 = \begin{pmatrix} \overline{R} & -C & P \end{pmatrix}$$

$$(3.48) \quad \overline{\mathbf{X}}_3 = \begin{pmatrix} Q & \overline{P} & -C \end{pmatrix}.$$

As for $\overline{\mathbf{X}}_3$ (3.45), the inner product $\overline{\mathbf{X}}_3 \mathbf{X}_3$ for (3.47) and (3.48) evaluates to zero using the URM3 dynamical equations (A5b) and (A5c) respectively.

It is important to keep in mind that the choice of $\overline{\mathbf{X}}_3$ will affect the resultant vector $\mathbf{X}^+$, as will be seen later. Since the only condition $\overline{\mathbf{X}}_3$ has to satisfy is that of orthogonality (3.21c), any $\overline{\mathbf{X}}_3$ lying in a plane with normal $\mathbf{X}_3$ will suffice. Since this is a 2D plane (in URM3), this means $\overline{\mathbf{X}}_3$ has two free parameters. The scaling coefficient $\bar{s}$ in $\mathbf{A}$ (2.25) is effectively absorbed into both of these parameters since it multiplies the entire vector $\overline{\mathbf{X}}_3$ in the top right of $\mathbf{A}$. However, there is also the third, free parameter s in the left column of $\mathbf{A}$. So, there are actually three, free parameters in this residual method presented here.

The particular choice of $\overline{\mathbf{X}}_3$, e.g. (3.45) above, introduces a dimensionality problem in that $\overline{\mathbf{X}}_3$ now has the units of the dynamical variables, whereas $\mathbf{X}_3$ has the units of coordinates. However, the arbitrary parameters s and $\bar{s}$ also play the role of conversion factors and are, consequently, not just dimensionless scalar factors but can have physical units. In this example, s has units of time T and $\bar{s}$ remains dimensionless. The goal of ascribing units to the scalars s and $\bar{s}$ is to make all elements of matrix $\mathbf{A}_3$ have consistent units, namely those of the URM3 unity root matrix $\mathbf{A}_3$, which is a velocity (J6) in the standard physical interpretation, Appendix (J); see also note (3.26a).

Keep in mind that the association of dynamical variables in $\overline{\mathbf{X}}_3$ (3.45) with velocity is only one such physical association, albeit a very useful one. There is also an alternative association, which is simply to reinterpret $\mathbf{X}$ (comprising 'coordinates' x, y, z) as having the same units as the dynamical variables, with no need for a conversion factor s. Since the standard physical interpretation is not the only interpretation, albeit very attractive, this alternative association is plausible. A consequence of this is that 'conversion' factors s and $\bar{s}$ are strictly no longer required, but given they are useful, arbitrary scaling (variational) parameters, with no effect on the results, then they are usefully retained as dimensionless scalars, i.e. with no physical units.

Lastly, another alternative solution is to use a form of vector $\overline{\mathbf{X}}_3$, which is of the same units as $\mathbf{X}$. Some such dimensionally consistent vectors, with the right orthogonal properties (3.21c), are

(3.49)

(3.49a) $\overline{\mathbf{X}}_3 = \begin{pmatrix} -y & +x & 0 \end{pmatrix}$

(3.49b) $\overline{\mathbf{X}}_3 = \begin{pmatrix} 0 & -z & +x \end{pmatrix}$

(3.49c) $\overline{\mathbf{X}}_3 = \begin{pmatrix} -z & 0 & +x \end{pmatrix}$

Indeed, the vector $\overline{\mathbf{X}}_3 = \mathbf{X}^{3-} = \begin{pmatrix} x & y & -z \end{pmatrix}$ (A35c) has already been introduced in Section (3-2), equation (3.21a). However, it is noted that, since (3.49) comprises acceleration coordinates x, y, z, they are still not of the same units as the matrix $\mathbf{A}_3$. Nevertheless, a simple scaling of both with s and $\overline{s}$, sharing common units of time (in this example), is sufficient to make every element of $\mathbf{A}$ have the same physical units, namely those of velocity. In the general, cubic polynomial case (which excludes the special quadratic case, Section (3-2)), the forms (3.49) do not give particularly neat solutions, and so they are not pursued further. The neatest form for $\overline{\mathbf{X}}_3$, in terms of the fewest terms in the elements of the solution for $\mathbf{X}^+$, seems to be obtained with those forms of $\overline{\mathbf{X}}_3$ that have dynamical variables as their elements, such as (3.45). However, it is the product $\overline{\mathbf{X}}_3\mathbf{A}_3$ that is desired, and another reason for choosing (3.45) for $\overline{\mathbf{X}}_3$ is that $\overline{\mathbf{X}}_3\mathbf{A}_3$ evaluates to some useful relations in the URM3 scale factors α_3, β_3 and γ_3 (I2), i.e. the elements of $\mathbf{X}^{3+}$ (A13).

Expanding $\overline{\mathbf{X}}_3\mathbf{A}_3$ in full, using (A1a) for $\mathbf{A}_3$ and (3.45) for $\overline{\mathbf{X}}_3$, gives

$$(3.50)\quad \overline{\mathbf{X}}_3\mathbf{A}_3 = \begin{pmatrix} -C & R & \overline{Q} \end{pmatrix}\begin{pmatrix} 0 & R & \overline{Q} \\ R & 0 & P \\ Q & \overline{P} & 0 \end{pmatrix},$$

which evaluates to

$$(3.51)\quad \overline{\mathbf{X}}_3\mathbf{A}_3 = \begin{pmatrix} R\overline{R} + Q\overline{Q}, & -CR + \overline{P}\overline{Q}, & -C\overline{Q} + RP \end{pmatrix}.$$

Using the DCE (A7) and the URM3 Potential V (A9), the first term $R\overline{R} + Q\overline{Q}$ in (3.51) is re-written as

(3.52a) $R\overline{R} + Q\overline{Q} = (C^2 - P\overline{P}) - V_3$,

and, using $(CR + \overline{PQ}) = \beta_3 x$ (A16c), the second term is re-written as

(3.52b) $- CR + \overline{PQ} = \beta_3 x - 2CR$.

Finally, using $(C\overline{Q} + RP) = \gamma_3 x$ (A16e), the third term is

(3.52c) $- C\overline{Q} + RP = \gamma_3 x - 2C\overline{Q}$.

Collating all three terms, then $\mathbf{X}_3\mathbf{A}_3$ (3.51) is now re-written in terms of the URM3 scale factors α_3, β_3 and γ_3 respectively as

(3.53) $\overline{\mathbf{X}}_3\mathbf{A}_3 = \left(\alpha_3 x - V_3, \quad \beta_3 x - 2CR, \quad \gamma_3 x - 2C\overline{Q}\right)$.

This completes the evaluation of the first term, $\overline{\mathbf{X}}_3\mathbf{A}_3$ in $\overline{\mathbf{Y}}$ (3.41).

The second term $2C\overline{\mathbf{X}}_3$ in $\overline{\mathbf{Y}}$ (3.41) is very simply given, using (3.45) for $\overline{\mathbf{X}}_3$, as

(3.54) $2C\overline{\mathbf{X}}_3 = \left(- 2C^2 \quad 2CR \quad 2C\overline{Q}\right)$,

and the last term $C\mathbf{X}^{3+}$ is also trivially given, using (A13) for $\mathbf{X}^{3+}$, as

(3.55) $C\mathbf{X}^{3+} = \left(C\alpha_3 \quad C\beta_3 \quad C\gamma_3\right)$.

The composite vector $\overline{\mathbf{Y}}$ (3.41) is now defined in terms of URM4 'scale factors' α, β and γ (I2).

URM4 Scale Factors

As analogs of the URM3 scale (or divisibility) factors α_3, β_3 and γ_3, the URM4 scale factors α, β and γ are defined in terms of their URM3 equivalents as follows:

(3.56)

(3.56a) $\alpha = \alpha_3(\overline{ss}x + C) - s\overline{s}(2C^2 + V_3)$

(3.56b) $\beta = \beta_3(\overline{ss}x + C)$

(3.56c) $\gamma = \gamma_3 (s\bar{s}x + C)$.

Using these definitions for α, β and γ, the row eigenvector $\overline{\mathbf{Y}}$, comprising $\overline{\mathbf{X}}_3 \mathbf{A}_3$ (3.53), $2C\overline{\mathbf{X}}_3$ (3.54) and $C\mathbf{X}^{3+}$ (3.55), is now the URM4 equivalent of URM3's vector $\mathbf{X}^{3+}$, and written as

(3.57) $\overline{\mathbf{Y}} = (\alpha \quad \beta \quad \gamma)$,

and $\mathbf{E}$ (3.43) is now written in outer product form as

(3.58) $\mathbf{E} = \begin{pmatrix} 0 \\ \mathbf{X}_3 \end{pmatrix} (s(2C^2 + V_3) \quad \alpha \quad \beta \quad \gamma)$.

Expanding out this product gives

(3.59) $\mathbf{E} = \left(s(2C^2 + V_3)\begin{pmatrix} 0 \\ \mathbf{X}_3 \end{pmatrix} \quad \alpha\begin{pmatrix} 0 \\ \mathbf{X}_3 \end{pmatrix} \quad \beta\begin{pmatrix} 0 \\ \mathbf{X}_3 \end{pmatrix} \quad \gamma\begin{pmatrix} 0 \\ \mathbf{X}_3 \end{pmatrix} \right)$.

With the URM4 residual matrix $\mathbf{E}$ written, once again, as the outer product of the four-vectors $\mathbf{X}$ (2.4) and $\mathbf{X}^+$ as in $\mathbf{E} = \mathbf{X}\mathbf{X}^+$ (3.3) then, finally, the four-vector $\mathbf{X}^+$ is thus

(3.60) $\mathbf{X}^+ = (s(2C^2 + V_3) \quad \alpha \quad \beta \quad \gamma)$.

For dimensional, consistency checks, the units are as follows, in accordance with appendix (J)

(J10a) $units(s) = \text{time} = T$

(J10b) $units(\bar{s}) = \text{none, when } \overline{\mathbf{X}}_3 = (-C \quad R \quad \overline{Q})$ (3.45)

(J3) $units(x, y, z) = units(\mathbf{X}, \mathbf{X}_3) = \text{acceleration} = LT^{-2}$

(J4) $units(C) = \text{velocity} = LT^{-1}$

(J7) $units(\mathbf{A}) = \text{velocity} = LT^{-1}$

(J14b) $units(V_3) = \text{energy per unit mass} = L^2T^{-2}$

(J15) $units(\alpha_3, \beta_3, \gamma_3) = units(\mathbf{X}^3) = \text{length} = L$

and the compound quantities are thus

(3.61a) $units(s\bar{s}x) = units(C) = \text{velocity} = LT^{-1}$

(3.61b) $units(\alpha) = units[\alpha_3(s\bar{s}x+C)] = L^2T^{-1}$

(3.61c) $units[s\bar{s}(2C^2 + V_3)]) = L^2T^{-1}$

so that, by (3.60),

(3.62) $units(\mathbf{X}^+) = units(\alpha) = L^2T^{-1}$.

Thus, the outer product $\mathbf{E}$ (3.59) has units of velocity cubed, as anticipated given it is a cubic matrix polynomial (3.2a) in $\mathbf{A}$ (units of velocity, LT^{-1}), i.e.

(J13) $units(\mathbf{E}) = units(\mathbf{XX}^+) = L^3T^{-3}$, using (J3) and (3.62).

Notes

The appearance of the x coordinate in the URM4 scale factors α, β, γ (3.56) is a consequence of using the chosen definition of $\overline{\mathbf{X}}_3$ (3.45). If one of the other forms of $\overline{\mathbf{X}}_3$ was used, i.e. (3.47) or (3.48), then equation (3.56) would be expressed in terms of y or z instead. For reference, the scale factors are parameterised in terms of each of the three coordinates, using all three forms, (3.45), (3.47) and (3.48), as follows:

The scale factors α, β, γ, parameterised in terms of x, are reproduced below from (3.56), for the case $\overline{\mathbf{X}}_3 = \left(-C \quad R \quad \overline{Q}\right)$, $\bar{x} = -C$, $\bar{y} = R$, $\bar{z} = \overline{Q}$ (3.45)

(3.56a) $\alpha = \alpha_3(s\bar{s}x+C) - s\bar{s}(V_3 + 2C^2)$

(3.56b) $\beta = \beta_3(s\bar{s}x+C)$

(3.56c) $\gamma = \gamma_3(s\bar{s}x+C)$.

The scale factors α, β, γ, parameterised in terms of y, are quoted below, without derivation, for the case $\overline{\mathbf{X}}_3 = \left(R \quad -C \quad P\right)$, $\bar{x} = R$, $\bar{y} = -C$, $\bar{z} = P$ (3.47)

(3.63a) $\alpha = \alpha_3(s\bar{s}y+C)$

(3.63b) $\beta = \beta_3(s\bar{s}y+C) - s\bar{s}(V_3 + 2C^2)$

(3.63c) $\gamma = \gamma_3 (s\bar{s}y + C)$.

The scale factors α, β, γ, parameterised in terms of z, are quoted below without derivation, for the case $\mathbf{X}_3 = \begin{pmatrix} Q & \bar{P} & -C \end{pmatrix}$, $\bar{x} = Q$, $\bar{y} = \bar{P}$, $\bar{z} = -C$ (3.48)

(3.64a) $\alpha = \alpha_3 (s\bar{s}z + C)$

(3.64b) $\beta = \beta_3 (s\bar{s}z + C)$

(3.64b) $\gamma = \gamma_3 (s\bar{s}z + C) - s\bar{s}(V_3 + 2C^2)$.

Each solution for the scale factors uses a form of orthogonal vector $\mathbf{\overline{X}}_3$, whereby the orthogonality relation $\mathbf{\overline{X}}_3 \mathbf{X}_3 = 0$ is a URM3 dynamical equation (A5). For example, the z parameterisation uses $\mathbf{\overline{X}}_3 = \begin{pmatrix} Q & \bar{P} & -C \end{pmatrix}$, and gives $\mathbf{\overline{X}}_3 \mathbf{X}_3 = xQ + \bar{P}y - Cz = 0$, which simply rearranges to the URM3 dynamical equation for z as in $Cz = Qx + \bar{P}y$ (A5c).

The solution for $\mathbf{X}^+$ (3.60) is a form of 'lifting' URM3 to URM4. So named because it uses an eigenvector solution $\mathbf{X}$ (2.4), which is really just a zero-padded, URM3 solution in x, y and z, but enables a full, 4D, non-trivial row eigenvector $\mathbf{X}^+$ to be obtained. The method of lifting is fundamental to any progress in obtaining solutions to URM4, URM5 and higher, which have the highly desirable physical property of an invariant, zero Potential (9.0). As such, the method of lifting is given a full treatment in Section (7), and used extensively in Part II.

(3-4) Determining the Zero Eigenvector $\mathbf{X}_0$

The same Residual Matrix method, used above to calculate $\mathbf{X}^+$, can also be used to calculate the eigenvectors $\mathbf{X}_0$ and $\mathbf{X}^0$. This is because the residual matrix $\mathbf{E}_0$ is the outer product of $\mathbf{X}_0$ and $\mathbf{X}^0$ as in

(3.70) $\mathbf{E}_0 = -\mathbf{X}_0 \mathbf{X}^0$ (the minus sign is a retained, URM3 legacy).

Using (3.2a) for $\mathbf{E}$, then $\mathbf{E}_0$ (3.2b) can be written equivalently in terms of $\mathbf{E}$ as

(3.71) $\mathbf{E}_0 = \mathbf{E} - C(\mathbf{A}^2 + C\mathbf{A} + V_3\mathbf{I})$.

The term $(\mathbf{A}^2 + C\mathbf{A} + V_3\mathbf{I})$ has already been evaluated (3.8), and is reproduced below,

$$(3.8) \quad \mathbf{A}^2 + C\mathbf{A} + V_3\mathbf{I} = \begin{pmatrix} V_3 & \bar{s}\overline{\mathbf{X}}_3\mathbf{A}_3 + \bar{s}C\overline{\mathbf{X}}_3 \\ 2s C\mathbf{X}_3 & s\bar{s}\mathbf{X}_3\overline{\mathbf{X}}_3 + \mathbf{X}_3\mathbf{X}^{3+} \end{pmatrix}.$$

Before evaluating the complete matrix $\mathbf{E}_0$ (3.71), a simple clue to the eigenvector $\mathbf{X}_0$ can be obtained from looking at the left column of $\mathbf{E}_0$ which, from (3.71), and using (3.8) and $\mathbf{E}$ (3.59), is given by

$$(3.72) \quad \mathbf{E}_0 = \left(s(2C^2 + V_3) \begin{pmatrix} 0 \\ \mathbf{X}_3 \end{pmatrix} \quad , \quad , \quad , \right) - C \begin{pmatrix} V_3 \\ 2sC\mathbf{X}_3 \end{pmatrix} \quad , \quad , \quad , \Bigg)$$

This evaluates to (first column only)

$$(3.73) \quad \mathbf{E}_0 = \left(V_3 \begin{pmatrix} -C \\ s\mathbf{X}_3 \end{pmatrix} \quad , \quad , \quad , \right).$$

Since $\mathbf{E}_0 = -\mathbf{X}_0\mathbf{X}^0$ (3.70), this implies that $\mathbf{X}_0$ is given by

$$(3.74) \quad \mathbf{X}_0 = \begin{pmatrix} +C \\ -s\mathbf{X}_3 \end{pmatrix},$$

and that the first element of $\mathbf{X}^0$ is V_3, as in

$$(3.75) \quad \mathbf{X}^0 = \begin{pmatrix} V_3 & , & , & , \end{pmatrix}.$$

The correctness of $\mathbf{X}_0$ (3.74) can be verified by evaluating its defining, eigenvector equation for $\lambda = 0$, i.e. $\mathbf{A}\mathbf{X}_0 = 0$, expanded in full as

$$(3.76) \quad \mathbf{A}\mathbf{X}_0 = \begin{pmatrix} -s\bar{s}\overline{\mathbf{X}}_3\mathbf{X}_3 \\ sC\mathbf{X}_3 - s\mathbf{A}_3\mathbf{X}_3 \end{pmatrix} = \begin{pmatrix} 0 \\ 0 \end{pmatrix},$$

where $\overline{\mathbf{X}}_3\mathbf{X}_3 = 0$ by the orthogonality condition (3.21c), and using $\mathbf{A}_3\mathbf{X}_3 = C\mathbf{X}_3$ (A4) gives zero in the bottom element.

With the eigenvector $\mathbf{X}_0$ determined, the residual matrix $\mathbf{E}_0$ is written in the following form, where the last three elements of $\mathbf{X}^0$ (3.75) are determined next.

(3.77) $\mathbf{E}_0 = -\mathbf{X}_0 \mathbf{X}^0 = \begin{pmatrix} +C \\ -s\mathbf{X}_3 \end{pmatrix} (V_3 \quad , \quad , \quad ,).$

To determine the remaining three elements of $\mathbf{X}^0$, i.e. complete the residual matrix $\mathbf{E}_0$ (3.71), the first step is to simplify the quadratic expression $\mathbf{A}^2 + C\mathbf{A} + V_3\mathbf{I}$ (3.8). As for the earlier computation of $\mathbf{E}$, a choice is made for orthogonal vector $\overline{\mathbf{X}}_3$ such it simplifies the algebra. Remaining with the choice of $\overline{\mathbf{X}}_3$ as $\overline{\mathbf{X}}_3 = \begin{pmatrix} -C & R & \overline{Q} \end{pmatrix}$ (3.45), then the expression $\overline{\mathbf{X}}_3 \mathbf{A}_3$ has already been calculated as

(3.53) $\overline{\mathbf{X}}_3 \mathbf{A}_3 = \begin{pmatrix} \alpha_3 x - V_3, & \beta_3 x - 2CR, & \gamma_3 x - 2C\overline{Q} \end{pmatrix},$

and so the top right element in (3.8) becomes

(3.78) $\overline{s}\overline{\mathbf{X}}_3 \mathbf{A}_3 + \overline{s}C\overline{\mathbf{X}}_3 = \overline{s}\begin{pmatrix} \alpha_3 x - V_3 - C^2, & \beta_3 x - CR, & \gamma_3 x - C\overline{Q} \end{pmatrix}.$

Each of the three elements in (3.78) contains a standard URM3 divisibility relation, see (A15) and (A16), and simplify significantly as follows:

(3.79)

(3.79a) $\alpha_3 x - C^2 = -P\overline{P}$, (A15a)

(3.79b) $\beta_3 x - CR = \overline{P}\overline{Q}$, (A16c)

(3.79c) $\gamma_3 x - C\overline{Q} = RP$, (A16e).

Thus, the top right element of (3.8), given by (3.78), reduces to

(3.80) $\overline{s}\overline{\mathbf{X}}_3 \mathbf{A}_3 + \overline{s}C\overline{\mathbf{X}}_3 = \overline{s}\begin{pmatrix} -V_3 - P\overline{P}, & \overline{P}\overline{Q}, & RP \end{pmatrix}.$

For the bottom right element $s\overline{s}\mathbf{X}_3\overline{\mathbf{X}}_3 + \mathbf{X}_3\mathbf{X}^{3+}$ of (3.8), this is re-written as the outer product of $\mathbf{X}_3$ and $s\overline{s}\overline{\mathbf{X}}_3 + \mathbf{X}^{3+}$, i.e.

(3.81) $s\overline{s}\mathbf{X}_3\overline{\mathbf{X}}_3 + \mathbf{X}_3\mathbf{X}^{3+} = \mathbf{X}_3\begin{pmatrix} s\overline{s}\overline{\mathbf{X}}_3 + \mathbf{X}^{3+} \end{pmatrix},$

and using the vector definitions for $\mathbf{X}_3$ (A3), $\overline{\mathbf{X}}_3$ (3.45) and $\mathbf{X}^{3+}$ (A13), the bracketed sum $\begin{pmatrix} s\overline{s}\overline{\mathbf{X}}_3 + \mathbf{X}^{3+} \end{pmatrix}$ is therefore

(3.82) $s\overline{s}\overline{\mathbf{X}}_3 + \mathbf{X}^{3+} = \begin{pmatrix} \alpha_3 - s\overline{s}C & \beta_3 + s\overline{s}R & \gamma_3 + s\overline{s}\overline{Q} \end{pmatrix}.$

Substituting (3.80) and (3.82) into (3.8) gives

(3.83)

$$\mathbf{A}^2 + C\mathbf{A} + V_3\mathbf{I} =$$
$$\begin{pmatrix} V_3 & -\bar{s}V_3 - \bar{s}P\bar{P} & \bar{s}P\bar{Q} & \bar{s}RP \\ 2s C\mathbf{X}_3 & (\alpha_3 - s\bar{s}C)\mathbf{X}_3 & (\beta_3 + s\bar{s}R)\mathbf{X}_3 & (\gamma_3 + s\bar{s}\bar{Q})\mathbf{X}_3 \end{pmatrix}.$$

Returning to the computation of $\mathbf{E}_0$ (3.71), using the above (3.83) and $\mathbf{E}$ (3.59), then $\mathbf{E}_0$ becomes (split into two parts purely for printing reasons)

$$(3.84)\ \mathbf{E}_0 = \begin{pmatrix} -V_3 C & -C\bar{s}(-V_3 - P\bar{P}) & - & - \\ V_3 s\mathbf{X}_3 & (\alpha - C(\alpha_3 - s\bar{s}C))\mathbf{X}_3 & - & - \end{pmatrix} +$$

$$\begin{pmatrix} - & - & -\bar{s}C P\bar{Q} & -\bar{s}CRP \\ - & - & (\beta - C(\beta_3 + s\bar{s}R))\mathbf{X}_3 & (\gamma - C(\gamma_3 + s\bar{s}\bar{Q}))\mathbf{X}_3 \end{pmatrix}.$$

Using the definitions (3.56) for the URM4 divisibility factors α, β, γ, the bottom row simplifies to give

(3.85)

$$\mathbf{E}_0 =$$
$$\begin{pmatrix} -V_3 C & -\bar{s}C(-V_3 - P\bar{P}) & -\bar{s}C P\bar{Q} & -\bar{s}CRP \\ V_3 s\mathbf{X}_3 & s\bar{s}(\alpha_3 x - C^2 - V_3)\mathbf{X}_3 & s\bar{s}(\beta_3 x - CR)\mathbf{X}_3 & s\bar{s}(\gamma_3 x - C\bar{Q})\mathbf{X}_3 \end{pmatrix}$$

Finally, the bottom row simplifies further using the same URM3 relations (3.79), as used earlier in (3.80), to give $\mathbf{E}_0$ as

$$(3.86)\ \mathbf{E}_0 = \begin{pmatrix} -V_3 C & -\bar{s}C(-V_3 - P\bar{P}) & -\bar{s}C P\bar{Q} & -\bar{s}CRP \\ V_3 s\mathbf{X}_3 & s\bar{s}(-V_3 - P\bar{P})\mathbf{X}_3 & -s\bar{s}P\bar{Q}\mathbf{X}_3 & -s\bar{s}RP\mathbf{X}_3 \end{pmatrix}$$

which is now re-written in a simpler form as follows, where a factor of -1 has been removed from the bracketed vector terms in anticipation that $\mathbf{E}_0 = -\mathbf{X}_0\mathbf{X}^0$ (3.70)

(3.87)

$$\mathbf{E}_0 =$$

$$-\left(V_3\begin{pmatrix} +C \\ -s\mathbf{X}_3 \end{pmatrix} \quad \bar{s}\left(-V_3 - P\bar{P}\right)\begin{pmatrix} +C \\ -s\mathbf{X}_3 \end{pmatrix} \quad \bar{s}\overline{PQ}\begin{pmatrix} +C \\ -s\mathbf{X}_3 \end{pmatrix} \quad \bar{s}RP\begin{pmatrix} +C \\ -s\mathbf{X}_3 \end{pmatrix}\right)$$

Thus, comparing this with the outer product form of $\mathbf{E}_0$ (3.70), and with $\mathbf{X}_0$ already evaluated as (3.74), the vector $\mathbf{X}^0$ is therefore

(3.88) $\mathbf{X}^0 = \left(V_3 \quad \bar{s}\left(-V_3 - P\bar{P}\right) \quad \bar{s}\overline{PQ} \quad \bar{s}RP \right).$

This is quite some simplification given the earlier forms for $\mathbf{E}_0$, above. Perhaps of most note is that the vector $\mathbf{X}^0$ is not a function of any coordinates x, y, z.

The correctness of $\mathbf{X}^0$ (3.88) can be verified by checking it satisfies its defining, eigenvector equation for $\lambda = 0$, i.e. $\mathbf{X}^0\mathbf{A} = 0$. Using $\mathbf{X}^0$ (3.88) and $\mathbf{A}$ (2.25), this product is

(3.89) $\mathbf{X}^0\mathbf{A} = \left(V_3 \quad \bar{s}\left(-V_3 - P\bar{P}\right) \quad \bar{s}\overline{PQ} \quad \bar{s}RP \right)\begin{pmatrix} 0 & \bar{s}\mathbf{X}_3 \\ s\mathbf{X}_3 & \mathbf{A}_3 \end{pmatrix}.$

Confirming the result $\mathbf{X}^0\mathbf{A} = 0$ involves quite a bit of algebra, nevertheless, for instructional purposes, it is given in full as follows:

Evaluating each column separately, then for the first column, using $\mathbf{X}_3$ (A3), the product is

(3.90) $\mathbf{X}^0\begin{pmatrix} 0 \\ \mathbf{X}_3 \end{pmatrix} = \bar{s}s\left(\left(-V_3 - P\bar{P}\right)x + \overline{PQ}y + RPz\right).$

This is actually very simply proven to be zero since it is equivalent to the inner product $\mathbf{X}^0\mathbf{X}$: given $\mathbf{X}$ is an eigenvector for eigenvalue C, and $\mathbf{X}^0$ is a row eigenvector for eigenvalue 0, then the two eigenvectors $\mathbf{X}^0$ and $\mathbf{X}$ are orthogonal by the rules of matrix algebra (F25), i.e.

(3.91) $\mathbf{X}^0\begin{pmatrix} 0 \\ \mathbf{X}_3 \end{pmatrix} = \mathbf{X}^0\mathbf{X} = 0.$

However, it is useful to independently verify the algebraic solution (3.88) for $\mathbf{X}^0$, which must also give the zero inner product as follows:

Substituting for V_3 (A9) into (3.90) gives the bracketed term on the right of (3.90) as

$$(3.92) \quad -PQRx - \overline{PQ}\,\overline{R}x - CP\overline{P}x + CP\overline{Q}y + CRPz,$$

Grouping the terms as follows:

$$(3.93) \quad PR(Cz - Qx) + \overline{PQ}(Cy - \overline{R}x) - CP\overline{P}x,$$

and using the two URM3 dynamical equations (A5b) and (A5c), this simplifies to

$$(3.94a) \quad P\overline{P}Ry + P\overline{P}\overline{Q}z - CP\overline{P}x,$$

and factoring $P\overline{P}$ gives

$$(3.94b) \quad P\overline{P}(Ry + \overline{Q}z - Cx).$$

But the bracket in (3.94b) is the rearranged form of the dynamical equation for x (A5a), which equates to zero, i.e.

$$(3.94c) \quad P\overline{P}(Ry + \overline{Q}z - Cx) = 0,$$

and so the entire bracket on the right of (3.90) is also zero, and therefore the inner product $\mathbf{X}^0\mathbf{X}$ (3.91) is zero.

For the right column of product (3.89),

$$(3.95) \quad \mathbf{X}^0\begin{pmatrix} \overline{s}\mathbf{X}_3 \\ \mathbf{A}_3 \end{pmatrix} = \overline{s}V_3\mathbf{X}_3 + \left(\left(-V_3 - P\overline{P}\right) \;\; \overline{PQ} \;\; RP\right)\!\overline{s}\mathbf{A}_3.$$

the first term on the right, using $\mathbf{X}_3 = \left(-C \;\; R \;\; \overline{Q}\right)$ (3.45), is simply

$$(3.96) \quad V_3\mathbf{X}_3 = \left(-V_3C \;\; V_3R \;\; V_3\overline{Q}\right),$$

where the overall multiplying factor $\overline{s}$ has been omitted, as it will be seen to cancel. Using $\mathbf{A}_3$ (A1a), the second term evaluates to

$$(3.97) \quad \left(PQR + \overline{PQ}\,\overline{R} \;\; -V_3R \;\; -V_3\overline{Q}\right).$$

Since the term $PQR + \overline{PQ}\,\overline{R} = V_3C$ by (A9), then this becomes

$$(3.98) \quad \left(V_3C \;\; -V_3R \;\; -V_3\overline{Q}\right),$$

and it is then easily seen that $V_3\overline{X}_3$ (3.96) cancels with this term (3.98) to leave zero, hence the second column is also verified as zero.

(3.99) $V_3\overline{X}_3 + \left(V_3 C \quad -V_3 R \quad -V_3\overline{Q}\right) = 0$.

This completes the proof that $X^0 A = 0$.

Once again, as mentioned earlier, the solution (3.88) for X^0 is dependent upon the form chosen for $\overline{X}_3$, here $\overline{X}_3 = \left(-C \quad R \quad \overline{Q}\right)$ (3.45). The solution X^0 for the other forms of $\overline{X}_3$, i.e. (3.47) and (3.48), has not been given, but suffice to say, solutions for X^0 will not be the same as (3.88) using $\overline{X}_3 = \left(-C \quad R \quad \overline{Q}\right)$ (3.45).

Having verified the inner product $X^0 X$ is zero then, for completeness, the inner product $X^+ X_0$ is also verified as zero as follows. As noted above, these two eigenvectors are orthogonal by (F25), and the algebraic verification acts more to verify the eigenvectors themselves have been correctly derived.

Using X^+ (3.60) and X_0 (3.74), the inner product $X^+ X_0$ is expanded in full as

(3.100) $X^+ X_0 = sC(2C^2 + V_3) - s(\alpha x + \beta y + \gamma z)$

From the definitions (3.56) of the URM4 divisibility factors α, β, γ in terms of their URM3 equivalents α_3, β_3 and γ_3, the term $\alpha x + \beta y + \gamma z$ expands to

(3.101)
$$
\begin{aligned}
&\alpha x + \beta y + \gamma z = \\
&s\overline{s}x(\alpha_3 x + \beta_3 y + \gamma_3 z) + \\
&C(\alpha_3 x + \beta_3 y + \gamma_3 z) + \\
&- s\overline{s}x(V_3 + 2C^2)
\end{aligned}
$$

However, the URM3 term $\alpha_3 x + \beta_3 y + \gamma_3 z$ is just the URM Potential equation (A17), i.e.

(3.102) $\alpha_3 x + \beta_3 y + \gamma_3 z = (2C^2 + V_3)$, and so the $s\bar{s}$ terms in (3.101) cancel and the entire expression reduces to

(3.103) $\alpha x + \beta y + \gamma z = C(\alpha_3 x + \beta_3 y + \gamma_3 z)$.

Using (3.102) again, this is written as

(3.104) $\alpha x + \beta y + \gamma z = C(2C^2 + V_3)$

The result (3.102) is a very neat and simple result linking the URM3 and URM4 scale factors and is used again, further below, when evaluating the inner product $\mathbf{X}^+\mathbf{X}$, known as the URM4 Potential equation by analogy with the URM3 equivalent (A17).

Continuing with evaluation of the inner product $\mathbf{X}^+\mathbf{X}_0$ (3.100), substituting the above result (3.104) into (3.100) shows it is zero, as expected by (F25), i.e.

(3.105) $\mathbf{X}^+\mathbf{X}_0 = 0$.

The inner product $\mathbf{X}^0\mathbf{X}_0$ is decomposed as follows, by splitting $\mathbf{X}_0$ (3.74) into two components:

(3.106) $\mathbf{X}^0 \begin{pmatrix} +C \\ -s\mathbf{X}_3 \end{pmatrix} = \mathbf{X}^0 \begin{pmatrix} 0 \\ -s\mathbf{X}_3 \end{pmatrix} + \mathbf{X}^0 \begin{pmatrix} +C \\ 0 \end{pmatrix}$.

The first term on the right evaluates to $-s\mathbf{X}^0\mathbf{X}_3$, which is zero since $\mathbf{X}^0\mathbf{X}_3 = 0$ (F5). For the second term on the right, using $\mathbf{X}^0$ (3.88), this product expands to

(3.107) $\mathbf{X}^0 \begin{pmatrix} +C \\ 0 \end{pmatrix} = \begin{pmatrix} V_3 & \bar{s}(-V_3 - P\bar{P}) & \bar{s}\overline{PQ} & \bar{s}RP \end{pmatrix} \begin{pmatrix} +C \\ 0 \end{pmatrix}$,

and since only the first element C of $\mathbf{X}_0$ contributes to the product, this gives a very simple result

(3.108) $\mathbf{X}^0\mathbf{X}_0 = CV_3$.

Note that this is not the same as the URM3 result $\mathbf{X}^{30}\mathbf{X}_{30} = +C^2$ (F3), which is actually the URM3 DCE under URM3 Pythagoras conditions. The reason being that this zero eigenvalue is a general consequence of the embedding method, and not due to URM4 Pythagoras conditions.

Note that the URM4 conditions include the URM3 conditions (A18) as a subset, and add a second, zero eigenvalue in the process; see Section (4) for more details.

(3-5) The URM4 Potential Equation

In URM3, one of the key conservation equations, namely the 'Potential Equation' (because it linked to the Potential V_3), is given by the inner product $\mathbf{X}^+\mathbf{X}$,

(A17) $\quad \mathbf{X}^+\mathbf{X} = \alpha_3 x + \beta_3 y + \gamma_3 z = 2C^2 + V_3,$ the URM3 Potential equation

Since the URM4 vector $\mathbf{X}^+$ has now been now analytically determined (3.60), this same inner product $\mathbf{X}^+\mathbf{X}$ can be evaluated for URM4. Using $\mathbf{X}$ (2.3) and $\mathbf{X}^+$ (3.60), then this product is given by the following matrix product

$$(3.110) \quad \mathbf{X}^+\mathbf{X} = \left(s(2C^2 + V_3) \quad \alpha \quad \beta \quad \gamma \right) \begin{pmatrix} 0 \\ x \\ y \\ z \end{pmatrix}.$$

Expanding in full, this becomes

(3.111) $\quad \mathbf{X}^+\mathbf{X} = \alpha x + \beta y + \gamma z,$

and using (3.104) this evaluates to

(3.112) $\quad \mathbf{X}^+\mathbf{X} = C(2C^2 + V_3),$

and it is seen that the URM4 inner product is simply a scalar multiple C of the URM3 equivalent, for non-zero V_3, i.e.

(3.113) $\quad \mathbf{X}^+\mathbf{X} = C\mathbf{X}^{3+}\mathbf{X}_{3+}.$

This extra factor of C actually arises because the URM4 residual matrix is a cubic matrix polynomial in the dynamical variables, whereas for URM3 it was a quadratic. From (3.112), if the Potential V_3 is zero then $\mathbf{X}^+\mathbf{X}$ is a constant $2C^3$, as compared to the URM3 quadratic value of $2C^2$. However, it was explained earlier in Section (3-2) that this extra factor of C is actually irrelevant when $V_3 = 0$ and $\overline{\mathbf{X}}_3 = \mathbf{X}^{3-}$

(3.21a), arising due to an unnecessary, additional evaluation of a cubic polynomial. The factor of C can be legitimately discarded, making the Potential equation in both URM3 and URM4 the same invariant quantity $2C^2$.

This completes the theory of the Residual Matrix Method in URM4, and this section now ends with a complete example.

(3-6) An Example of the Residual Method

This example is an embedding of the standard cubic example, first given in Appendix C of [1],#1,. It is purposefully chosen such that it does not satisfy any Pythagoras conditions, given the majority of the rest of this book focuses on them. Before giving the full URM4 treatment, the original URM3 solution, as in [1],#1, is reproduced below.

The URM3 coordinates x, y, z

(3.120) $x = 9$, $y = 31$, $z = 70$.

The URM3 coordinate vector $\mathbf{X}_3$ (A3)

(3.121) $\mathbf{X}_3 = \begin{pmatrix} 9 \\ 31 \\ 70 \end{pmatrix}$.

The URM3 dynamical variables P, Q, R and $\overline{P}, \overline{Q}, \overline{R}$

(3.122)

(3.122a) $P = -2$, $Q = -6$, $R = -11$

(3.122b) $\overline{P} = +4$, $\overline{Q} = +5$, $\overline{R} = +19$.

The URM3 unity root matrix $\mathbf{A}_3$

(3.123) $\mathbf{A}_3 = \begin{pmatrix} 0 & -11 & +5 \\ +19 & 0 & -2 \\ -6 & +4 & 0 \end{pmatrix}$.

The single, unity eigenvalue

(3.124) $\lambda = C = 1$

The URM3 Potential V_3

$$(3.125)\ V_3 = \frac{PQR + \overline{P}\,\overline{Q}\,\overline{R}}{C} = +248$$

The URM3 dynamical equations $\mathbf{A}_3\mathbf{X}_3 = C\mathbf{X}_3$ (A4) are verified using $\mathbf{A}_3$ (3.123), C (3.124), and $\mathbf{X}_3$ (3.121)

$$(3.126)\ \begin{pmatrix} 9 \\ 31 \\ 70 \end{pmatrix} = \begin{pmatrix} 0 & -11 & +5 \\ +19 & 0 & -2 \\ -6 & +4 & 0 \end{pmatrix}\begin{pmatrix} 9 \\ 31 \\ 70 \end{pmatrix}.$$

The URM3 orthogonal vector $\overline{\mathbf{X}}_3 = \begin{pmatrix} -C & R & \overline{Q} \end{pmatrix}$ (3.45)

$$(3.127)\ \overline{\mathbf{X}}_3 = \begin{pmatrix} -1 & -11 & +5 \end{pmatrix}$$

The URM3 Orthogonality condition (3.21c) $\overline{\mathbf{X}}_3\mathbf{X}_3 = 0$

$$(3.128)\ \begin{pmatrix} -1 & -11 & +5 \end{pmatrix}\begin{pmatrix} 9 \\ 31 \\ 70 \end{pmatrix} = 0.$$

The URM3 divisibility factors α_3, β_3 and γ_3 are calculated from (A15) using the dynamical variables (3.122) and unity eigenvalue $C = 1$ (3.124)

$$(3.129)\ \alpha_3 = +1,\ \beta_3 = +1,\ \gamma_3 = +3.$$

The URM3 row eigenvector of scale (or divisibility) factors $\mathbf{X}^{3+}$ (A13)

$$(3.130)\ \mathbf{X}^{3+} = \begin{pmatrix} 1 & 1 & 3 \end{pmatrix}.$$

The URM3 conjugate dynamical equations $\mathbf{X}^{3+}\mathbf{A}_3 = C\mathbf{X}^{3+}$ (A14) are verified using $\mathbf{A}_3$, C and $\mathbf{X}^{3+}$

$$(3.131)\ \begin{pmatrix} 1 & 1 & 3 \end{pmatrix} = \begin{pmatrix} 1 & 1 & 3 \end{pmatrix}\begin{pmatrix} 0 & -11 & +5 \\ +19 & 0 & -2 \\ -6 & +4 & 0 \end{pmatrix}.$$

This completes the reproduction of the URM3 solution from [1],#1. The solution is now embedded in URM4, and the residual method used to determine the URM4 matrix $\mathbf{A}$ and eigenvectors $\mathbf{X}$, $\mathbf{X}^+$, $\mathbf{X}_0$ and $\mathbf{X}^0$.

The URM4 coordinates w, x, y, z

(3.132) $w = 0$, $x = 9$, $y = 31$, $z = 70$.

The URM4 coordinate vector $\mathbf{X}$ (2.3)

$$(3.133) \; \mathbf{X} = \begin{pmatrix} 0 \\ 9 \\ 31 \\ 70 \end{pmatrix}.$$

The URM4 orthogonal vector $\overline{\mathbf{X}}$ (2.12), using $\mathbf{X}_3$ (3.127)

(3.134) $\overline{\mathbf{X}} = \begin{pmatrix} 0 & -1 & -11 & +5 \end{pmatrix}$.

The orthogonality (2.9) is trivially verified since, by (3.128),

$$(3.135) \; \overline{\mathbf{X}}\mathbf{X} = \begin{pmatrix} 0 & -1 & -11 & +5 \end{pmatrix} \begin{pmatrix} 0 \\ 9 \\ 31 \\ 70 \end{pmatrix} = 0.$$

The conversion factors s and $\bar{s}$ are temporarily set to unity for simplicity

(3.136) $s = 1$, $\bar{s} = 1$.

From the URM4 matrix $\mathbf{A}$ (2.25) in block matrix form, using s, $\bar{s}$, $\mathbf{X}_3$ (3.121), $\overline{\mathbf{X}}_3$ (3.127) and $\mathbf{A}_3$ (3.123), then the URM4 unity root matrix $\mathbf{A}$ is, expanded in full as

$$(3.137) \; \mathbf{A} = \begin{pmatrix} 0 & -1 & -11 & +5 \\ +9 & 0 & -11 & +5 \\ +31 & +19 & 0 & -2 \\ +70 & -6 & +4 & 0 \end{pmatrix}.$$

Determination of $\mathbf{X}^+$ from $\mathbf{E} = \mathbf{X}\mathbf{X}^+$ (3.3), using the matrix polynomial $\mathbf{E} = \mathbf{A}(\mathbf{A}^2 + C\mathbf{A} + V_3\mathbf{I})$ (3.2a).

$$(3.138) \ \mathbf{A}^2 = \begin{pmatrix} 0 & -239 & +31 & +17 \\ +9 & -248 & -79 & +67 \\ +31 & -19 & -558 & +250 \\ +70 & +6 & -704 & +312 \end{pmatrix}, \ C = 1, \ V_3 = 248$$

$$(3.139) \ \mathbf{A}^2 + C\mathbf{A} + V_3\mathbf{I} = \begin{pmatrix} +248 & -240 & +20 & +22 \\ +18 & 0 & -90 & +72 \\ +62 & 0 & -310 & +248 \\ +140 & 0 & -700 & +560 \end{pmatrix}.$$

The URM3 sub-matrix $\mathbf{E}_3$ is extracted from the bottom right 3x3 matrix of (3.139)

$$(3.140) \ \mathbf{E}_3 = \begin{pmatrix} 0 & -90 & +72 \\ 0 & -310 & +248 \\ 0 & -700 & +560 \end{pmatrix}$$

and using $\mathbf{E}_3 = \mathbf{X}_3\mathbf{X}^{3+}$ [1],#2 and $\mathbf{X}_3$ (3.121) implies that $\mathbf{X}^{3+}$ is

$$(3.141) \ \mathbf{X}^{3+} = \begin{pmatrix} 0 & -10 & 8 \end{pmatrix}$$

i.e.

$$(3.142) \ \mathbf{E}_3 = \mathbf{X}_3\mathbf{X}^{3+} = \begin{pmatrix} 9 \\ 31 \\ 70 \end{pmatrix} \begin{pmatrix} 0 & -10 & 8 \end{pmatrix}$$

The URM3 Potential equation (A17)

$$(3.143) \ \mathbf{X}^{3+}\mathbf{X}_3 = 2C^2 + V_3 = 250, \ C = 1, \ V_3 = 248$$

$$(3.144) \ \mathbf{X}^{3+}\mathbf{X}_3 = \begin{pmatrix} 0 & -10 & 8 \end{pmatrix} \begin{pmatrix} 9 \\ 31 \\ 70 \end{pmatrix} = 250$$

Continuing with URM4, the residual matrix $\mathbf{E}$ (3.2a) is

(3.145)

$$E = A(A^2 + CA + V_3 I) = \begin{pmatrix} 0 & 0 & 0 & 0 \\ 2250 & -2160 & 90 & 270 \\ 7750 & -7440 & 310 & 930 \\ 17500 & -16800 & 700 & 2100 \end{pmatrix}$$

which implies, using $E = XX^+$ (3.3) and X (3.133), that

(3.146) $X^+ = \begin{pmatrix} 250 & -240 & 10 & 30 \end{pmatrix}$

The analytic form of X^+ (3.60) is confirmed as correct and equal to (3.146) by calculation of the URM4 scale factors α, β, γ (3.56) as follows, using $(\alpha_3, \beta_3, \gamma_3) = (1,1,3)$ (3.129), $s = 1$, $\bar{s} = 1$ (3.136), $x = 9$ (3.132), $C = 1$ (3.124), $V_3 = 248$ (3.125)

(3.147) $\alpha = -240$, $\beta = 10$, $\gamma = 30$

The URM4 Potential equation X^+X (3.112) can be used as a check on eigenvectors X (3.133) and X^+ (3.146), which, for $C = 1$, equates to the URM3 equivalent $(2C^2 + V_3)$ (A17) as follows:

(3.148) $X^+X = \begin{pmatrix} 250 & -240 & 10 & 30 \end{pmatrix} \begin{pmatrix} 0 \\ 9 \\ 31 \\ 70 \end{pmatrix} = 250$.

Moving on to the zero eigenvalue case, $\lambda = 0$. Using the analytic form (3.74) for X_0, with $s = 1$, $C = 1$ and X_3 (3.121), then

(3.149) $X_0 = \begin{pmatrix} +1 \\ -9 \\ -31 \\ -70 \end{pmatrix}$.

From the analytic form (3.88) for X^0, with $\bar{s} = 1$, $V_3 = 248$, $P = -2$, $\bar{P} = +4$, $\bar{Q} = +5$, $R = -11$, then

(3.150) $X^0 = \begin{pmatrix} 248 & -240 & 20 & 22 \end{pmatrix}$.

Verifying the orthogonality relations $\mathbf{X}^0\mathbf{X}$ (3.91) and $\mathbf{X}^+\mathbf{X}_0$ (3.105)

$$(3.151)\quad \mathbf{X}^0\mathbf{X} = \begin{pmatrix} 248 & -240 & 20 & 22 \end{pmatrix}\begin{pmatrix} 0 \\ 9 \\ 31 \\ 70 \end{pmatrix} = 0$$

$$(3.152)\quad \mathbf{X}^+\mathbf{X}_0 = \begin{pmatrix} 250 & -240 & 10 & 30 \end{pmatrix}\begin{pmatrix} +1 \\ -9 \\ -31 \\ -70 \end{pmatrix} = 0.$$

The residual matrix $\mathbf{E}_0$ is calculated from the matrix polynomial definition (3.2b), using $\mathbf{A}^2 + C\mathbf{A} + V_3\mathbf{I}$ (3.139) and calculating $\mathbf{A} - C\mathbf{I}$ as follows:

$$(3.153)\quad \mathbf{A} - C\mathbf{I} = \begin{pmatrix} -1 & -1 & -11 & +5 \\ +9 & -1 & -11 & +5 \\ +31 & +19 & -1 & -2 \\ +70 & -6 & +4 & -1 \end{pmatrix},$$

to give for $\mathbf{E}_0$

$$(3.154)\quad \mathbf{E}_0 = \begin{pmatrix} -248 & +240 & -20 & -22 \\ +2232 & -2160 & +180 & +198 \\ +7688 & -7440 & +620 & +682 \\ +17360 & -16800 & +1400 & +1540 \end{pmatrix}.$$

Using $\mathbf{X}_0$ (3.149) and $\mathbf{X}^0$ (3.150), the residual matrix $\mathbf{E}_0$ is verified as the outer product $\mathbf{E}_0 = -\mathbf{X}_0\mathbf{X}^0$.

$$\mathbf{E}_0 = -\begin{pmatrix} +1 \\ -9 \\ -31 \\ -70 \end{pmatrix}\begin{pmatrix} 248 & -240 & 20 & 22 \end{pmatrix}.$$

Verifying the inner product $\mathbf{X}^0\mathbf{X}_0 = CV_3$ (3.108) using $\mathbf{X}_0$ (3.149) and $\mathbf{X}^0$ (3.150), with $C = 1$, $V_3 = 248$,

$$(3.155) \quad \mathbf{X}^0\mathbf{X}_0 = \begin{pmatrix} 248 & -240 & 20 & 22 \end{pmatrix} \begin{pmatrix} +1 \\ -9 \\ -31 \\ -70 \end{pmatrix} = 248.$$

Repeating some of the calculations for $s = 1$, $\bar{s} = 2$.

The URM4 scale factors α, β, γ (3.56)

$(3.156) \quad \alpha = -481$, $\beta = 19$, $\gamma = 57$.

Using these scale factors and $C = 1$, $V_3 = 248$, gives $\mathbf{X}^+$ (3.60) as

$(3.157) \quad \mathbf{X}^+ = \begin{pmatrix} 250 & -481 & 19 & 57 \end{pmatrix}$

Since the URM3 Potential V_3 (A9) and eigenvalue C evidently have no functional dependence on s and $\bar{s}$, then the quantity $C(2C^2 + V_3)$ in the Potential equation (3.112) is invariant to s and $\bar{s}$, and thus $\mathbf{X}^+\mathbf{X}$ is the same as (3.148), as verified by

$$(3.158) \quad \mathbf{X}^+\mathbf{X} = \begin{pmatrix} 250 & -481 & 19 & 57 \end{pmatrix} \begin{pmatrix} 0 \\ 9 \\ 31 \\ 70 \end{pmatrix} = 250.$$

With $\bar{s} = 2$, then $\mathbf{X}^0$ (3.88) is

$(3.159) \quad \mathbf{X}^0 = \begin{pmatrix} 248 & -480 & 40 & 44 \end{pmatrix}$

The inner product $\mathbf{X}^0\mathbf{X}_0 = CV_3$ (3.108) is also invariant to s and $\bar{s}$, by the same reasoning as for $\mathbf{X}^+\mathbf{X}$ (3.158), and thus the same as (3.155), confirmed using $\mathbf{X}^0$ (3.159) and $\mathbf{X}_0$ (3.149) by

$$(3.160) \quad \mathbf{X}^0 \mathbf{X}_0 = \begin{pmatrix} 248 & -480 & 40 & 44 \end{pmatrix} \begin{pmatrix} +1 \\ -9 \\ -31 \\ -70 \end{pmatrix} = 248.$$

Repeating the calculations for $s = 2$, $\bar{s} = 1$.

The URM4 scale factors α, β, γ (3.56) are actually the same as the case $s = 1$, $\bar{s} = 2$, since they are symmetric upon interchange of s and $\bar{s}$, so too is the invariant quantity $(2C^2 + V_3)$, therefore they are as per above (3.156). Likewise, $\mathbf{X}^+$ is the same as in (3.157).

The $\mathbf{X}_0$ eigenvector (3.74) becomes, for $s = 2$, $C = 1$ and $\mathbf{X}_3$ (3.121),

$$(3.161) \quad \mathbf{X}_0 = \begin{pmatrix} +1 \\ -18 \\ -62 \\ -140 \end{pmatrix}.$$

The $\mathbf{X}^0$ eigenvector is also invariant since it is only a function of $\bar{s}$, so $\mathbf{X}^0$ is as for (3.150), i.e. $\mathbf{X}^0 = \begin{pmatrix} 248 & -240 & 20 & 22 \end{pmatrix}$.

With $\mathbf{X}^0$ and $\mathbf{X}$ invariant to s then, unsurprisingly, the orthogonality relation $\mathbf{X}^0 \mathbf{X}$ (3.91) holds for arbitrary s. The inner product $\mathbf{X}^0 \mathbf{X}_0 = CV_3$ (3.108) is also invariant, as mentioned above for (3.160), and confirmed using $\mathbf{X}^0$ (3.150) and $\mathbf{X}_0$ (3.161) by

$$(3.162) \quad \mathbf{X}^0 \mathbf{X}_0 = \begin{pmatrix} 248 & -240 & 20 & 22 \end{pmatrix} \begin{pmatrix} -1 \\ -18 \\ -62 \\ -140 \end{pmatrix} = -248.$$

For the interested reader, the following URM3 calculations, used in the above example, are provided here for reference, reproduced from [1],#1. Otherwise, this completes the example and section.

In all equations the eigenvalue C is unity, i.e. $C = 1$.

(A42a) $(C^2 - P\overline{P}) = \alpha_3 x$

(3.170a) $(1 - (-2)(+4)) = 9\alpha_3$, $\alpha_3 = +1$.

(A42b) $(C^2 - Q\overline{Q}) = \beta_3 y$

(3.170b) $(1 - (-6)(+5)) = 31\beta_3$, $\beta_3 = +1$.

(A42c) $(C^2 - R\overline{R}) = \gamma_3 z$

(3.170c) $(1 - (-11)(+19)) = 70\gamma_3$, $\gamma_3 = +3$.

The kinetic term K

(3.171) $K = P\overline{P} + Q\overline{Q} + R\overline{R} = -247$

The Potential V_3

(3.172) $V_3 = PQR + \overline{P}\,\overline{Q}\,\overline{R} = +248$

Verifying the conservation equation

(A12) $C^2 = K + V_3$

(3.173) $1 = (-247) + (+248)$.

Verifying the Potential equation (A17)

(A17) $2C^2 + V_3 = \alpha_3 x + \beta_3 y + \gamma_3 z$

(3.174) $250 = (+1)(+9) + (+1)(+31) + (+3)(+70)$.

The coordinate equation

(2.21) $0 = x^n + y^n - z^n + xyz.k(x, y, z)$

(3.175a) $0 = 9^3 + 31^3 - 70^3 + 9.31.70.k(x, y, z)$

(3.175b) $k(x, y, z) = +16$

The dynamical variables, which are also cubic unity roots since $C = 1$ and exponent $n = 3$, satisfy the following congruence relations by their definitions (A2).

(A2a) $P^n \equiv +1 \pmod{x}$

(3.176a) $-2^3 \equiv +1 \pmod 9$.

(A2b) $Q^n \equiv +1 \pmod{y}$

(3.176b) $-6^3 \equiv +1 \pmod{31}$.

(A2c) $R^n \equiv -1 \pmod{z}$

(3.176c) $-11^3 \equiv -1 \pmod{70}$.

(A2d) $\overline{P}^n \equiv +1 \pmod{x}$

(3.176d) $+4^3 \equiv +1 \pmod 9$.

(A2e) $\overline{Q}^n \equiv +1 \pmod{y}$

(3.176e) $5^3 \equiv +1 \pmod{31}$.

(A2f) $\overline{R}^n \equiv -1 \pmod{z}$

(3.176f) $+19^3 \equiv -1 \pmod{70}$.

(A2g) $\overline{P} \equiv P^{n-1} \pmod{x}$

(3.176g) $+4 \equiv (-2)^2 \pmod 9$.

(A2h) $\overline{Q} \equiv Q^{n-1} \pmod{y}$

(3.176h) $+5 \equiv (-6)^2 \pmod{31}$.

(A2i) $\overline{R} \equiv -R^{n-1} \pmod{z}$

(3.176i) $+19 \equiv -(-11)^2 \pmod{70}$.

(4) URM4 Pythagoras Conditions

As first discussed in Section (1), obtaining eigenvector solutions to the general, 4x4 unity root matrix **A** (1.3a) invariably requires simplifying conditions. This section elaborates upon the third case, (1.20c) 'URM4 Pythagoras Conditions', of the three simplifying cases considered in (1.20). Note that this case is independent of the 'embedding' methods given in cases (1.20a) and (1.20b), as detailed further in Section (2). Nevertheless, the merging of embedding and URM4 Pythagoras conditions, under the auspices of 'lifting' solutions, eventually forms the bedrock of all subsequent n-dimensional work, and is covered later in Sections (7), (11) and all of Part II.

In an analogous way to URM3, the URM4 Pythagoras conditions are such that the **X** eigenvector, for eigenvalue $\lambda = C$, is a Pythagorean quadruple, denoted by $\mathbf{X}_+$ ($\sim \mathbf{X}_{4+}$ URM4). Additionally, the conditions generate a second eigenvector for eigenvalue $\lambda = -C$, denoted by $\mathbf{X}_-$ ($\sim \mathbf{X}_{4-}$ URM4), which is also a Pythagorean quadruple. The conditions on the dynamical variables that give these two Pythagorean quadruples can be determined by inspection and analogy with the URM3 equivalent conditions, and leads to the following URM4 'Pythagoras conditions':

(4.1)

(4.1a) $\overline{S} = -S$, $\overline{T} = -T$, $\overline{U} = U$

(4.1b) $\overline{P} = P$, $\overline{Q} = Q$, $\overline{R} = -R$.

It is of note that the partial conditions (4.1b) are the same as the complete URM3 Pythagoras conditions (A18), which means that the URM3 Potential V_3 is always zero (A24), and not the general URM4 Potential V (1.17). This latter point is only made to avoid any confusion between the two Potentials, V and V_3. In general, the method of embedding, Section (2), references the URM3 Potential V_3, whereas it is the URM4 Potential V that is of interest in this section, which may or may not be zero.

Substituting the URM4 conditions (4.1) into the $\mathbf{A}$ matrix (1.3a), gives

$$(4.2)\ \mathbf{A} = \begin{pmatrix} 0 & S & T & U \\ -S & 0 & R & Q \\ -T & -R & 0 & P \\ U & Q & P & 0 \end{pmatrix}.$$

The full DCE is derived from the characteristic equation $\det(\mathbf{A} - \lambda\mathbf{I}) = 0$ (1.8) for eigenvalue $\lambda = C$, and under URM4 Pythagoras conditions (4.1) becomes

(4.3a) the DCE

$$C^2 = P^2 + Q^2 + U^2 - \left(R^2 + S^2 + T^2\right) - \frac{[QT - (PS + RU)]^2}{C^2}.$$

For advance information, justified later in this section, the DCE (4.3a) splits nicely into a kinetic term K

$$(4.3b)\ K = P^2 + Q^2 + U^2 - \left(R^2 + S^2 + T^2\right),$$

derived from (1.18) under conditions (4.1), and a Potential term V, which is basically all the remaining terms, and later derived in this section as

$$(4.3c)\ V = \frac{[QT - (PS + RU)]^2}{C^2}.$$

With the eigenvector $\mathbf{X}_+$ written in the standard, four-coordinate $'w, x, y, z'$ form (1.5), then the dynamical equations $\mathbf{AX} = C\mathbf{X}$ (1.7), for eigenvalue C, are

(4.4)

(4.4a) $Cw = Sx + Ty + Uz$

(4.4b) $Cx = -Sw + Ry + Qz$

(4.4c) $Cy = -Tw - Rx + Pz$

(4.4d) $Cz = Uw + Qx + Py,$

and the coordinates satisfy the 4-dimensional form of the Pythagoras equation

(4.5) $0 = w^2 + x^2 + y^2 - z^2$.

This Pythagorean relation amongst the elements of $\mathbf{X}_+$ (1.5) can be verified by multiplying (4.4a) by w, (4.4b) by x, (4.4c) by y and (4.4d) by z, and then summing appropriately. In fact, using these same arguments, the elements of any eigenvector, for a non-zero eigenvalue, also form a Pythagorean quadruple. This includes the eigenvector $\mathbf{X}_-$ (below), for eigenvalue $-C$ (see further (4.10)), and all complex eigenvectors for any complex eigenvalues that may arise in URMT, see also (4.14b) further below and also Section (2-6).

It can also be verified that for eigenvalue $-C$, with eigenvector equation $\mathbf{AX} = -C\mathbf{X}$, there is another eigenvector $\mathbf{X}_-$:

(4.6) $\mathbf{X}_- = \begin{pmatrix} a \\ b \\ c \\ -d \end{pmatrix}$,

which also satisfies the 4-dimensional form of the Pythagoras equation

(4.7) $d^2 = a^2 + b^2 + c^2$.

The minus in front of the fourth element, i.e. $-d$, is intentional and done in hindsight with URM3 in mind.

Returning to the general characteristic equation (1.9), reproduced below,

(1.9) $\lambda^4 + a_2 \lambda^2 + a_1 \lambda + a_0 = 0$,

the coefficients a_2, a_1, a_0, under Pythagoras conditions (4.1), reduce to the following

(4.8)

(4.8a) $a_0 = -[QT - (PS + RU)]^2$,

(4.8b) $a_1 = 0$,

(4.8c) $a_2 = R^2 + S^2 + T^2 - (P^2 + Q^2 + U^2)$.

With $a_1 = 0$, the characteristic equation simplifies to the following quartic, which is now quadratic in the eigenvalue λ, a fact mentioned earlier in Section (1-4) with Pythagoras in mind.

(1.12) $\lambda^4 + a_2\lambda^2 + a_0 = 0$,

To recap Section (1-4), defining the coefficients in terms of the kinetic and Potential terms, K and V, with $a_1 = 0$ (4.8b),

(1.20c) $a_0 = -VC^2$

(1.16) $a_2 = -K$,

then the characteristic equation (1.12) becomes

(4.9) $\lambda^4 - K\lambda^2 - VC^2 = 0$.

Note that the Potential V here is a true URM4 Potential, not to be confused with the URM3 Potential V_3 (A9).

Since $\lambda = +C$ is a known eigenvalue, and (4.9) is a quadratic in λ^2, then it can be inferred that $\lambda = -C$ is also an eigenvalue, hence two symmetric eigenvalues are now known, i.e.

(4.10) $\lambda = \pm C$.

Upon substitution of either eigenvalue the characteristic equation reduces to

(4.11) $C^4 - KC^2 - VC^2 = 0$,

and dividing throughout by C^2 returns the familiar DCE $C^2 = K + V$ (1.14).

If, instead, the DCE is mandated, and keeping with the coefficient definitions for a_0 (1.20c) and a_2 (1.16), then substituting for the kinetic energy K, in terms of V and C from (1.14), into the characteristic equation (4.9), gives

(4.12) $\lambda^4 - (C^2 - V)\lambda^2 - VC^2 = 0$,

which has a very simple solution

(4.13a) $\lambda^2 = C^2$ or $\lambda^2 = -V$

i.e.

(4.13b) $\lambda = \pm C$ or $\lambda = \pm\sqrt{-V}$.

The first two eigenvalues $\lambda = \pm C$ are already given by (4.10), but the second two eigenvalues are new. Equating the definition $a_0 = -VC^2$ (1.20c) with the expression for the coefficient a_0 (4.8a) gives the Potential in terms of the dynamical variables as per (4.3c). It is clear from (4.3c) that the Potential is a perfect square, and thus always zero or positive. Hence the other two eigenvalues, according to (4.13b), are either zero or complex conjugates of each other, i.e.

(4.14a) $\lambda = \pm C, 0, 0$, $V = 0$, $a_0 = 0$,

or

(4.14b) $\lambda = \pm C, \pm i\sqrt{V}$, $V > 0$.

This is a slightly different situation to URM3 in that the URM4 Potential (4.3c) is always zero or positive here, whereas it could be zero, positive or negative in URM3. Most importantly though, when under URM3 Pythagoras conditions (A18), the URM3 Potential is always zero (A24), whereas under URM4 Pythagoras conditions (4.1) it need not be. In other words, a zero Potential solution is not implicit under URM4 Pythagoras conditions and, instead, an additional criterion is needed: by (1.20c), if $V = 0$ then

(4.15) $a_0 = 0$,

which, by (4.8a), implies that

(4.16) $QT = (PS + RU)$,

and thus the eigenvalues are given by (4.14a).

By (4.14b), if the Potential V is non-zero then the other two eigenvalues are always complex since the potential is always greater than zero by (4.3c).

In all cases, under the Pythagoras conditions (4.1), the Potential controls the eigenvalues, which is also true in URM3.

The inclusion of complex eigenvalues was not considered under URM3, primarily because it was not particularly necessary. Nevertheless, complex eigenvalues can be incorporated into URM3 since it is perfectly possible to generate a complex eigenvalue by application of a local, URM3 variational transformation. Doing so can move the Potential away from zero in either direction, positive or negative. It must be stressed that this does destroy the URM3 Pythagoras conditions, which keep $V_3 = 0$ by definition, and radically changes the URM3 eigenvector landscape. In effect, only the single eigenvector $\mathbf{X}_3$ ($\sim \mathbf{X}_+$ or $\mathbf{X}_{3+}$) and its transpose conjugate $\mathbf{X}^-$ ($\sim \mathbf{X}^{3-}$) remain Pythagorean triples when $V_3 \neq 0$, instead of both $\mathbf{X}_3$, $\mathbf{X}_-$ and their conjugates $\mathbf{X}^-$, $\mathbf{X}^+$, when $V_3 = 0$.

Returning to the subject of URM4 under Pythagoras conditions, a possible, zero-Potential is now examined in the following section.

(5) The Zero Potential $PS+RU$ Solution

As first elaborated in [1], and restated in the introduction, a constant Potential solution is desirable as regards physical properties within URMT since it represents the trajectory of a particle with no force acting upon it. It is therefore of constant kinetic energy, $K = C^2$, by the DCE (1.14).

Before proceeding, it is emphasized that this solution is only valid under URM4 Pythagoras conditions (4.1), as outlined in the previous section. As also mentioned in Section (4), URM4 Pythagoras conditions (4.1) alone are not sufficient to obtain a zero, URM4 Potential (4.3c), and there is an additional condition (4.16) on the dynamical variables, namely,

(4.16) $QT = PS + RU \Rightarrow V = 0$.

(5-1) Eigenvalues

With a zero Potential, under Pythagoras conditions (4.1), the eigenvalues are

(4.14a) $\lambda = \pm C$, $\lambda = 0$ (repeated).

It is fully anticipated, from considerations of URM3, that the repeated $\lambda = 0$ eigenvalue solution will give two eigenvectors, $\mathbf{X}_{0A}$ and $\mathbf{X}_{0B}$, with elements comprising solely of dynamical variables, i.e. no coordinates w, x, y, z (1.5) or scale factors α, β, γ (3.56). This stems from the fact that, under URM3 Pythagoras conditions (A18), there is a single zero eigenvalue $\lambda = 0$, with eigenvector $\mathbf{X}_{30}$ (A33b), comprising only the dynamical variables P, Q, R. Indeed, this will also be seen to be the case in URM4 for the two eigenvectors $\mathbf{X}_{0A}$ and $\mathbf{X}_{0B}$. Ideally then, it would be convenient if the number of dynamical variables were

75

reduced to four, corresponding to the four elements of a URM4, zero eigenvector. This desire fits in nicely with the zero Potential condition expressed as $QT = PS + RU$ in (4.16), which is satisfied by the following choice of dynamical variables P,Q,R,S,T,U

(5.1)

(5.1a) $Q = 0$

(5.1b) $T = 0$

(5.1c) $PS + RU = 0$.

Superficially, this only gives three independent dynamical variables since there are six in total P,Q,R,S,T,U, with the three constraints above. Nevertheless, it is shown later in Section (9), when studying variational methods and the eigenvector evolution, that the $T = 0$ constraint no longer holds true, and T becomes non-zero for arbitrary variations. Therefore, with T unconstrained, there are now only two constraints (5.1a) and (5.1c), giving four of six independent dynamical variables. The DCE, (5.2a) below, is an extra constraint on these four, but with the eigenvalue C acting as an initial, arbitrary parameter, this constraint is effectively removed.

Because of constraint (5.1c), the solution arising from the zero Potential conditions (4.16) is thus termed the 'PS plus RU zero' solution, or 'PS+RU' for short.

With conditions (5.1), the Kinetic K (4.3b) and Potential V (4.3c) terms reduce to

(5.2a) $K = C^2 = (P^2 + U^2) - (R^2 + S^2)$, the DCE.

(5.2b) $V = 0$

Also, with the above conditions (5.1), the unity root matrix reduces to the following simple form, each row having only two non-zero elements

(5.3) $\mathbf{A} = \begin{pmatrix} 0 & S & 0 & U \\ -S & 0 & R & 0 \\ 0 & -R & 0 & P \\ U & 0 & P & 0 \end{pmatrix}$.

Given the four eigenvalues (4.14a), there are four dynamical equations in the eigenvectors $\mathbf{X}_+$, $\mathbf{X}_-$, $\mathbf{X}_{0A}$, $\mathbf{X}_{0B}$,

(5.4a) $\mathbf{AX}_+ = +C\mathbf{X}_+$

(5.4b) $\mathbf{AX}_- = -C\mathbf{X}_-$

(5.4c) $\mathbf{AX}_{0A} = 0$

(5.4d) $\mathbf{AX}_{0B} = 0$.

(5-2) To solve for the eigenvectors

For a general eigenvalue λ then, as outlined in [1]#1, there are several methods to obtain the eigenvectors $\mathbf{X}_\lambda$ of matrix $\mathbf{A}$ (5.3), i.e.

(5.10) $\mathbf{AX}_\lambda = \lambda\mathbf{X}_\lambda$

The standard algebraic method is one such way to obtain a solution for $\mathbf{X}_\lambda$, and is simply a matter of rearranging the dynamical equations (below) to solve for, say, x, y, z in terms of w, the latter acting as a free-parameter.

(5.11)

(5.11a) $\lambda w = Sx + Uz$

(5.11b) $\lambda x = -Sw + Ry$

(5.11c) $\lambda y = -Rx + Pz$

(5.11d) $\lambda z = Uw + Py$

(5.11e) $Q = 0$ (5.1a) and $T = 0$ (5.1b).

The dynamical variables are not so free in that they must satisfy the zero Potential constraint (5.1c) and the DCE (5.2a). It was explained in Section (4) that, for non-zero eigenvalues, the two eigenvectors $\mathbf{X}_+$ and $\mathbf{X}_-$ of the $\mathbf{A}$ matrix (4.2) are naturally Pythagorean quadruples. Therefore, by solving for these eigenvectors $\mathbf{X}_+$ and $\mathbf{X}_-$ by direct algebraic manipulation, the solution obtained will be a parameterisation of Pythagorean quadruples given in terms of the dynamical variables. Indeed, such a parameterisation will emerge but, instead of algebraically solving as above, a simple solution for the dynamical variables, that satisfies the zero Potential conditions (5.1), will first be obtained, and

77

then the residual matrix method, Section (3), used to obtain the eigenvector solutions. It is done this way because, in practice, it is simple and easily generalizes to an arbitrary number of dimensions, where algebraic methods become very clumsy.

To solve specifically for the dynamical variables P, R, S, U, with $Q = 0$, $T = 0$ and $PS + RU = 0$ in accordance with (5.1), the non-zero dynamical variables P, R, S, U are parameterised in terms of four parameters w, x, y, z. These parameters will be found to be the same as those coordinates used in the eigenvector $\mathbf{X}_+$ (1.5). However, until this is demonstrated, w, x, y, z should, for now, be considered as parameters only. Note that when originally developing the work, parameters a, b, c, d were used instead and it was later seen that they formed the eigenvector $\mathbf{X}_+$, where a, b, c, d related to w, x, y, z via simple linear relations $w = c$, $x = -a$, $y = -b$ and $z = d$. It was then decided to re-label all expressions in a, b, c, d using w, x, y, z such that the eigenvector $\mathbf{X}_+$ came out exactly as in (1.5).

A simple solution for the dynamical variables, that satisfies the zero Potential conditions (5.1), is parameterised in terms w, x, y, z as follows:

(5.12)

(5.12a) $P = yz$,

(5.12b) $Q = 0$, (5.1a)

(5.12c) $R = xy$,

(5.12d) $S = -wx$,

(5.12e) $T = 0$, (5.1b)

(5.12f) $U = wz$

where w, x, y, z are subject to the usual co-primality criterion

(1.6) $\gcd(w, x, y, z) = 1$.

Note that this solution was obtained empirically, based upon solving $PS = -RU$ (4.16) as simply and non-trivially as possible. Ultimately its justification is manifest in the results it provides, i.e. the form of the eigenvectors, as will be seen shortly in Section (5-3).

Using the parameterisation (5.12), the DCE (5.2a) becomes

(5.13) $C^2 = (z^2 - x^2)(w^2 + y^2)$

The two, bracketed factors on the right are chosen to each equal the eigenvalue, i.e.

(5.14)

(5.14a) $C = (z^2 - x^2)$

(5.14b) $C = (w^2 + y^2)$.

These conditions mean that the eigenvalue C is both the sum and difference of two squares.

The choice is made such that C is positive by URMT convention.

(5.15) $C > 0 \Rightarrow |z| > |x|$, by (5.14a)

If $|z| = |x|$ then four repeated zero eigenvalues, $\lambda = 0$, are obtained. This solution is not studied further as it is considered too simplistic.

Note that equating the two in (5.14) to eliminate the eigenvalue C gives

(5.16) $z^2 = w^2 + x^2 + y^2$,

and so the ordered quadruple (w, x, y, z) naturally forms a Pythagorean quadruple, justifying the choice of factorisation (5.14). Given $\mathbf{X}_+$ and $\mathbf{X}_-$ are also Pythagorean quadruples, these parameters w, x, y, z will also be seen to form the eigenvectors.

With w, x, y, z forming a Pythagorean quadruple, the smallest, non-trivial quadruple is known to be, to within a sign of the elements,

(5.17) $(w, x, y, z) = (1,2,2,3)$.

With this solution, the smallest value of the eigenvalue C is five since, using (5.14),

(5.18) $x = 2$, $z = 3 \Rightarrow C = 5$.

That w, x, y, z satisfy the 4D Pythagoras equation (5.16) is actually a constraint on these parameters, so they are not now all entirely arbitrary.

Using (5.14a), the eigenvalue factors as follows

(5.19) $C = (z - x)(z + x)$.

The parameters x and z are now, themselves, given in terms of two new parameters f and g as follows

(5.20)

(5.20a) $f = z + x$, $f \in \mathbb{Z}$.

(5.20b) $g = z - x$, $g \in \mathbb{Z}$.

such that

(5.21)

(5.21a) $fg = (z^2 - x^2)$,

and by Pythagoras (5.16) this also implies

(5.21b) $fg = (w^2 + y^2)$.

Rearranging (5.20) gives x and z in terms of f and g as

(5.22)

(5.22a) $x = (f - g)/2$

(5.22b) $z = (g + f)/2$.

The Diophantine equation $fg = (w^2 + y^2)$ (5.21b) has already been solved for URM3 (Appendix (A) in [1],#2); the solution is reproduced below in terms of four new integer parameters i, j, s, t as follows, where a factor of two has been added to each term so as to clear the denominator of 2 in x (5.22a) and z (5.22b)

(5.23)

(5.23a) $f = 2(t^2 + s^2)$, $t, s \in \mathbb{Z}$.

(5.23b) $g = 2(i^2 + j^2)$, $i, j \in \mathbb{Z}$.

(5.24)

(5.24a) $w = 2(is + jt)$

(5.24b) $y = 2(it - js)$.

Finally, eliminating parameters f and g gives x and z in terms of arbitrary parameters i, j, s, t.

(5.24c) $x = (t^2 + s^2) - (i^2 + j^2)$

(5.24d) $z = (t^2 + s^2) + (i^2 + j^2)$.

Equations (5.24) constitute the four-parameter solution for the coordinates w, x, y, z, and are a form of parameterisation for Pythagorean quadruples known as the 'Lebesgue Identity'. This is a better parameterisation than the more usual, three-parameter variant (below), which cannot give all primitive quads, see Section (7-2). To get the three-parameter variant set $t = 0$, which gives

(5.25a) $w = 2is$

(5.25b) $x = s^2 - i^2 - j^2$

(5.25c) $y = -2js$

(5.25d) $z = s^2 + i^2 + j^2$.

With $C > 0$, and hence $|z| > |x|$ by (5.15), this means at least one of i, j is non-zero

(5.26) $|z| > |x| \Rightarrow (i, j) \neq (0,0)$.

The above parameterisation (5.24) only gives positive $z > 0$; for $z < 0$ there is another parameterisation that can be obtained by multiplying each i, j, s, t by the complex number i, but since symbol i (and j) is already used as one of the parameters, writing this out in full using i ambiguously will be confusing. Thus, the solution is simply stated as follows, which is just a sign reversal in each term

(5.27a) $w = -2(is + jt)$

(5.27b) $x = (i^2 + j^2) - (t^2 + s^2)$

(5.27c) $y = 2(js - it)$

(5.27d) $z = -(t^2 + s^2) - (i^2 + j^2)$.

Of course, whilst z is now unconditionally zero or less, by varying the four parameters i, j, s, t suitably, the signs of w, x, y can be varied as desired.

The dynamical variables P, R, S, U are also implicitly specified in terms of parameters i, j, s, t via their explicit parameterisation (5.12) in terms of w, x, y, z.

The solution (5.24) completely solves for the coordinates w, x, y, z and dynamical variables P, R, S, U, with $Q = 0$ and $T = 0$ as pre-conditions. It also hints that the parameters w, x, y, z are elements of the eigenvectors $\mathbf{X}_+$ and $\mathbf{X}_-$ since they satisfy the Pythagoras equation. However, most importantly, it doesn't yet give the actual eigenvectors. This requires further work using the Residual Matrix method, which now follows.

(5-3) Residual matrix method

As outlined in Section (3), the residual matrix method is used to obtain the standard eigenvectors and their reciprocal, row eigenvectors by calculation of a matrix, which is effectively the outer product of a vector and its reciprocal, as in

(5.30a) $\mathbf{E}_+ = \mathbf{X}_+ \mathbf{X}^+$

(5.30b) $\mathbf{E}_0 = -\mathbf{X}_0 \mathbf{X}^0$

(5.30c) $\mathbf{E}_- = \mathbf{X}_- \mathbf{X}^-$.

The residual matrices $\mathbf{E}_+$, $\mathbf{E}_0$, $\mathbf{E}_-$, for eigenvalues $\lambda = C$, $\lambda = 0$ and $\lambda = -C$ respectively, are defined as follows, in accordance with the quadratic factorisation discussed in section (3-2):

(5.31a) $\mathbf{E}_+ = \mathbf{A}(\mathbf{A} + C\mathbf{I})$

(5.31b) $\mathbf{E}_0 = (\mathbf{A}^2 - C^2 \mathbf{I})$

(5.31c) $\mathbf{E}_- = \mathbf{A}(\mathbf{A} - C\mathbf{I})$.

(5.32) To recap the earlier explanation for using a quadratic polynomial, e.g. $\mathbf{E}_+ = \mathbf{A}(\mathbf{A} + C\mathbf{I})$, instead of the full, cubic form, e.g. $\mathbf{E}_+ = \mathbf{A}^2(\mathbf{A} + C\mathbf{I})$: the reason is that a repeated multiplication by $\mathbf{A}$ of $\mathbf{E}_+$ and $\mathbf{E}_-$ only actually produces a new matrix that is just a scalar multiple (eigenvalue C) of the original matrices, $\mathbf{E}_+$ and $\mathbf{E}_-$. This is because the quadratic expressions for $\mathbf{E}_+$ and $\mathbf{E}_-$ produce matrices whose columns are already eigenvectors $\mathbf{X}_+$ and $\mathbf{X}_-$ of $\mathbf{A}$. Hence, repeated multiplication of $\mathbf{E}_+$ and $\mathbf{E}_-$ by $\mathbf{A}$ only scales them by a factor C on each multiplication, without changing the eigenvectors embedded within them.

In the case of $\mathbf{E}_0$, a repeated multiplication by $\mathbf{A}$ to give a cubic, as in $\mathbf{E}_0 = \mathbf{A}(\mathbf{A}^2 - C^2\mathbf{I})$, will actually annihilate the matrix, i.e. return the zero matrix, since the eigenvalue is zero - see the subject of 'minimal', 'minimum' or 'reduced' matrix polynomials in linear algebra texts. There is a downside to not using the full cubic expression for $\mathbf{E}_+$ and $\mathbf{E}_-$, and that is the extra factor, eigenvalue C, is no longer included in the residual matrix. Ultimately, this means that one of the eigenvectors is, strictly speaking, missing a factor C. This is discussed again further below.

Using the solution (5.12) for all six dynamical variables, then the $\mathbf{A}$ matrix (5.3) is written in terms of w, x, y, z as

$$(5.33) \quad \mathbf{A} = \begin{pmatrix} 0 & -wx & 0 & +wz \\ +wx & 0 & +xy & 0 \\ 0 & -xy & 0 & +yz \\ +wz & 0 & +yz & 0 \end{pmatrix}.$$

Using this form for $\mathbf{A}$, the factors $(\mathbf{A}+C\mathbf{I})$ and $(\mathbf{A}-C\mathbf{I})$ are

$$(5.34) \quad \mathbf{A} \pm C\mathbf{I} = \begin{pmatrix} \pm C & -wx & 0 & +wz \\ +wx & \pm C & +xy & 0 \\ 0 & -xy & \pm C & +yz \\ +wz & 0 & +yz & \pm C \end{pmatrix},$$

and the square of $\mathbf{A}$, used later for $\mathbf{E}_0$, is

$$(5.35) \quad \mathbf{A}^2 = \begin{pmatrix} +Cww & 0 & +Cyw & 0 \\ 0 & -Cxx & 0 & +Cxz \\ +Cwy & 0 & +Cyy & 0 \\ 0 & -Cxz & 0 & +Czz \end{pmatrix}.$$

Note that the two relations in (5.14) are used several times in simplifying the matrix elements. This remark also applies to many of the matrix calculations that follow.

Multiplying (5.33) by $(\mathbf{A}+C\mathbf{I})$ (5.34) gives $\mathbf{E}_+$ (5.31a) as follows:

$$(5.36) \ \mathbf{E}_+ = \begin{pmatrix} +Cww & -Cwx & +Cwy & +Cwz \\ +Cxw & -Cxx & +Cxy & +Cxz \\ +Cyw & -Cyx & +Cyy & +Cyz \\ +Czw & -Czx & +Czy & +Czz \end{pmatrix}.$$

Comparing $\mathbf{E}_+$ above with its outer product form $\mathbf{E}_+ = \mathbf{X}_+\mathbf{X}^+$ (5.30a), gives the eigenvectors $\mathbf{X}_+$ and $\mathbf{X}^+$ as

$$(5.37) \ \mathbf{X}_+ = \begin{pmatrix} w \\ x \\ y \\ z \end{pmatrix},$$

$$(5.38) \ \mathbf{X}^+ = \begin{pmatrix} +Cw & -Cx & +Cy & +Cz \end{pmatrix}.$$

It is now confirmed that parameters w, x, y, z are, indeed, the same as the coordinate elements (w, x, y, z) of $\mathbf{X}_+$ (1.5).

With $\mathbf{X}_+$ as above (5.37), the dynamical equation $\mathbf{A}\mathbf{X}_+ = C\mathbf{X}_+$ (5.4a) can be verified, with the help of relations (5.14). Likewise, the reciprocal row eigenvector equation $\mathbf{X}^+\mathbf{A} = C\mathbf{X}^+$ is also verified.

(5.39) Notice that the eigenvalue C has been assigned to the $\mathbf{X}^+$ row eigenvector, and the $\mathbf{X}_+$ vector has been left in an unscaled, primitive form, according to $\gcd(w, x, y, z) = 1$ (1.6). This gcd condition on the elements $\mathbf{X}_+$ continues the practice followed in URM3 where the $\mathbf{X}_+$ eigenvector is kept primitive. In fact, by inspection of the eigenvector equation (5.4a), any common factor in w, x, y, z, and hence in $\mathbf{X}_+$, could actually be absorbed into the eigenvalue C and the matrix $\mathbf{A}$ (5.3), i.e. absorbed into dynamical variables P, R, S, U. For example, if $\gcd(w, x, y, z) = k$ for some integer constant k, then removing k from $\mathbf{X}_+$ so that $\mathbf{X}_+$ is now primitive again, i.e. $\gcd(w, x, y, z) = 1$, would give an eigenvector equation $(\mathbf{A}k)\mathbf{X}_+ = (Ck)\mathbf{X}_+$. The eigenvalue is now Ck for a transformed (scaled) matrix Ak. This common scale factor in Ak can actually be transformed away using variational methods; see [1],#6, for more information on the general subject of non-unity eigenvalues and the related subject of scaling dynamical variables by a common factor.

Returning to the determination of the eigenvectors X_- and X^-, the residual matrix E_- is obtained by multiplying (5.33) by $(A - CI)$ (5.34) to give

$$(5.40) \quad E_- = \begin{pmatrix} +Cww & +Cwx & +Cwy & -Cwz \\ -Cxw & -Cxx & -Cxy & +Cxz \\ +Cyw & +Cyx & +Cyy & -Cyz \\ -Czw & -Czx & -Czy & +Czz \end{pmatrix}.$$

Again, the two relations (5.14) have been used several times in simplifying the matrix elements of E_-.

Comparing E_- (5.40) with its outer-product form $E_- = X_-X^-$ (5.31c), gives the eigenvectors X_- and X^- as

$$(5.41) \quad X_- = \begin{pmatrix} +Cw \\ -Cx \\ +Cy \\ -Cz \end{pmatrix},$$

$$(5.42) \quad X^- = \begin{pmatrix} +w & +x & +y & -z \end{pmatrix}.$$

The elements of X_+ and X_- are seen to be almost identical, i.e. both contain w, x, y, z in the same position, barring a sign and scaling by the eigenvalue C (see below). This is in contrast to the URM3 zero Potential solution, where X_{3+} (A3) and X^{3-} (A35c) comprise coordinates x, y, z and scale factors $\alpha_3, \beta_3, \gamma_3$, which are not the same quantities, albeit related (A15). In fact, this similarity between the elements of X_+ and X_- is a URM4 feature unique to this solution constraining both Q and T to zero, and when the subjects of invariance transformations and eigenvector evolution are studied (Sections (8) and (9)), the two vectors will no longer share the same elements.

85

In the definitions of $\mathbf{X}_-$ (5.41) and $\mathbf{X}^-$ (5.42), the eigenvalue C scales the $\mathbf{X}_-$ vector, and the $\mathbf{X}^-$ remains primitive, i.e. unscaled by C. This is the opposite to the vectors $\mathbf{X}_+$ (5.37) and $\mathbf{X}^+$ (5.38), where $\mathbf{X}_+$ remains primitive but $\mathbf{X}^+$ is scaled, as mentioned above. This raises the immediate question as to why the eigenvector $\mathbf{X}_-$ has an embedded factor of the eigenvalue C whereas $\mathbf{X}_+$ does not? Similarly, why does $\mathbf{X}^+$ have an embedded scale factor of C, that is not present in $\mathbf{X}^-$? This omission of a scale factor C has been mentioned earlier (5.32), and arises due to using a quadratic, not cubic, polynomial form for the residual matrices $\mathbf{E}_+$ (5.31a) and $\mathbf{E}_-$ (5.31c). Given, for now, that this is the case, the explanation for the partitioning of C amongst the eigenvectors is that it is actually done to be consistent with conjugate (reciprocal) relations; see (A38) for URM3 and Appendix (E) for general n-dimensional, URMT. The detailed explanation is as follows: the conjugate relation (E5) relates $\mathbf{X}^-$, i.e. $\overline{\mathbf{X}}_+$, to $\mathbf{X}_+$ by the $\mathbf{T}$ operator transformation

(E5) $\mathbf{X}^- = \overline{\mathbf{X}}_+ = (\mathbf{T}\mathbf{X}_+)^T$.

Using the transpose rule $(AB)^T = B^T A^T$ for matrices, and since $\mathbf{T}^T = \mathbf{T}$ (E4), then this becomes

(5.43) $\mathbf{X}^- = \overline{\mathbf{X}}_+ = \mathbf{X}_+^T \mathbf{T}$.

Using (5.37) for $\mathbf{X}_+$ and (E3) for $\mathbf{T}$, this becomes

(5.44) $\mathbf{X}^- = \overline{\mathbf{X}}_+ = (+w \quad +x \quad +y \quad -z)$

Thus, looking at (5.44), if $\mathbf{X}^-$ had an embedded scale factor C then the conjugate transform (5.43) would also have to contain the scale factor, i.e. $\mathbf{X}^- = \overline{\mathbf{X}}_+ = (C\mathbf{T}\mathbf{X}_+)^T$, since $\mathbf{X}_+$ is primitive with no common factor C.

Likewise, the conjugate transformation (E7) relates $\mathbf{X}^+$, i.e. $\overline{\mathbf{X}}_-$, to $\mathbf{X}_-$ by the $\mathbf{T}$ operator transformation,

(E7) $\mathbf{X}^+ = \overline{\mathbf{X}}_- = \mathbf{X}_-^T \mathbf{T}$

Using (5.41) for $\mathbf{X}_-$ and (E3) for $\mathbf{T}$, this becomes

(5.45) $\mathbf{X}^+ = \overline{\mathbf{X}}_- = \left(+Cw \quad -Cx \quad +Cy \quad +Cz \right)$

And, by similar arguments to $\mathbf{X}^-$ (5.44), if $\mathbf{X}_-$ didn't have the scale factor C then the conjugate transform (5.45) would also have to contain the scale factor, i.e. $\mathbf{X}^+ = \overline{\mathbf{X}}_- = (C\mathbf{T}\mathbf{X}_-)^T$.

In brief, the distribution of the scale factor C amongst the four eigenvectors $\mathbf{X}_+$, $\mathbf{X}_-$, $\mathbf{X}^+$ and $\mathbf{X}^-$ is to obtain consistent conjugate transformations in accordance with Appendix (E). Having said all this, it would be simpler if the extra factor C was included in the first place by using a cubic polynomial form for the residual matrices instead of a quadratic form, see (5.32). The only disadvantage would be that the eigenvectors $\mathbf{X}_+$ and $\mathbf{X}^-$ would no longer be primitive, i.e. they would not satisfy the gcd condition (1.6), but now have a gcd of C instead. Of course, if $C = 1$, this would make no difference, except it is never unity in this particular PS+RU solution. Omitting the factor C also raises an additional issue with consistency between the elements of an eigenvector and their conjugates, e.g. w, x, y, z and $\overline{w}, \overline{x}, \overline{y}, \overline{z}$ in $\mathbf{X}_+$ and $\overline{\mathbf{X}}_+$ respectively - see Appendix (A39) (URM3 only) about conjugation of the eigenvector elements under Pythagoras conditions. However, all these problematic issues are largely artificial, arising due to simplifications and/or definitions, and it suffices here to make the reader aware of them and move on, especially since this PS+RU solution is supplied more as an educational example rather than a platform for further development.

(5-3) Derivation of the Zero Eigenvectors

To obtain the eigenvectors $\mathbf{X}_{0A}$ and $\mathbf{X}_{0B}$ for the repeated eigenvalue $\lambda = 0$, $\mathbf{E}_0$ is calculated from (5.31b) using $\mathbf{A}^2$ (5.35) and relations (5.14) to give

(5.50) $\mathbf{E}_0 = \begin{pmatrix} -Cyy & 0 & +Cyw & 0 \\ 0 & -Czz & 0 & +Czx \\ +Cwy & 0 & -Cww & 0 \\ 0 & -Cxz & 0 & +Cxx \end{pmatrix}.$

This residual matrix $\mathbf{E}_0$ is then further decomposed as the sum of two sub-matrices, designated $\mathbf{E}_{0A}$ and $\mathbf{E}_{0B}$, as follows

(5.51) $\mathbf{E}_0 = \mathbf{E}_{0A} + \mathbf{E}_{0B}$

(5.52) $\mathbf{E}_{0A} = \begin{pmatrix} -Cyy & 0 & +Cyw & 0 \\ 0 & 0 & 0 & 0 \\ +Cwy & 0 & -Cww & 0 \\ 0 & 0 & 0 & 0 \end{pmatrix}$

(5.53) $\mathbf{E}_{0B} = \begin{pmatrix} 0 & 0 & 0 & 0 \\ 0 & -Czz & 0 & +Czx \\ 0 & 0 & 0 & 0 \\ 0 & -Cxz & 0 & +Cxx \end{pmatrix}$

By comparing $\mathbf{E}_{0A}$ and $\mathbf{E}_{0B}$ with their outer-product forms, $\mathbf{E}_{0A} = -\mathbf{X}_{0A}\mathbf{X}^{0A}$ and $\mathbf{E}_{0B} = -\mathbf{X}_{0B}\mathbf{X}^{0B}$ respectively, then eigenvectors $\mathbf{X}_{0A}$, $\mathbf{X}_{0B}$ and their reciprocal row eigenvectors $\mathbf{X}^{0A}$, $\mathbf{X}^{0B}$ are assigned as follows:

(5.54a) $\mathbf{X}_{0A} = \begin{pmatrix} -y \\ 0 \\ +w \\ 0 \end{pmatrix}$ (5.54b) $\mathbf{X}_{0B} = \begin{pmatrix} 0 \\ +z \\ 0 \\ +x \end{pmatrix}$

(5.54c) $\mathbf{X}^{0A} = \begin{pmatrix} -Cy & 0 & +Cw & 0 \end{pmatrix}$

(5.54d) $\mathbf{X}^{0B} = \begin{pmatrix} 0 & +Cz & 0 & -Cx \end{pmatrix}$.

This completes the determination of the eigenvectors for each eigenvalue.

(5-4) Vector Inner Products

Having obtained all four eigenvectors and their conjugates, the full suite of inner products can be evaluated, and will be seen to give the familiar URM3 invariants and conservation equations, Appendix (F). Most importantly, the quadratic nature of the invariant quantity, i.e. $+C^2$, is retained. This identity with URM3 is highly desirable when it comes to the geometric and physical considerations in both URM3 and URM4.

For ease of reference, the four eigenvectors and their conjugates are reproduced below,

$$(5.55) \quad \mathbf{X}_+ = \begin{pmatrix} w \\ x \\ y \\ z \end{pmatrix}, \quad \mathbf{X}_{0A} = \begin{pmatrix} -y \\ 0 \\ +w \\ 0 \end{pmatrix}, \quad \mathbf{X}_{0B} = \begin{pmatrix} 0 \\ +z \\ 0 \\ +x \end{pmatrix}, \quad \mathbf{X}_- = \begin{pmatrix} +Cw \\ -Cx \\ +Cy \\ -Cz \end{pmatrix}$$

(5.56)

$$\mathbf{X}^+ = (+Cw \quad -Cx \quad +Cy \quad +Cz)$$

$$\mathbf{X}^{0A} = (-Cy \quad 0 \quad +Cw \quad 0)$$

$$\mathbf{X}^{0B} = (0 \quad +Cz \quad 0 \quad -Cx)$$

$$\mathbf{X}^- = (+w \quad +x \quad +y \quad -z).$$

The inner products are calculated as follows, using the above, (5.55) and (5.56), and relations (5.14) :

(5.57)

$$\mathbf{X}^+\mathbf{X}_+ = C(w^2 - x^2 + y^2 + z^2) = 2C^2$$

$$\mathbf{X}^-\mathbf{X}_- = C(w^2 - x^2 + y^2 + z^2) = 2C^2$$

$$\mathbf{X}^-\mathbf{X}_+ = w^2 + x^2 + y^2 - z^2 = 0$$

$$\mathbf{X}^+\mathbf{X}_- = C^2(w^2 + x^2 + y^2 - z^2) = 0$$

$$\mathbf{X}^{0B}\mathbf{X}_{0A} = 0, \quad \mathbf{X}^{0A}\mathbf{X}_{0B} = 0$$

$$\mathbf{X}^{0A}\mathbf{X}_{0A} = C(w^2 + y^2) = C^2$$

$$\mathbf{X}^{0B}\mathbf{X}_{0B} = C(z^2 - x^2) = C^2$$

$$\mathbf{X}^{0A}\mathbf{X}_+ = 0, \quad \mathbf{X}^+\mathbf{X}_{0A} = 0$$

$$\mathbf{X}^{0B}\mathbf{X}_+ = 0, \quad \mathbf{X}^+\mathbf{X}_{0B} = 0$$

$$\mathbf{X}^{0A}\mathbf{X}_- = 0, \quad \mathbf{X}^-\mathbf{X}_{0A} = 0$$

$$\mathbf{X}^{0B}\mathbf{X}_- = 0, \ \mathbf{X}^-\mathbf{X}_{0B} = 0$$

(5-5) Dimensional Considerations

In this URM4 solution the eigenvector $\mathbf{X}_-$ comprises coordinates w, x, y, z, just like the $\mathbf{X}_+$ vector. This is contrary to URM3 whereby the $\mathbf{X}_-$ solution comprised unique, scale factors, $\alpha_3, \beta_3, \gamma_3$, and only $\mathbf{X}_+$ comprised coordinates x, y, z. Likewise, in this URM4 solution, the vectors $\mathbf{X}_{0A}$ and $\mathbf{X}_{0B}$ also both comprise coordinates whereas, in URM3, they were exclusive functions of the dynamical variables. In fact, all four eigenvectors $\mathbf{X}_+$, $\mathbf{X}_-$, $\mathbf{X}_{0A}$ and $\mathbf{X}_{0B}$ are linear in the coordinates. Were it not for the scaling by the eigenvalue C in $\mathbf{X}_-$, all four standard forms would have the same physical units, i.e. those of the coordinates. This is most definitely not the case in URM3 where each is usefully associated with a different physical quantity under the standard physical interpretation, Appendix (J). This dimensionality issue is another reason to discount this PS+RU solution on physical grounds, i.e. not enough variety in the physical quantities involved.

Although there are a few minor issues, such as the above dimensionality and the uneven distribution of the eigenvalue C, as mentioned earlier in Section (5-2), the main concern over this PS+RU solution is its non-unity eigenvalue. Given [1] showed that URMT need only be studied for unity eigenvalues, and all other cases could be derived from it, it would be nicer if such a simple unity eigenvalue solution could be obtained for URM4 (and URM5). In fact, there is a general resolution to obtain unity eigenvalue solutions by 'lifting' URM3 solutions, detailed in Section (7). This lifting method, which is really just a form of embedding (Section (2)), actually resolves many issues, such as obtaining an invariant, zero Potential solution, and will form the basis of Part II and higher dimensional URMT in general. Beforehand, however, another simple solution is given for a unity eigenvalue case.

(6) URM4 Unity Eigenvalue 2a2p1 Solution

This solution is essentially a variant of the PS+RU zero solution in the previous section, but for a unity eigenvalue,

(6.1a) $C = 1$

(6.1b) $\mathbf{AX} = \mathbf{X}$ (1.7) for $C = 1$.

The aforementioned PS+RU solution suffered from having a non-unity eigenvalue, with a minimum value $C = 5$ (5.18). However, there is a desire in URMT, inherited from URM3, for unity eigenvalue solutions, since in [1],#6 it is shown that URM3 can be solved for any arbitrary eigenvalue using just the unity eigenvalue solution. This extends to URM4, and URMT in general, as the arguments in [1] apply to any formulation URMn, $n \geq 3$.

(6-1) Analytic Solution

The Potential V (4.3c) is zero, as for the previous PS+RU solution, but in this 2a2p1 case, instead of constraints (5.1), where $Q = 0$, $T = 0$ and $(PS + RU) = 0$, only the single dynamical variable U is set to zero, with the only other constraint being $QT - PS = 0$, i.e.

(6.2)

(6.2a) $U = 0$

(6.2b) $QT - PS = 0$,

With these settings, the URM4 Potential (4.3c) is zero, i.e.

(6.3) $V = \dfrac{\left[QT - (PS + RU)\right]^2}{C^2} = 0$.

By reducing the three conditions of the PS+RU solution to two, this, in practice, allows for an arbitrary eigenvalue, but still permits a zero Potential. As a consequence, the eigenvalue C is chosen to be unity (6.1a) so as to restore URM3 compatibility.

Because the Potential is zero, and the eigenvalue unity, the DCE (1.14) is simply the unity kinetic term

(6.4) $K = 1$, the DCE.

To satisfy constraint (6.2b), the dynamical variables P, Q, S, T are assigned in a similar but not identical fashion to the PS+RU solution (5.12), in terms of four integer parameters a, b, c, d, i.e.

(6.5a) $P = ab$, $Q = -bc$, $R = P = ab$, $S = -2cd$, $T = 2ad$, $U = 0$,

(6.5b) $a, b, c, d \in \mathbb{Z}$.

Note that the earlier PS+RU solution (5.12) used w, x, y, z as parameters since the solution for eigenvector $\mathbf{X}_+$ (5.37) had these very same parameters as elements of the vector. However, in this unity-eigenvalue case, this is not so and general parameters a, b, c, d are used instead.

Parameters c, d are defined as follows, which effectively eliminates two of the four parameters leaving just a and b

(6.6) $c = 1$, $d = a$.

Inserting these values for c, d into (6.5a) gives for the dynamical variables

(6.7) $P = ab$, $Q = -b$, $R = P = ab$, $S = -2a$, $T = 2a^2$, $U = 0$.

The full DCE (4.3a), when under URM4 Pythagoras conditions (4.1), and with a zero Potential (6.3), comprises only the Kinetic term K (4.3b). Furthermore, using the fact that $C = 1$, $U = 0$ and $R = P$, then the DCE (4.3a) becomes

(6.8) $1 = Q^2 - S^2 - T^2$, the DCE $\sim K = 1$ (6.4).

This form will be used again shortly as an alternate justification for the parametric solution (6.7). For now, using (6.7), then after some rearrangement the DCE (6.8) becomes

(6.9) $b^2 = 4a^4 + 4a^2 + 1 = (2a^2 + 1)^2$,

so that parameter b is now given in terms of a as

(6.10) $b = \pm(2a^2 + 1)$.

From here onward only the positive form of (6.10) is used, i.e.

(6.11) $b = 2a^2 + 1$.

This parameterisation is termed '2a2p1' as an abbreviation of its 'two a-squared plus one' form, and hence the solution is also coined this name.

With this expression (6.11) for b in terms of a, there now remains only one free parameter, i.e. a. However, for simplifying the notation, b is often retained in expressions even though it is not an independent or free parameter.

With only one parameter a specifying what is anticipated to be a Pythagorean quadruple as an eigenvector, which would normally be specified with three, free parameters, this is obviously already a very restricted solution. For the purposes of this 2a2p1 solution, and as an example of the URM4 unity eigenvalue problem, this is good enough, and satisfactory, like the previous PS+RU solution, for illustrating properties of URM4, including further advancement on variational methods, Sections (7) to (11).

Returning to the DCE (6.8), on rearrangement this form makes the quadruple $(1, S, T, Q)$ a Pythagorean quadruplet as in

(6.12) $Q^2 = 1 + S^2 + T^2$,

This equation is a solved problem, as given by the parameterisation, (5.24) or (5.27), for the Pythagorean quadruple (w, x, y, z) in terms of four arbitrary integer parameters i, j, s, t. Using the following settings with $t = 0$ in (5.27),

(6.13) $s = i = a$, $j = 1$, $t = 0$,

and equating (w, x, y, z) in (5.27) with the ordered quadruple $(S, 1, T, Q)$, using $b = 2a^2 + 1$ (6.11), then the three dynamical variables S, T, Q are parameterised as follows

(6.14) $S = x = -2a$, $T = y = 2a^2$, $Q = z = -b$.

Finally, using constraint $QT - PS = 0$ (6.2b), then P is obtained as follows, which agrees with (6.7) and is also equal to R by definition (6.5a)

(6.15) $P = ab$.

This completes the parametric determination of the 2a2p1 solution for a unity eigenvalue, zero Potential and conserved kinetic quantity $K = 1$ (6.4).

Although completely arbitrary, the parameter a is not an evolutionary parameter unlike, for example, the parameter s in Section (2-3), or the commonly used URM3 parameter 'm' (or δ) throughout [1] and Part I of this book. The point is made because evolutionary parameters play an important role in the physical, time-evolution aspects of URMT, especially in Part II of this book, and so it is necessary to outline when a free parameter is or isn't of a temporal nature.

For zero Potential (6.3), and unity eigenvalue (6.1a), the full set of four eigenvalues is, by (4.14a),

(6.16) $\lambda = \pm 1, 0, 0$.

Determination of the eigenvector $\mathbf{X}_+$, for eigenvalue $C = 1$, is best solved analytically for reasons of simplicity, and then the residual matrix method is used to determine the eigenvector $\mathbf{X}^+$.

Substituting for dynamical variables S, T, U, P, Q, R (6.7) into $\mathbf{A}$ (4.2) gives the unity root matrix as

$$(6.17) \ \mathbf{A} = \begin{pmatrix} 0 & -2a & +2a^2 & 0 \\ +2a & 0 & +ab & -b \\ -2a^2 & -ab & 0 & +ab \\ 0 & -b & +ab & 0 \end{pmatrix}.$$

Solving the unity eigenvector equation (6.1b) by algebraic methods gives the analytic solution for $\mathbf{X}_+$ (1.5) as

(6.18)

(6.18a) $w = 2a$

(6.18b) $x = b - 2$

(6.18c) $y = 2a$ (note $y = w$)

94

(6.18d) $z = b$.

$$(6.19) \ \mathbf{X}_+ = \begin{pmatrix} 2a \\ b-2 \\ 2a \\ b \end{pmatrix}$$

The row eigenvector $\mathbf{X}^+$ is calculated from the quadratic expression (5.31a) for residual matrix $\mathbf{E}_+$ and unity eigenvalue (6.1a) as

(6.20)

$$\mathbf{E}_+ = \mathbf{A}(\mathbf{A}+\mathbf{I}) =$$

$$\begin{pmatrix} 0 & -2a & +2a^2 & 0 \\ +2a & 0 & +ab & -b \\ -2a^2 & -ab & 0 & +ab \\ 0 & -b & +ab & 0 \end{pmatrix} \begin{pmatrix} +1 & -2a & +2a^2 & 0 \\ +2a & +1 & +ab & -b \\ -2a^2 & -ab & +1 & +ab \\ 0 & -b & +ab & +1 \end{pmatrix}.$$

There is no need to calculate all elements of $\mathbf{E}_+$, and a single row is sufficient when already given the vector $\mathbf{X}_+$. In this particular example, by inspection of (6.20), it is relatively easy to calculate the bottom row of $\mathbf{E}_+$. If $\mathbf{X}^+$ is denoted by

$$(6.21) \ \mathbf{X}^+ = \begin{pmatrix} w^+ & x^+ & y^+ & z^+ \end{pmatrix},$$

and using $\mathbf{X}_+$ (6.19) with a fourth element z, where $z = b$, then the fourth, bottom row of $\mathbf{E}_+$, as denoted by $\mathbf{E}_+(4,-)$ and equivalent to $\mathbf{X}_+\mathbf{X}^+(4,-)$, is simply

$$(6.22) \ \mathbf{E}_+(4,-) = b\mathbf{X}^+.$$

Alternatively, calculating the fourth row of (6.20) by the usual matrix multiplication, and dividing throughout by b, gives $\mathbf{X}^+$ as

$$(6.23) \ \mathbf{X}^+ = \begin{pmatrix} -a(b+1) & -(a^2b+1) & -a(b-1) & +b(a^2+1) \end{pmatrix}.$$

The $\mathbf{T}$ operator relation (E7), for $n = 4$, is used to derive $\mathbf{X}^-$ from $\mathbf{X}_+$, and $\mathbf{X}_-$ from $\mathbf{X}^+$, to give

$$(6.24) \ \mathbf{X}^- = \begin{pmatrix} 2a & b-2 & 2a & -b \end{pmatrix}$$

$$(6.25)\ \mathbf{X}_- = \begin{pmatrix} -a(b+1) \\ -(a^2b+1) \\ -a(b-1) \\ -b(a^2+1) \end{pmatrix}.$$

Note that although the analytic solution for $\mathbf{X}_+$ (6.18) is fairly simple and limited, in so far as $w = y$ and $x = z - 2$, the $\mathbf{X}^+$ (6.21) solution is less trivial with four distinct elements.

Moving on to the two eigenvectors $\mathbf{X}_{0A}$ and $\mathbf{X}_{0B}$ and their conjugates. Because there are two repeated eigenvalues of zero, using the residual method to determine the eigenvectors is not straightforward, and the analytic solution to the eigenvector equations $\mathbf{AX}_0 = 0$ is more simply solved by direct algebraic manipulation. Doing just so gives the following two eigenvectors $\mathbf{X}_{0A}$ and $\mathbf{X}_{0B}$

$$(6.26)\ \mathbf{X}_{0A} = \begin{pmatrix} b \\ 0 \\ 0 \\ 2a \end{pmatrix},\ \mathbf{X}_{0B} = \begin{pmatrix} 0 \\ a \\ 1 \\ a \end{pmatrix}.$$

Likewise, solving $\mathbf{X}^0 \mathbf{A} = 0$ algebraically gives $\mathbf{X}^{0A}$ and $\mathbf{X}^{0B}$ as follows:

$$(6.27)\ \mathbf{X}^{0A} = \begin{pmatrix} b & 0 & 0 & -2a \end{pmatrix},\ \mathbf{X}^{0B} = \begin{pmatrix} 0 & a & 1 & -a \end{pmatrix}.$$

Notice that (6.26) and (6.27) are related, as expected, by the $\mathbf{T}$ operator transformations (E6).

By the rules of matrix algebra (F25), $\mathbf{X}^{0A}$ and $\mathbf{X}^{0B}$ are both orthogonal to $\mathbf{X}_+$ and $\mathbf{X}_-$, and the following inner products are thus zero:

(6.28)

$$\mathbf{X}^+ \mathbf{X}_{0A} = \mathbf{X}^{0A} \mathbf{X}_+ = 0$$

$$\mathbf{X}^+ \mathbf{X}_{0B} = \mathbf{X}^{0B} \mathbf{X}_+ = 0$$

$$\mathbf{X}^- \mathbf{X}_{0A} = \mathbf{X}^{0A} \mathbf{X}_- = 0$$

$$\mathbf{X}^- \mathbf{X}_{0B} = \mathbf{X}^{0B} \mathbf{X}_- = 0.$$

However, $\mathbf{X}^{0A}$ and $\mathbf{X}^{0B}$ are not orthogonal to either of their conjugates, $\mathbf{X}_{0A}$ and $\mathbf{X}_{0B}$ respectively, which is actually contrary to the PS+RU solution. Given they share the same eigenvalue (zero), this is not surprising since the matrix orthogonality rule (F25) only applies to eigenvectors of different eigenvalues. Suffice to say, orthogonality between the zero eigenvectors is a nicety but not a necessity. However, a complete set of orthogonal eigenvectors is preferable, and only those solutions with a natural set of orthogonal eigenvectors are generally advanced in URMT development, e.g. the general solution in Part II, Section (14-9).

The non-zero inner products of $\mathbf{X}_{0A}$ and $\mathbf{X}_{0B}$ are as follows:

(6.29)

$$\mathbf{X}^{0A}\mathbf{X}_{0B} = -2a^2$$

$$\mathbf{X}^{0B}\mathbf{X}_{0A} = -2a^2$$

$$\mathbf{X}^{0A}\mathbf{X}_{0A} = b^2 - 4a^2, = 4a^4 + 1 \text{ using (6.11)}$$

$$\mathbf{X}^{0B}\mathbf{X}_{0B} = 1.$$

Because this 2a2p1 solution, like the PS+RU solution, is only an example of a URM4 eigenvalue problem, under Pythagoras conditions, further work on the 2a2p1 eigenvectors, $\mathbf{X}^{0A}$ and $\mathbf{X}^{0B}$, is not pursued. However, before completing this example, a few numeric examples of eigenvectors $\mathbf{X}_{+}$ and $\mathbf{X}_{-}$ are given to illustrate how the eigenvector complexity parametrically increases with a.

(6-2) Example Case 1 $a = 0 \Rightarrow b = 1$

This case gives a trivial solution, i.e. one or more elements are zero, where non-trivial means that all four elements are non-zero, see (I10).

(6.30)

$$\mathbf{X}_{+} = \begin{pmatrix} 0 \\ -1 \\ 0 \\ +1 \end{pmatrix}, \mathbf{X}_{-} = \begin{pmatrix} 0 \\ -1 \\ 0 \\ -1 \end{pmatrix}$$

$$\mathbf{X}^{+} = \begin{pmatrix} 0 & -1 & 0 & +1 \end{pmatrix},$$

$$\mathbf{X}^- = \begin{pmatrix} 0 & -1 & 0 & -1 \end{pmatrix}$$

$$\mathbf{X}^+\mathbf{X}_+ = \mathbf{X}^-\mathbf{X}_- = 2C^2 = 2$$

$$\mathbf{X}^+\mathbf{X}_- = \mathbf{X}^-\mathbf{X}_+ = 0$$

(6-3) Example Case 2 $a = 1 \Rightarrow b = 3$

This case gives the smallest magnitude, non-trivial solution, i.e. all four eigenvector elements are non-zero

(6.31)

$$\mathbf{X}_+ = \begin{pmatrix} 2 \\ 1 \\ 2 \\ 3 \end{pmatrix}, \quad \mathbf{X}_- = \begin{pmatrix} -4 \\ -4 \\ -2 \\ -6 \end{pmatrix}$$

$$\mathbf{X}^+ = \begin{pmatrix} -4 & -4 & -2 & +6 \end{pmatrix}$$

$$\mathbf{X}^- = \begin{pmatrix} 2 & 1 & 2 & -3 \end{pmatrix}$$

$$\mathbf{X}^+\mathbf{X}_+ = \mathbf{X}^-\mathbf{X}_- = 2C^2 = 2$$

$$\mathbf{X}^+\mathbf{X}_- = \mathbf{X}^-\mathbf{X}_+ = 0$$

In this example, barring a scale factor of -2, $\mathbf{X}_+$ and $\mathbf{X}_-$ are similar but not identical, i.e. they represent what is essentially the same Pythagorean quadruple $(1,2,2,3)$. Nevertheless, $\mathbf{X}_+$ and $\mathbf{X}_-$ are linearly independent, as expected, for unique eigenvalues $+C$ and $-C$ respectively.

(6-4) Example Case 3 $a = 2 \Rightarrow b = 9$

Lastly, for the case $a = 2$, both eigenvectors are unique (not the same Pythagorean quadruple)

(6.33)

$$\mathbf{X}_+ = \begin{pmatrix} 4 \\ 7 \\ 4 \\ 9 \end{pmatrix}, \quad \mathbf{X}_- = \begin{pmatrix} -20 \\ -37 \\ -16 \\ -45 \end{pmatrix}$$

$$\mathbf{X}^+ = \begin{pmatrix} -20 & -37 & -16 & +45 \end{pmatrix}$$

$$\mathbf{X}^- = \begin{pmatrix} 4 & 7 & 4 & -9 \end{pmatrix}$$

$$\mathbf{X}^+\mathbf{X}_+ = \mathbf{X}^-\mathbf{X}_- = 2C^2 = 2$$

$$\mathbf{X}^+\mathbf{X}_- = \mathbf{X}^-\mathbf{X}_+ = 0$$

A look at the inner products in all three examples $\mathbf{X}^+\mathbf{X}_+$ and $\mathbf{X}^-\mathbf{X}_-$ (6.33e) shows they are all equal to $+2C^2$, i.e. $+2$ when $C=1$ (6.1a); see also Appendix (F). This is algebraically confirmed for arbitrary parameter a and means that this zero Potential, 2a2p1 solution, once again, has the same quadratic invariants as per URM3 for a zero Potential (6.3).

(7) Lifting Solutions

(7-1) Introduction

Lifting, in the context of URMT, is the process of generating eigenvector solutions for an $(n+1)\times(n+1)$ matrix $\mathbf{A}_{n+1}$, $n\geq 2$, using a single eigenvector solution $\mathbf{X}_n$ to an $n\times n$ matrix $\mathbf{A}_n$, for eigenvalue C. The matrix $\mathbf{A}_n$ is embedded in $\mathbf{A}_{n+1}$, and an eigenvector solution $\mathbf{X}_n$, to $\mathbf{A}_n$, is also an eigenvector to $\mathbf{A}_{n+1}$, with appropriate zero padding.

For example, if

(7.1) $\mathbf{A}_n\mathbf{X}_n = C\mathbf{X}_n$,

and defining $\mathbf{X}_{n+1}$ as the $n+1$ dimensional, zero-padded form of $\mathbf{X}_n$, and the matrix $\mathbf{A}_{n+1}$ in terms of $\mathbf{A}_n$, as follows

(7.2a) $\mathbf{X}_{n+1} = \begin{pmatrix} 0 \\ \mathbf{X}_n \end{pmatrix}$

(7.2b) $\mathbf{A}_{n+1} = \begin{pmatrix} 0 & s\overline{\mathbf{X}}_n \\ s\mathbf{X}_n & \mathbf{A}_n \end{pmatrix}$,

where $\overline{\mathbf{X}}_n$ is defined as orthogonal to $\mathbf{X}_n$, i.e.

(7.2c) $\overline{\mathbf{X}}_n \cdot \mathbf{X}_n = 0$, orthogonality cf. (2.9)

then the lifted solution $\mathbf{X}_{n+1}$ satisfies

(7.2d) $\mathbf{A}_{n+1}\mathbf{X}_{n+1} = C\mathbf{X}_{n+1}$.

This, by itself, is evidently no big deal since, firstly, it is the same as embedding, Section (2), but generalised here from four dimensions to n-dimensions. Secondly, as for embedding, the eigenvector $\mathbf{X}_{n+1}$ is really just a zero-padded version of $\mathbf{X}_n$. However, the main advantage of lifting is that it is done under Pythagoras conditions, and the complete set of $n+1$ dimensional eigenvectors for $\mathbf{A}_{n+1}$ can be obtained from just the n-dimensional eigenvector $\mathbf{X}_{n+}$, and related row eigenvector $\mathbf{X}^{n-} \sim \overline{\mathbf{X}}_{n+}$ (E8a), of matrix $\mathbf{A}_n$.

Focussing on the particular case of lifting a URM3 solution $\mathbf{X}_3$ to URM4, i.e. the $n=3$ case here, then lifting is a special form of embedding done under both URM4 and URM3 Pythagoras conditions, where the conjugate vector $\overline{\mathbf{X}}_3$, embedded in the 4x4, URM4 matrix $\mathbf{A}$ (2.25), is actually the Pythagorean triple $\mathbf{X}^{3-}$ (A35c), conjugate to $\mathbf{X}_3$ (A3), i.e.

(A37a) $\overline{\mathbf{X}}_3 = \mathbf{X}^{3-} = \begin{pmatrix} x & y & -z \end{pmatrix}$, URM3 Pythagoras conditions (A18).

By the definition of $\mathbf{X}^{3-}$, the orthogonality condition (2.9) is automatically satisfied using Pythagoras (F1), i.e.

(7.3) $\overline{\mathbf{X}}_3 \mathbf{X}_3 = \mathbf{X}^{3-} \mathbf{X}_3 = 0$.

Furthermore, the embedding matrix $\mathbf{A}$ (2.25) has the additional constraint on the arbitrary parameters s and $\overline{s}$

(7.4) $\overline{s} = -s$.

Using this and $\overline{\mathbf{X}}_3 = \mathbf{X}^{3-}$ (A37a), the matrix $\mathbf{A}_{40}$ ($\sim \mathbf{A}$ under URM4 Pythagoras conditions (4.1)), becomes

(7.5) $\mathbf{A}_{40} = \begin{pmatrix} 0 & -s\mathbf{X}^{3-} \\ s\mathbf{X}_3 & \mathbf{A}_{30} \end{pmatrix}$.

With the 4D eigenvector $\mathbf{X}_{4+}$ ($\sim \mathbf{X}_4$ under URM4 Pythagoras conditions), defined as per embedding (2.4)

(2.4) $\mathbf{X}_{4+} = \begin{pmatrix} 0 \\ \mathbf{X}_3 \end{pmatrix}$, then the standard dynamical equations (2.7) are seen to be satisfied

$$(7.6) \quad \mathbf{A}_{40}\mathbf{X}_{4+} = \begin{pmatrix} \mathbf{X}^{3-}\mathbf{X}_3 \\ \mathbf{A}_{30}\mathbf{X}_3 \end{pmatrix} = \begin{pmatrix} 0 \\ C\mathbf{X}_3 \end{pmatrix} = C\mathbf{X}_{4+},$$

since $\mathbf{A}_3\mathbf{X}_3 = C\mathbf{X}_3$ (A4) and $\mathbf{X}^{3-}\mathbf{X}_3 = 0$ by orthogonality (7.3). Note that $\mathbf{A}_{30} \sim \mathbf{A}_3$ under URM3 Pythagoras conditions (A18).

Furthermore, the reciprocal vector $\mathbf{X}^{4-}$ is defined as the following embedding of $\mathbf{X}^{3-}$:

$$(7.7) \quad \mathbf{X}^{4-} = \begin{pmatrix} 0 & \mathbf{X}^{3-} \end{pmatrix},$$

and related to the reciprocal vector by the usual relation

$$(7.8) \quad \mathbf{X}^{4-} = (\mathbf{TX}_{4+})^T, \text{ cf. (A34e) for URM3 and (E5) URM} n,$$

also satisfies its URM4 dynamical equation for eigenvalue $-C$, i.e.

$$(7.9) \quad \mathbf{X}^{4-}\mathbf{A}_{40} = \begin{pmatrix} 0 & \mathbf{X}^{3-}\mathbf{A}_{30} \end{pmatrix} = -C\mathbf{X}^{4-}, \text{ by (A36c).}$$

The matrix $\mathbf{A}_{40}$ (7.5) is expanded in full using $\mathbf{X}_3$ (A3), $\mathbf{X}^{3-}$ (A37a) and $\mathbf{A}_{30}$ (A19), as follows:

$$(7.10) \quad \mathbf{A}_{40} = \begin{pmatrix} 0 & -sx & -sy & sz \\ sx & 0 & R & Q \\ sy & -R & 0 & P \\ sz & Q & P & 0 \end{pmatrix}.$$

Comparing this with the general form of $\mathbf{A}_{40}$ (1.3a), then the URM4-specific dynamical variables S,T,U are now

(7.11)

$(7.11a) \quad \bar{S} = -S = sx$

$(7.11b) \quad \bar{T} = -T = sy$

$(7.11c) \quad \bar{U} = U = sz,$

The URM4 Potential V (4.3c), under Pythagoras conditions (4.1), can be verified as zero by substituting for the dynamical variables S,T,U

from (7.11) into the denominator term $QT - (PS + RU)$ in (4.3c), to give

(7.12) $QT - (PS + RU) = s(Px - Qy - Rz)$.

The bracketed term on the right, $Px - Qy - Rz$, is the URM3 'delta' conservation equation (F5), and equates to zero under URM3 Pythagoras conditions (A18).

(F5) $Px - Qy - Rz = 0$.

Hence the URM4 Potential V (4.3c) is also zero for arbitrary s, i.e.

(7.13) $V = \dfrac{[QT - (PS + RU)]^2}{C^2} = 0 \ \forall \ s$.

The invariance of this zero Potential to variations will be revisited again in Section (11).

The analytic solution for $\mathbf{X}^+$ has already been derived in Section (3), equation (3.24), and reproduced below,

(3.24) $\mathbf{X}^+ = \left(2sC \quad s\bar{s}\mathbf{X}^{3-} + \mathbf{X}^{3+}\right)$.

Substituting $\bar{s} = -s$ from (7.4), and using $\mathbf{X}^{3-}$ (A37a) and $\mathbf{X}^{3+}$ (A35a), then (3.24) becomes

(7.14) $\mathbf{X}^+ = \left(2sC \quad \alpha_3 - s^2 x \quad \beta_3 - s^2 y \quad \gamma_3 + s^2 z\right)$.

Some examples of lifting are given after the following notes.

Notes

In general, lifting is applicable to any conjugate vector $\overline{\mathbf{X}}_{n+}$ $(= \mathbf{X}^{n-})$, for URMn, embedded in the $(n+1) \times (n+1)$ matrix $\mathbf{A}_{n+1}$, where $\mathbf{X}^{n-} = \left(\mathbf{T}\mathbf{X}_{n+}\right)^T$ (E5), both $\mathbf{X}^{n-}$ and $\mathbf{X}_{n+}$ being Pythagorean n-tuples,

(E8a) $\overline{\mathbf{X}}_{n+} = \mathbf{X}^{n-}$

(E8c) $\overline{\mathbf{X}}_{n-} = \mathbf{X}^{n+}$

(F7) $\overline{\mathbf{X}}_{n\pm}\mathbf{X}_{n\pm} = \mathbf{X}^{n\mp}\mathbf{X}_{n\pm} = 0$.

The dimension n can also be the, hitherto unmentioned, case $n = 2$, i.e. 'URM2'. Indeed, this is the first example given further below; see also Appendix (H).

The guaranteed existence of a reciprocal (conjugate) vector $\mathbf{X}^{n-}$ only occurs under URM3 Pythagoras conditions, where $\mathbf{X}^{n-} = \left(\mathbf{TX}_{n+} \right)^T$ (E5). In the general method of embedding, when not under Pythagoras conditions, there is no such easily obtainable vector and no such $\mathbf{T}$ operator relation.

Under Pythagoras conditions, complete sets of eigenvectors are usually obtained for all eigenvalues, whereas the method of embedding usually restricts to just eigenvectors $\mathbf{X}$ and $\mathbf{X}_0$ for eigenvalues $\lambda = C$ and $\lambda = 0$ respectively, since these are the only guaranteed eigenvalues in general embedding, Section (2).

The general solution to obtain eigenvectors, in the special case when $\overline{\mathbf{X}}_3 = \mathbf{X}^{3-}$ (A37a), has already been mentioned in Section (3-2), using a quadratic, and not a cubic, residual matrix polynomial. This special quadratic form of polynomial, and the full solution to obtain all eigenvectors, is used throughout the remainder of this book, which now just concentrates on lifting solutions under Pythagoras conditions.

The term 'lifting' originates from the number theoretic method of lifting solutions for exponent n to $n+1$, see a number theory text, e.g. Ref. [6]. Whilst URMT lifting is not identical, the concept is similar.

(7-2) Example 1. Generation of Pythagorean Triples from (1,1)

This example shows how to lift a URM2 solution (Appendix (H)) to URM3. The lifting to URM4 and URM5 is then shown in the second and third examples.

The goal of this example is not just to demonstrate lifting in URMT but that, in essence, every (a cautionary note follows) Pythagorean triple, quadruple, quintuple or n-tuple can be derived from the simplest ordered pair $(1,1)$. Note that 'every' Pythagorean n-tuple means as given by the standard parameterisation, i.e. (7.22), further below, in the case of quadruples. Strictly speaking, this parameterisation has several limitations: 1) it cannot generate every primitive triple (I12); 2) it cannot generate all sign combinations; and 3), it cannot generate all rearrangements of the same quadruple. Basically, to use the term 'every', one has to accept that, for example, (1,2,2,3) is the same as (4,-2,-4,6), i.e. by disregarding the common factor 2, rearranging the elements and changing the signs, they are the same quadruple. This may seem like a lot of limitations but, nevertheless, generating 'every' Pythagorean n-tuple from (1,1) is quite some result because, although trivial, the pair (1,1) is a legitimate Pythagorean pair, i.e. $1^2 = 1^2$, and shows consistency of the Pythagorean treatment in all forms of URMT, from URM2 to 3, 4, 5 and beyond. The reader is advised to search the web on 'Pythagorean Quadruples' for more information on this and the 'Lebesgue Identity', (5.24) and (5.27), which resolves some issues.

URM2 is discussed in Appendix (H). However, to recap, the ordered pair (1,1) is represented by the URM2 vector $\mathbf{X}_{2+}$

$$\text{(H9)} \quad \mathbf{X}_{2+} = \begin{pmatrix} 1 \\ 1 \end{pmatrix},$$

and with a unity root matrix $\mathbf{A}_2$, eigenvalue C (remember only when $C = 1$ is matrix $\mathbf{A}_2$ a true 'unity' root matrix)

$$\text{(H8)} \quad \mathbf{A}_2 = \begin{pmatrix} 0 & C \\ C & 0 \end{pmatrix}, \quad \mathbf{A}_2 \sim \mathbf{A}_{20}$$

then the usual URM2 dynamical equations are seen to be satisfied:

$$\text{(H4b)} \quad \mathbf{A}_2 \mathbf{X}_{2+} = C \mathbf{X}_{2+}$$

The conjugate vector $\overline{\mathbf{X}}_2$ ($\sim \overline{\mathbf{X}}_{2+}$) is derived from $\mathbf{X}_{2+}$ (H9) in the usual way, i.e.

$$\text{(H14)} \quad \overline{\mathbf{X}}_{2+} = \mathbf{X}^{2-} = \left(\mathbf{T} \mathbf{X}_{2+} \right)^T,$$

and thus

$$\text{(H11b)} \quad \mathbf{X}^{2-} = \begin{pmatrix} 1 & -1 \end{pmatrix},$$

which satisfies the URM2 eigenvector equation

$$\mathbf{X}^{2-} \mathbf{A}_2 = -C \mathbf{X}^{2-},$$

and the orthogonality relation

(H15) $\mathbf{X}^{2-}\mathbf{X}_{2+} = \overline{\mathbf{X}}_{2+}\mathbf{X}_{2+} = 0$.

That's enough of URM2; proceeding on to URM3, the solution $\mathbf{X}_{2+}$ (H9) and $\mathbf{X}^{2-}$ (H11b) is then embedded in the URM3 vectors $\mathbf{X}_{3+}$ and $\mathbf{X}^{3-}$ respectively, as follows:

(7.20a) $\mathbf{X}_{3+} = \begin{pmatrix} 0 \\ \mathbf{X}_{2+} \end{pmatrix}$, $\mathbf{X}^{3-} = \begin{pmatrix} 0 & \mathbf{X}^{2-} \end{pmatrix}$

i.e.

(7.20b) $\mathbf{X}_{3+} = \begin{pmatrix} 0 \\ 1 \\ 1 \end{pmatrix}$, $\mathbf{X}^{3-} = \begin{pmatrix} 0 & +1 & -1 \end{pmatrix}$.

Of course, $\mathbf{X}_{3+}$ is now a trivial Pythagorean triple (as is $\mathbf{X}^{3-}$), i.e. $0^2 + 1^2 = 1^2$.

The $\mathbf{A}_2$ matrix is embedded into a URM3 $\mathbf{A}_{30}$ matrix in a similar fashion to (7.5), i.e.

(7.21) $\mathbf{A}_{30} = \begin{pmatrix} 0 & -s\mathbf{X}^{2-} \\ s\mathbf{X}_{2+} & \mathbf{A}_2 \end{pmatrix}$

For advance information, both s and eigenvalue C will appear in the standard parameterisation of Pythagorean triples as in

(7.22) $x = 2sC$, $y = C^2 - s^2$, $z = C^2 + s^2$.

Note, as mentioned at the start of this Section (7-2), this parameterisation (7.22) does not give all primitive triples.

Matrix $\mathbf{A}_{30}$ (7.21) is expanded in full as follows, using $\mathbf{X}^{2-}$ and $\mathbf{X}_{2+}$

(7.23) $\mathbf{A}_{30} = \begin{pmatrix} 0 & -s & s \\ s & 0 & C \\ s & C & 0 \end{pmatrix}$.

Given the block matrix notation is rather heavy-handed for 2x2 and 3x3 matrices, this expanded 3x3 form will be now used throughout; the original intention of using block matrix form being to highlight the relation to the earlier theory of lifting (embedding), which is best expressed in block matrix form for the $n \times n$ matrices of URM n.

Comparing $\mathbf{A}_{30}$ (7.23) with the standard URM3 matrix form $\mathbf{A}$ (A1a), the dynamical variables are thus

(7.24)

(7.24a) $P = C$, $Q = s$, $R = -s$

(7.24b) $\overline{P} = C$, $\overline{Q} = s$, $\overline{R} = s$,

and it is noted that they satisfy the URM3 Pythagoras conditions (A18)

(A18) $\overline{P} = P$, $\overline{Q} = Q$, $\overline{R} = -R$.

As a consequence, the URM3 Potential V_3 (A9) is zero

(7.25) $V_3 = \dfrac{(PQR + \overline{P}\,\overline{Q}\,\overline{R})}{C} = 0$,

and the kinetic term $K = P\overline{P} + Q\overline{Q} + R\overline{R}$ (A8) is just the DCE for zero Potential, as also expected under URM3 Pythagoras conditions, i.e.

(7.26) $K = C^2$

The eigenvalues are also as per Pythagoras (I14)

(7.27) $\lambda = +C$, $\lambda = 0$, $\lambda = -C$,

and so the Cayley-Hamilton polynomial is as follows, which will subsequently be used to obtain the eigenvectors using the residual matrix method, Section (3),

(7.28) $0 = \mathbf{A}_{30}(\mathbf{A}_{30} + C\mathbf{I}_3)(\mathbf{A}_{30} - C\mathbf{I}_3)$,

where $\mathbf{I}_3$ is the 3×3 identity matrix.

To produce a Pythagorean triple eigenvector $\mathbf{X}^{3+}$ from the above definitions, the residual matrix $\mathbf{E}_+ = \mathbf{X}^{3+}\mathbf{X}_{3+}$, cf. (3.3), is obtained by factoring the Cayley-Hamilton polynomial and removing linear factor $(\mathbf{A}_{30} - C\mathbf{I}_3)$, for eigenvalue $\lambda = +C$, to give

(7.29) $\mathbf{E}_{3+} = \mathbf{A}_{30}(\mathbf{A}_{30} + C\mathbf{I}_3) = \mathbf{A}_{30}^2 + C\mathbf{A}_{30}$

The square of matrix $\mathbf{A}_{30}$ (7.23) is

$$(7.30) \quad \mathbf{A}_{30}^2 = \begin{pmatrix} 0 & Cs & -Cs \\ Cs & C^2 - s^2 & s^2 \\ Cs & -s^2 & C^2 + s^2 \end{pmatrix},$$

and using $\mathbf{A}_{30}$ (7.23) again for $C\mathbf{A}_{30}$, then $\mathbf{E}_{3+}$ is evaluated from (7.29) as

$$(7.31) \quad \mathbf{E}_{3+} = \begin{pmatrix} 0 & 0 & 0 \\ 2sC & C^2 - s^2 & C^2 + s^2 \\ 2sC & C^2 - s^2 & C^2 + s^2 \end{pmatrix}.$$

Defining $\mathbf{X}^{3+}$ as the 1x3 row vector, where $\mathbf{X}^{3+}\mathbf{A}_{30} = C\mathbf{X}^{3+}$ by definition (A36a)

$$(7.32) \quad \mathbf{X}^{3+} = \begin{pmatrix} 2sC & C^2 - s^2 & C^2 + s^2 \end{pmatrix},$$

then $\mathbf{E}_{3+}$ is the 3x3 outer product of the 3x1 vector $\mathbf{X}_{3+}$ with the 1x3, row vector $\mathbf{X}^{3+}$, as in $\mathbf{E}_+ = \mathbf{X}\mathbf{X}^+$ (3.3)

$$(7.33) \quad \mathbf{E}_{3+} = \begin{pmatrix} 0 \\ 1 \\ 1 \end{pmatrix} \begin{pmatrix} 2sC & C^2 - s^2 & C^2 + s^2 \end{pmatrix}.$$

Hence all Pythagorean triples covered by the standard parameterisation (7.22) can be generated from the trivial triple $(0,1,1)$ by the method of lifting.

Of course, this may seem a cheat in so far as the $\mathbf{X}_{3+}$ triple $(0,1,1)$ has no parameterisation in s, C, which only appears in the row vector $\mathbf{X}^{3+}$. However, the two vectors, $\mathbf{X}_{3+}$ and are $\mathbf{X}^{3+}$, are intrinsically linked, not just together in the outer product forming $\mathbf{E}_{3+}$ but, in fact, the vector $\mathbf{X}^{3+}$ is actually related to the column vector $\mathbf{X}_{3-}$ defined as follows,

$$(7.34) \ \mathbf{X}_{3-} = \begin{pmatrix} 2sC \\ C^2 - s^2 \\ -(C^2 + s^2) \end{pmatrix},$$

and formed from $\mathbf{X}^{3+}$ by relation $\mathbf{X}_{3-} = (\mathbf{X}^{3+}\mathbf{T}_3)^T$ (A34b).

Both the column vectors $\mathbf{X}_{3+}$ and $\mathbf{X}_{3-}$ are Pythagorean triples, as are $\mathbf{X}^{3+}$ and $\mathbf{X}^{3-}$, albeit only $\mathbf{X}_{3-}$ and $\mathbf{X}^{3+}$ contain three non-zero elements, i.e. what would be classed as non-trivial Pythagorean triples.

Before proceeding, this solution for $\mathbf{X}_{3+}$ is called the 'almost trivial solution' in Appendix C of [1],#3, for parameter $l = 1$ (this parameter l has not been used here). It ($\mathbf{X}_{3+}$) is basically the smallest possible solution to URM3 under Pythagoras conditions, and is very useful for checking any URM3 methods and results when under these conditions.

Given that eigenvectors $\mathbf{X}_{3+}$ and $\mathbf{X}_{3-}$, for eigenvalues $\lambda = +C$ and $\lambda = -C$ respectively, have been obtained, it remains to find the eigenvector $\mathbf{X}_{30}$ for $\lambda = 0$. This eigenvector $\mathbf{X}_{30}$ and row eigenvector $\mathbf{X}^{30}$ are obtained from the residual matrix $\mathbf{E}_{30}$, which is equivalent to the outer product $-\mathbf{X}^{30}\mathbf{X}_{30}$ (3.70), and is calculated from the matrix polynomial (5.31b), reproduced below,

$$(5.31b) \ \mathbf{E}_{30} = \mathbf{A}_{30}^2 - C^2\mathbf{I}, \ \mathbf{E}_{30} \sim \mathbf{E}_3 = -\mathbf{X}_{30}\mathbf{X}^{30}.$$

Using $\mathbf{A}_3^2$ (7.30), then $\mathbf{E}_{30}$ is calculated as

$$(7.35) \ \mathbf{E}_{30} = \mathbf{A}_{30}^2 - C^2\mathbf{I} = \begin{pmatrix} -C^2 & Cs & -Cs \\ Cs & -s^2 & s^2 \\ Cs & -s^2 & s^2 \end{pmatrix}.$$

Comparing (7.35) with the outer product form of $\mathbf{E}_{30} = -\mathbf{X}_{30}\mathbf{X}^{30}$, then the zero eigenvalue has eigenvectors $\mathbf{X}_{30}$ or $\mathbf{X}^{30}$ given by

$$(7.36) \ \mathbf{X}_{30} = \begin{pmatrix} C \\ -s \\ -s \end{pmatrix}$$

(7.37) $\mathbf{X}^{30} = \begin{pmatrix} C & -s & +s \end{pmatrix}$.

Although not evaluated here, the complete set of URM3 inner products, when under URM3 Pythagoras conditions, are given in Appendix (F), and the reader may wish to verify that these are met using the above eigenvector solution.

(7-3) Example 2. Generation of Pythagorean Quadruples from (0,1,1)

Having obtained a URM3 solution in Example 1 by lifting a solution from UMR2, the next natural step is to lift a URM3 solution to URM4. Once again, the solution will be obtained under Pythagoras conditions, with Pythagorean triples as two of the four possible eigenvectors in URM4.

The URM3 solution to be lifted is chosen to be $\mathbf{X}_{3+}$ and $\mathbf{X}^{3-}$, as in Example 1, and so the lifted URM4 eigenvectors $\mathbf{X}_{4+}$ and $\mathbf{X}^{4-}$ are, very simply,

$$(7.40)\ \mathbf{X}_{4+} = \begin{pmatrix} 0 \\ 0 \\ 1 \\ 1 \end{pmatrix}, \quad \mathbf{X}^{4-} = \begin{pmatrix} 0 & 0 & +1 & -1 \end{pmatrix}.$$

Since $\mathbf{X}_{4+}$ is based on $\mathbf{X}_{3+}$ (7.20b) which itself is based on $\mathbf{X}_{2+}$ (H9), this example can also be thought of as an embedding of URM2 in URM4, or as a lifting of a URM2 solution to URM4.

In an identical manner to Example 1 in Section (6-2), the URM3 matrix $\mathbf{A}_3$ (7.23) is embedded in a URM4 matrix, represented in block matrix form by (7.41) below, where an extra, integer, parameter t has been introduced. Note that the block form shows URM2 matrices and vectors, and not URM3 vectors. However, it can be seen that the bottom right, 3x3 matrix block is that of $\mathbf{A}_{30}$ (7.23), and the top row is the lifted conjugate vector $\mathbf{X}^{3-} = \begin{pmatrix} 0 & t\mathbf{X}^{2-} \end{pmatrix}$.

$$(7.41)\ \mathbf{A}_{40} = \begin{pmatrix} 0 & 0 & -t\mathbf{X}^{2-} \\ 0 & 0 & -s\mathbf{X}^{2-} \\ t\mathbf{X}_{2+} & s\mathbf{X}_{2+} & \mathbf{A}_2 \end{pmatrix}, \quad s,t \in \mathbb{Z},\ \mathbf{A}_2 \sim \mathbf{A}_{20}\ (\text{H8}).$$

Expanding in full, using $\mathbf{A}_2$ (H8), $\mathbf{X}_{2+}$ (H9) and $\mathbf{X}^{2-}$ (H11b), matrix $\mathbf{A}_{40}$ becomes

$$(7.42) \quad \mathbf{A}_{40} = \begin{pmatrix} 0 & 0 & -t & t \\ 0 & 0 & -s & s \\ t & s & 0 & C \\ t & s & C & 0 \end{pmatrix}.$$

Comparing $\mathbf{A}_{40}$ (7.42) with the standard URM4 matrix (1.3a) gives the dynamical variables as follows:

(7.43)

(7.43a) $S = 0$, $T = -t$, $U = t$

(7.43b) $\bar{S} = 0$, $\bar{T} = t$, $\bar{U} = t$

(7.43c) $P = C$, $Q = s$, $R = -s$

(7.43d) $\bar{P} = C$, $\bar{Q} = s$, $\bar{R} = s$,

and it is noted that they satisfy the URM4 Pythagoras conditions (4.1)

(4.1a) $\bar{S} = -S$, $\bar{T} = -T$, $\bar{U} = U$

(4.1b) $\bar{P} = P$, $\bar{Q} = Q$, $\bar{R} = -R$.

A zero Potential solution is not implicit under the above, URM4 Pythagoras conditions but, in this specific case, it is zero, i.e.

$$(7.44) \quad V = \frac{[QT - (PS + RU)]^2}{C^2} = 0,$$

since $QT = -st$, $PS = 0$ and $RU = -st$ by (7.43).

The kinetic term K (4.3b) can be verified as the constant energy term C^2, as would be expected for the DCE (4.3a) for a zero Potential, i.e.

(7.45) $K = C^2$, the DCE.

Using $V = 0$ and $K = C^2$, then the characteristic equation (4.9) becomes

(7.46) $\lambda^4 - C^2 \lambda^2 = 0$,

and so the eigenvalues are

(7.47) $\lambda = \pm C$, $\lambda = 0,0$.

With these eigenvalues, the Cayley-Hamilton polynomial is thus

(7.48) $0 = \mathbf{A}_{40}^2 (\mathbf{A}_{40} + C\mathbf{I}_4)(\mathbf{A}_{40} - C\mathbf{I}_4)$,

where $\mathbf{I}_4$ is the 4x4 identity matrix.

For reasons detailed in Section (3-2), the residual matrix $\mathbf{E}_{4+}$ is constructed from the same quadratic, matrix polynomial expression as in the 3x3 case (7.29), namely

(7.49) $\mathbf{E}_{4+} = \mathbf{A}_{40}^2 + C\mathbf{A}_{40}$.

The fact that the 3x3 and 4x4 matrix polynomial expression is the same quadratic in both cases is key to why this residual matrix method, for generation of Pythagorean triples, works for any order n, and is due to having only two non-zero eigenvalues.

Using $\mathbf{A}_{40}$ (7.42) then $\mathbf{A}_{40}^2$ is calculated as

(7.50) $\mathbf{A}_{40}^2 = \begin{pmatrix} 0 & 0 & tC & -tC \\ 0 & 0 & sC & -sC \\ tC & sC & C^2 - s^2 - t^2 & s^2 + t^2 \\ tC & sC & -s^2 - t^2 & C^2 + s^2 + t^2 \end{pmatrix}$,

and $\mathbf{E}_{4+}$ (7.49) is evaluated as

(7.51) $\mathbf{E}_{4+} = \begin{pmatrix} 0 & 0 & 0 & 0 \\ 0 & 0 & 0 & 0 \\ 2tC & 2sC & C^2 - s^2 - t^2 & C^2 + s^2 + t^2 \\ 2tC & 2sC & C^2 - s^2 - t^2 & C^2 + s^2 + t^2 \end{pmatrix}$.

Defining the 1x4 row-vector $\mathbf{X}^{4+}$ as

(7.52) $\mathbf{X}^{4+} = \begin{pmatrix} 2tC & 2sC & C^2 - s^2 - t^2 & C^2 + s^2 + t^2 \end{pmatrix}$,

then, once again, $\mathbf{E}_{4+}$ is the 4x4 outer product of the 4x1 column vector $\mathbf{X}_{4+}$ (7.40) with the 1x4 row vector $\mathbf{X}^{4+}$ (7.52)

$$(7.53) \quad \mathbf{E}_{4+} = \begin{pmatrix} 0 \\ 0 \\ 1 \\ 1 \end{pmatrix} \begin{pmatrix} 2tC & 2sC & C^2 - s^2 - t^2 & C^2 + s^2 + t^2 \end{pmatrix}.$$

The row eigenvector $\mathbf{X}^{4+}$ is seen to be a Pythagorean quadruplet with the standard parameterisation

(7.54)

$(7.54a) \quad w = 2tC$

$(7.54b) \quad x = 2sC$

$(7.54c) \quad y = C^2 - s^2 - t^2$

$(7.54d) \quad z = C^2 + s^2 + t^2$,

and so the elements of $\mathbf{X}^{4+}$ satisfy the Pythagoras equation

$(7.55) \quad z^2 = w^2 + x^2 + y^2$.

See the beginning of Section (7-2) for a cautionary note on the parameterisation (7.54).

Because there are two zero eigenvalues, there are two associated zero eigenvectors, denoted by $\mathbf{X}_{40A}$ and $\mathbf{X}_{40B}$, and obtained from the residual matrix $\mathbf{E}_{40}$ given by

$(7.57) \quad \mathbf{E}_{40} = \mathbf{A}_{40}^2 - C^2 \mathbf{I}_4$.

Using $\mathbf{A}_{40}^2$ (7.50) gives for $\mathbf{E}_{40}$

$$(7.58) \quad \mathbf{E}_{40} = \begin{pmatrix} -C^2 & 0 & tC & -tC \\ 0 & -C^2 & sC & -sC \\ tC & sC & -s^2 - t^2 & s^2 + t^2 \\ tC & sC & -s^2 - t^2 & s^2 + t^2 \end{pmatrix}.$$

This is nicer separated into two parts, $\mathbf{E}_{40A}$ and $\mathbf{E}_{40B}$, one for parameter s and one for t, whereby

$(7.59) \quad \mathbf{E}_{40} = \mathbf{E}_{40A} + \mathbf{E}_{40B}$

$$(7.60) \quad \mathbf{E}_{40A} = \begin{pmatrix} 0 & 0 & 0 & 0 \\ 0 & -C^2 & sC & -sC \\ 0 & sC & -s^2 & s^2 \\ 0 & sC & -s^2 & s^2 \end{pmatrix},$$

$$\mathbf{E}_{40B} = \begin{pmatrix} -C^2 & 0 & tC & -tC \\ 0 & 0 & 0 & 0 \\ tC & 0 & -t^2 & t^2 \\ tC & 0 & -t^2 & t^2 \end{pmatrix}.$$

Using $\mathbf{E}_{40A} = -\mathbf{X}_{40A}\mathbf{X}^{40A}$ and $\mathbf{E}_{40B} = -\mathbf{X}_{40B}\mathbf{X}^{40B}$ (3.70), then the eigenvectors are found to be

$$(7.61) \quad \mathbf{X}_{40A} = \begin{pmatrix} 0 \\ +C \\ -s \\ -s \end{pmatrix}, \quad \mathbf{X}^{40A} = \begin{pmatrix} 0 & +C & -s & +s \end{pmatrix}$$

$$(7.62) \quad \mathbf{X}_{40B} = \begin{pmatrix} +C \\ 0 \\ -t \\ -t \end{pmatrix}, \quad \mathbf{X}^{40B} = \begin{pmatrix} +C & 0 & -t & +t \end{pmatrix}.$$

(7-4) Example 3. Generation of Pythagorean Quintuples and beyond from (0,0,1,1)

This third example lifts the previous example's URM4 solution $\mathbf{X}_{4+}$ (7.40) to URM5, once again, under Pythagoras conditions. By now the method should be clear from the previous two examples, so this third example is kept relatively brief.

The URM5 lifted eigenvector $\mathbf{X}_{5+}$ is defined as

$$(7.70) \quad \mathbf{X}_{5+} = \begin{pmatrix} 0 \\ 0 \\ 0 \\ 1 \\ 1 \end{pmatrix}.$$

As for the previous example, the vector $\mathbf{X}_{5+}$ is basically a lifting of a URM2 solution $\mathbf{X}_{2+}$ (H9) to URM5, even though it is also considered as a lifting of a URM4 solution $\mathbf{X}_{4+}$ (7.40). Likewise, it can be considered as a lift of a URM3 solution $\mathbf{X}_{3+}$ (7.20b) to URM5.

The URM5 $\mathbf{A}_{50}$ matrix is defined as follows, embedding the URM4 $\mathbf{A}_{40}$ matrix (7.42), and adding another integer parameter u to the existing set of s, t and C:

$$(7.71) \quad \mathbf{A}_{50} = \begin{pmatrix} 0 & 0 & 0 & -u & u \\ 0 & 0 & 0 & -t & t \\ 0 & 0 & 0 & -s & s \\ u & t & s & 0 & C \\ u & t & s & C & 0 \end{pmatrix}, \quad s,t,u,C \in \mathbb{Z}.$$

The residual matrix $\mathbf{E}_{5+}$ is then calculated using the same matrix polynomial expression as in the 3x3 and 4x4 cases, i.e.

$$(7.72) \quad \mathbf{E}_{5+} = \mathbf{A}_{50}^2 + C\mathbf{A}_{50}.$$

Doing so, using (7.71) for $\mathbf{A}_{50}$, shows that $\mathbf{E}_{5+}$ is expressed as the outer product of $\mathbf{X}_{5+}$ (7.70) with the 1x5 row-vector $\mathbf{X}^{5+}$ defined by

(7.73)

$$\mathbf{X}^{5+} = \begin{pmatrix} 2uC & 2tC & 2sC & C^2 - s^2 - t^2 - u^2 & C^2 + s^2 + t^2 + u^2 \end{pmatrix}.$$

The row eigenvector $\mathbf{X}^{5+}$ is seen to be a Pythagorean quintuplet with the standard parameterisation

(7.74a) $v = 2uC$,

(7.74b) $w = 2tC$,

(7.74c) $x = 2sC$,

(7.74d) $y = C^2 - s^2 - t^2 - u^2$,

(7.74a) $z = C^2 + s^2 + t^2 + u^2$

and so the elements of $\mathbf{X}^{5+}$ satisfy the Pythagoras equation

(7.75) $z^2 = v^2 + w^2 + x^2 + y^2$.

This completes the first three examples, which are essentially a lift of the URM2 solution $\mathbf{X}_{2+}$ (H9) to URM3, 4 and 5. The next and last example moves on to a slightly less basic solution.

(7-5) Example 4. Lifting $(0,4,3,5)$

This is the URM3 $(4,3,5)$ solution embedded in URM4.

Under URM3 Pythagoras conditions (A18), the URM3 unity root matrix $\mathbf{A}_{30}$ is

$$(7.80)\ \mathbf{A}_{30} = \begin{pmatrix} 0 & -2 & +2 \\ +2 & 0 & -1 \\ +2 & -1 & 0 \end{pmatrix},$$

with eigenvalues

(7.81) $\lambda = +1, 0, -1$, i.e. $C = 1$.

Comparing $\mathbf{A}_{30}$ with $\mathbf{A}_3$ (A1a), then the URM3 dynamical variables P, Q, R and $\overline{P}, \overline{Q}, \overline{R}$ are

(7.82)

(7.82a) $P = -1$, $Q = +2$, $R = -2$

(7.82b) $\overline{P} = -1$, $\overline{Q} = +2$, $\overline{R} = +2$.

By definition, these conditions give a zero, URM3 Potential V_3 (A9).

The eigenvector $\mathbf{X}_{3+}$ $(\lambda = +1)$, and conjugate eigenvector $\overline{\mathbf{X}}_3 = \mathbf{X}^{3-}$ (A37a) $(\lambda = -1)$, are:

$$(7.83) \quad \mathbf{X}_{3+} = \begin{pmatrix} 4 \\ 3 \\ 5 \end{pmatrix}, \quad x = 4, \ y = 4, \ z = 5.$$

$$(7.84) \quad \overline{\mathbf{X}}_3 = \mathbf{X}^{3-} = \begin{pmatrix} 4 & 3 & -5 \end{pmatrix},$$

and satisfy the orthogonality condition $\mathbf{X}^{3-}\mathbf{X}_3 = 0$ (7.3).

The solution $\mathbf{X}_{3+}$ (7.83) is embedded in URM4 $\mathbf{X}_{4+}$ as follows, according to (2.4),

$$(7.85) \quad \mathbf{X}_{4+} = \begin{pmatrix} 0 \\ 4 \\ 3 \\ 5 \end{pmatrix}.$$

with scalars s and $\overline{s}$ (7.4), defined as

(7.86) $\overline{s} = -s = -1$.

Using s, $\mathbf{X}_{3+}$ (7.83), $\mathbf{X}^{3-}$ (7.84), and $\mathbf{A}_{30}$ (7.80), then the URM4 matrix $\mathbf{A}_{40}$ (7.5) is

$$(7.87) \quad \mathbf{A}_{40} = \begin{pmatrix} 0 & -4 & -3 & +5 \\ +4 & 0 & -2 & +2 \\ +3 & +2 & 0 & -1 \\ +5 & +2 & -1 & 0 \end{pmatrix},$$

and the URM4 dynamical variables S, T, U and $\bar{S}, \bar{T}, \bar{U}$ are, by comparison of (7.87) with (1.3a),

(7.88a) $S = -4$, $T = -3$, $U = +5$

(7.88b) $\bar{S} = 4$, $\bar{T} = 3$, $\bar{U} = +5$.

Most importantly, these satisfy the URM4 Pythagoras conditions (4.1).

Evaluating the quadratic matrix expression for $\mathbf{E}_{4+}$ (7.49) shows that the top row remains all zero, as before,

$$(7.89)\ \mathbf{E}_{4+} = \begin{pmatrix} 0 & 0 & 0 & 0 \\ +8 & -16 & -16 & +24 \\ +6 & -12 & -12 & +18 \\ +10 & -20 & -20 & +30 \end{pmatrix}.$$

Writing $\mathbf{E}_{4+}$ in outer product form, using $\mathbf{X}_{4+}$ (7.85),

$$(7.90)\ \mathbf{E}_{4+} = \mathbf{X}_{4+}\mathbf{X}^{4+} = \begin{pmatrix} 0 \\ 4 \\ 3 \\ 5 \end{pmatrix}(+2 \quad -4 \quad -4 \quad +6),$$

then $\mathbf{X}^{4+}$ is

(7.91) $\mathbf{X}^{4+} = (+2 \quad -4 \quad -4 \quad +6)$.

This Pythagorean quadruple $(2, -4, -4, +6)$ is not primitive, but twice the primitive solution $(1, -2, -2, +3)$. To within a sign, this primitive solution appears numerous times within URM4, under Pythagoras conditions, and is the simplest, non-trivial quadruple.

The URM3 scale factors $\alpha_3, \beta_3, \gamma_3$ are obtained from the divisibility relations (A42), using the values for the dynamical variables $P, Q, R, \bar{P}, \bar{Q}, \bar{R}$ (7.82) and x, y, z (7.83), to give

(7.92) $\alpha_3 = 0$, $\beta_3 = -1$, $\gamma_3 = +1$.

The analytic solution (3.24) for $\mathbf{X}^{4+}$ $(\sim \mathbf{X}^+)$ can be used to verify (7.91). By substituting the values for $\alpha_3, \beta_3, \gamma_3$, $C = 1$, $\bar{s} = -s = -1$ (7.86) and x, y, z (7.83) into (3.24), the solution

$\mathbf{X}^{4+} = \begin{pmatrix} 2 & -4 & -4 & 6 \end{pmatrix}$ is obtained, matching (7.91). Furthermore, leaving s variable gives $\mathbf{X}^{4+}$ as the following, one-parameter solution:

(7.93) $\mathbf{X}^{4+} = \begin{pmatrix} 2s & -4s^2 & -3s^2-1 & 5s^2+1 \end{pmatrix}$,

which can be verified as a Pythagorean quadruple. This, then, is just a specific example of 'lifting' the classic Pythagorean triple $\begin{pmatrix} 4 & 3 & 5 \end{pmatrix}$ to produce an infinite set of Pythagorean quadruples $\mathbf{X}^{4+}$ (7.93) parameterised by the single parameter s.

In primitive form, the first few quadruples for $s = 1..3$ are

(7.94a) $\mathbf{X}^{4+} = 2\begin{pmatrix} 1 & -2 & -2 & 3 \end{pmatrix}$, $s = 1$

(7.94b) $\mathbf{X}^{4+} = \begin{pmatrix} 4 & -16 & -13 & 21 \end{pmatrix}$, $s = 2$

(7.94c) $\mathbf{X}^{4+} = 2\begin{pmatrix} 3 & -18 & -14 & 23 \end{pmatrix}$, $s = 3$.

The computation of the zero eigenvectors, $\mathbf{X}_{40A}$, $\mathbf{X}_{40B}$ $\mathbf{X}^{40A}$ and $\mathbf{X}^{40B}$, has been skipped in this section and left for when the topic of lifting solutions, using variational methods, is discussed in Section (11).

(8) Variational Methods

(8-1) The General Variational Method and URM3

The variational method in URMT entails defining a variable, square, $n \times n$ matrix Δ that annihilates the eigenvector X, for eigenvalue C, i.e.

(8.1) $\Delta X = 0$.

Defining the matrix A' as the initial, unvaried matrix A, and Δ as a 'variational' matrix that is a function of one or more variational parameters (to be defined), then the variational form of the $n \times n$ unity root matrix A is written as

(8.2) $A = A' + \Delta$.

Using the annihilation definition of Δ (8.1), then the transformation represented by A (8.2) leaves the coordinate eigenvector X invariant since

(8.3) $AX = (A' + \Delta)X = A'X + \Delta X = A'X = CX$.

As a reminder of URM3 variational methods in [1],#1, the variational matrix Δ_3 ($\sim \Delta$ in [1]) is defined in terms of three arbitrary, integer variations $\eta, \delta, \varepsilon$

(8.4) $\Delta_3 = \begin{pmatrix} 0 & +\eta z & -\eta y \\ -\delta z & 0 & +\delta x \\ +\varepsilon y & -\varepsilon x & 0 \end{pmatrix}$, $\eta, \delta, \varepsilon \in \mathbb{Z}$,

and, with coordinate eigenvector X_3 (A3), it can be verified that the annihilation condition (8.1) is satisfied, i.e.

(8.5) $\Delta_3 X_3 = 0$,

and so too, therefore, the transformation invariance of $\mathbf{X}_3$ to linear mappings defined by (8.3).

Note that definition (8.2) was reversed in [1] and written as $\mathbf{A}' = \mathbf{A} + \mathbf{\Delta}$, where $\mathbf{A}'$ represented the variational form, and not the initial form of $\mathbf{A}$. This notation was inconsistent in [1] since the primed superscript is generally used in URMT to denote an initial, constant value, and not a variational term. By redefining $\mathbf{A}'$ in (8.1) to represent the initial, constant component of matrix $\mathbf{A}$, this inconsistency is now rectified within this book.

Given that the URM3 unity root matrix $\mathbf{A}_3$ is defined in terms of the three dynamical variables P, Q, R and their conjugate forms $\overline{P}, \overline{Q}, \overline{R}$ as in

$$(\text{A1a}) \quad \mathbf{A}_3 = \begin{pmatrix} 0 & R & \overline{Q} \\ \overline{R} & 0 & P \\ Q & \overline{P} & 0 \end{pmatrix},$$

then (8.1) represents the following coordinate transformation in the dynamical variables:

(8.6)

(8.6a) $P \rightarrow P + \delta x$

(8.6b) $Q \rightarrow Q + \varepsilon y$

(8.6c) $R \rightarrow R + \eta z$

(8.6d) $\overline{P} \rightarrow \overline{P} - \varepsilon x$

(8.6e) $\overline{Q} \rightarrow \overline{Q} - \eta y$

(8.6f) $\overline{R} \rightarrow \overline{R} - \delta z$.

(8.7) These transformations invoke an **Invariance Principle**

The dynamical equations and their solutions are invariant

to a coordinate transformation in the dynamical variables.

This principle is algebraically stated by (8.3), and given in URM3 notation, as

$$(8.8) \quad (\mathbf{A}_3 + \mathbf{\Delta}_3)\mathbf{X}_3 = \mathbf{A}_3\mathbf{X}_3 + \mathbf{\Delta}_3\mathbf{X}_3 = \mathbf{A}_3\mathbf{X}_3 = C\mathbf{X}_3.$$

By applying transformations (8.6) to each of the terms in the URM3 DCE (A12), and asserting the invariance principle, this implies that every term with $\eta, \delta, \varepsilon$ as a coefficient must vanish, i.e. all equations must be independent of any one of the variations $\eta, \delta, \varepsilon$. Asserting this condition actually gives both the URM3 dynamical equations (A5) and their solution, i.e. the eigenvector $\mathbf{X}_3$ (A3). It is quite something that both equations and solutions arise from this variational method, which is considered original to URMT. Normally, variational calculus will only give the equations of motion from such a principle, e.g. The Principle of Least Action, and the resulting equations then have to be solved to give the solution.

As an example, consider the following: since all equations should be invariant to any variation, suppose $\delta = 0$ and $\varepsilon = 0$ such that only η is left to vary, then the transformations in the dynamical variables affect only R and $\overline{Q}$ as follows, by (8.6),

(8.9)

(8.9a) $R \rightarrow R + \eta z$

(8.9b) $\overline{Q} \rightarrow \overline{Q} - \eta y$.

Looking at $\mathbf{A}_3$ (A1a), only the dynamical variables in the first row are affected, hence such a variation is termed a 'first row variation' in URMT parlance.

Applying these two variations (8.9) to the URM3 DCE, reproduced in full below,

(8.10) $+C^2 = P\overline{P} + Q\overline{Q} + R\overline{R} + (PQR + \overline{P}\,\overline{Q}\,\overline{R})/C$,

give the following terms in η

(8.11) $Q(\overline{Q} - \eta y) + (R + \eta z)\overline{R} + \dfrac{PQ(R + \eta z) + \overline{P}(\overline{Q} - \eta y)\overline{R}}{C}$.

Since the conserved quantity remains invariant with a value C^2, and all other terms stay constant, then this variational term should also be the same as before, i.e.

(8.12)
$$Q\overline{Q} + R\overline{R} + (PQR + \overline{P}\overline{Q}\overline{R}) =$$
$$Q(\overline{Q} - \eta y) + (R + \eta z)\overline{R} + \frac{PQ(R + \eta z) + \overline{P}(\overline{Q} - \eta y)\overline{R}}{C}$$

Therefore all variational terms must equate to zero, which implies

(8.13) $\quad 0 = -\eta y Q + \eta z \overline{R} + \dfrac{\eta z P Q - \eta y \overline{P}\overline{R}}{C}$.

Factoring η and multiplying throughout by eigenvalue C tidies this to

(8.14) $\quad 0 = \eta(-yCQ + zC\overline{R} + zPQ - y\overline{P}\overline{R})$.

For an arbitrary variation in η, the bracketed term on right must always equate to zero. Hence collecting terms in y and z gives the following solution linking the two variables

(8.15) $\quad y(CQ + \overline{P}\overline{R}) = z(C\overline{R} + PQ)$.

Note that this is a solution to the dynamical equations (A5), and not a dynamical equation in itself, because it only involves two of the three coordinates x, y and z, i.e. it expresses y in terms of z .

Other solutions can be obtained similarly by varying δ and ε one at a time, with $\eta = 0$. Furthermore, by varying two at a time, e.g. δ and ε together with $\eta = 0$, gives both quadratic variational terms and linear terms, i.e. terms in $\delta\varepsilon$, δ and ε in this example. Note that applying all three variations in $\eta, \delta, \varepsilon$ simultaneously gives two additional cubic terms, which cancel leaving three quadratic and three linear terms. Rather neatly, the three quadratic terms give the three dynamical equations (A5), and the three linear terms give the eigenvector solutions; see [1],#4 for a full list of the URM3 eigenvector solutions.

This completes the overview of URM3 variational methods; for URM4 and beyond, the definitions and principles are identical. However, whereas URM3 has three 'local' variations $\eta, \delta, \varepsilon$, URM4 has far more, up to twelve in fact (eight minimum). The full URM4 variational matrix Δ is defined as follows, for arbitrary 'variational' integers δ_{ij}, where the subscripted i, j denotes the ith row and jth column of Δ, and $\delta_{ij} = 0$ for $i = j$, or generally non-zero when $i \neq j$. Note that the lead diagonal is all zero, as always in URMT, and so Δ has twelve non-zero variational coefficients.

(8.16)

$$\Delta = \begin{pmatrix} 0 & \delta_{12}y + \delta_{14}z & -\delta_{12}x + \delta_{13}z & -\delta_{14}x - \delta_{13}y \\ \delta_{21}y + \delta_{24}z & 0 & -\delta_{21}w + \delta_{23}z & -\delta_{24}w - \delta_{23}y \\ \delta_{31}x + \delta_{34}z & -\delta_{31}w + \delta_{32}z & 0 & -\delta_{34}w - \delta_{32}x \\ \delta_{41}x + \delta_{43}y & -\delta_{41}w + \delta_{42}y & -\delta_{43}w - \delta_{42}x & 0 \end{pmatrix}$$

$\delta_{ij} \in \mathbb{Z}, i, j = 1 \ldots n$.

The twelve local variational parameters are distributed three per row. As will be seen later, for URM5 there are a sizeable, thirty, local variational parameters. In both these cases this is actually an excess amount over the minimum required number of parameters, i.e. not all variational parameters are linearly independent. For URM4, only eight parameters are strictly required, and only fifteen for URM5. Like other areas of mathematics, the excess is nicer to work with and gives a much simpler form of Δ to understand. The following explanations show how the twelve parameters in URM4 originate, with the minimum number explained shortly thereafter.

There are many ways to view Δ, this section highlights some of them. First and foremost, the general URMT annihilator definition $\Delta X = 0$ (8.1) means that every row of Δ is a row eigenvector, say $\overline{X}$, orthogonal to the coordinate vector X. In URM4, with four dimensions, $\overline{X}$ lies in the three-dimensional hyperplane orthogonal to X. Such a plane can be spanned by three linearly independent, but not necessarily mutually orthogonal vectors, call them X^s, X^t and X^u such that

(8.17) $X^s X = 0$, $X^t X = 0$, $X^u X = 0$.

Then, for some arbitrary, integer parameters s, t, u, a row eigenvector $\overline{X}$ can be constructed as a linear sum of the vectors X^s, X^t and X^u, orthogonal to the coordinate vector X, i.e.

(8.18) $\overline{X} = sX^s + tX^t + uX^u$, $s, t, u \in \mathbb{Z}$.

and

(8.19) $\overline{\mathbf{X}}\mathbf{X} = 0$.

Such a vector $\mathbf{X}$ is then a perfectly adequate row-vector of matrix $\mathbf{\Delta}$. Given there are four rows, then four separate vectors can be created, one for each row, using three different parameters for each row, and making twelve parameters in total. But there is a catch, which is that the vectors $\mathbf{X}^s$, $\mathbf{X}^t$ and $\mathbf{X}^u$ are not all linearly independent, as will be shown shortly. However, beforehand, taking $\overline{\mathbf{X}}$ (8.18) as the top row of $\mathbf{\Delta}$, then the following choice for the three vectors $\mathbf{X}^s$, $\mathbf{X}^t$ and $\mathbf{X}^u$, and parameters s,t,u, satisfies (8.18)

(8.20a) $\mathbf{X}^s = \begin{pmatrix} 0 & +y & -x & 0 \end{pmatrix}$, $s = \delta_{12}$

(8.20b) $\mathbf{X}^t = \begin{pmatrix} 0 & +z & 0 & -x \end{pmatrix}$, $t = \delta_{14}$

(8.20c) $\mathbf{X}^u = \begin{pmatrix} 0 & 0 & +z & -y \end{pmatrix}$, $u = \delta_{13}$.

As mentioned above, this would be fine except the three vectors $\mathbf{X}^s$, $\mathbf{X}^t$ and $\mathbf{X}^u$ are not actually linearly independent. Note too that they all have a zero first element and are really vectors in a 3D subspace of coordinates x, y, z, with $w = 0$. If they are not linearly independent then the following vector equation can be solved for some non-zero integers a, b, c

(8.21) $0 = a\mathbf{X}^s + b\mathbf{X}^t + c\mathbf{X}^u$, $a, b, c \in \mathbb{Z}$.

Writing in matrix form this becomes

(8.22) $\begin{pmatrix} 0 & +z & +y \\ +z & 0 & -x \\ -y & -x & 0 \end{pmatrix} \begin{pmatrix} c \\ b \\ a \end{pmatrix} = \begin{pmatrix} 0 \\ 0 \\ 0 \end{pmatrix}$,

which will have a non-zero solution for a, b, c if the usual non-singular condition is satisfied, i.e.

(8.23) $\det \begin{pmatrix} 0 & +z & +y \\ +z & 0 & -x \\ -y & -x & 0 \end{pmatrix} = 0$.

Indeed, the determinant evaluates to zero and so three integers $(a, b, c) \neq (0, 0, 0)$ can be found such that (8.21) is satisfied, and the vectors $\mathbf{X}^s$, $\mathbf{X}^t$ and $\mathbf{X}^u$ are, therefore, not linearly independent. Consequently, the top row of $\boldsymbol{\Delta}$ (8.16) can be constructed from just two vectors, orthogonal to $\mathbf{X}$, and requiring just two parameters. For four rows then, a minimum of eight parameters are required.

In fact, from URM3, it is known that under URM3 Pythagoras conditions there are two linearly independent row eigenvectors, $\mathbf{X}^{3-}$ ($\lambda = -C$) and $\mathbf{X}^{30}$ ($\lambda = -C$), which, by the orthogonality property of eigenvectors for distinct eigenvalues (F25), are both orthogonal to $\mathbf{X}_3$ ($\sim \mathbf{X}_+$ or $\mathbf{X}_{3+}$). This means that $\overline{\mathbf{X}}_3$ can be expanded as a linear sum of $\mathbf{X}^{3-}$ and $\mathbf{X}^{30}$ for parameters s, t (8.18), i.e.

(8.24) $\overline{\mathbf{X}}_3 = s\mathbf{X}^{3-} + t\mathbf{X}^{30}$,

and using $\mathbf{X}^{3-}\mathbf{X}_3 = 0$ (F1) and $\mathbf{X}^{30}\mathbf{X}_3 = 0$ (F5), then $\overline{\mathbf{X}}_3$ is thus orthogonal to $\mathbf{X}_3$, i.e.

(8.25) $\overline{\mathbf{X}}_3\mathbf{X}_3 = 0$.

The equivalent, four-element vector $\overline{\mathbf{X}}$, orthogonal to $\mathbf{X}$, is thus

(8.26) $\overline{\mathbf{X}} = \begin{pmatrix} 0 & \overline{\mathbf{X}}_3 \end{pmatrix}$.

With $\mathbf{X}$ given by

(8.27) $\mathbf{X} = \begin{pmatrix} w \\ \mathbf{X}_3 \end{pmatrix}$,

then the orthogonality is simply seen to be satisfied

(8.28) $\overline{\mathbf{X}}\mathbf{X} = \begin{pmatrix} 0 & \overline{\mathbf{X}}_3 \end{pmatrix}\begin{pmatrix} w \\ \mathbf{X}_3 \end{pmatrix} = 0$.

Concluding all this, whilst only eight variational parameters are strictly required, the twelve-parameter form of the variational matrix $\boldsymbol{\Delta}$ is preferred on the grounds of simplicity and symmetry. It is also very easy to extend to URMn, in particular URM5.

From the above arguments, the minimum number of variational parameters is $n-2$ per row, where $n-1$ vectors are required but only $n-2$ are linearly independent. For n rows, the minimum number of variational parameters is thus

(8.29) $n(n-2)$,

which gives the following minimum values for the URMT theories under consideration

(8.30)

(8.30a) $n = 2$, zero variational parameters

(8.30b) $n = 3$, three variational parameters

(8.30c) $n = 4$, eight variational parameters

(8.30d) $n = 5$, fifteen variational parameters.

In general, using the excess form of Δ (8.16), the number of free variational parameters is given for URMn by the following cubic polynomial, explained further below,

(8.31) excess number of variational parameters $=$

$$ {}^{n}c_{2}n = \frac{n}{2}(n-1)(n-2), \ n \geq 2, $$

which gives the following values for the specific URMT theories under consideration

(8.32)

(8.32a) $n = 2$, zero variational parameters

(8.32b) $n = 3$, three variational parameters

(8.32c) $n = 4$, twelve variational parameters

(8.32d) $n = 5$, thirty variational parameters.

From (8.30) and (8.32) it is seen that, for URM2 and URM3, both the minimum and excess number of parameters is the same, i.e. zero and three respectively. Thereafter, the excess formula largely overestimates the figure since it is given by a cubic polynomial in n, rather than a quadratic for the minimum. However, this is all academic in that most variational work reduces all the variations to usually one global variation for each dimension, from three upward, when under Pythagoras conditions. For URM3, there is one global 'delta' variation, and for URM4 or higher, Part II of this book adds an extra variational parameter for each excess dimension (I4), i.e. each dimension, four or higher.

There is also a combinatoric explanation for the 'excess' formula (8.31) as follows, which is best understood with an example, given afterward. For each row in Δ, one of the elements is always zero, i.e. that on the leading diagonal and, hence, there are $n-1$ elements per row that can be varied. The number of unique, linear variational terms for all elements in a row is then simply the number of pairs of coordinates (see further below 'cancelling in pairs') that can be formed from the $n-1$ available, which is given by the combinatorial symbol $^{n-1}c_2$. Given that there are n rows, then there are a total of $^{n-1}c_2 n$ variational terms, each requiring a coefficient. Since $^{n-1}c_2 = (n-1)(n-2)/2$ then multiplying by n gives the polynomial (8.31).

For example, starting with the top row of Δ (8.16) for URM4, $n = 4$, the first element of the first row is zero ($\Delta_{11} = 0$) as it is on the lead diagonal. Since it multiplies coordinate w in X (8.27), the product is always zero. The full, first row matrix product ΔX is $\Delta_{11}w + \Delta_{12}x + \Delta_{13}y + \Delta_{14}z$, and since $\Delta_{11} = 0$, then this product reduces to $\Delta_{12}x + \Delta_{13}y + \Delta_{14}z = 0$ and, given all coordinates w, x, y, z are linearly independent, then no other term Δ_{12}, Δ_{13} or Δ_{14} to the right of Δ_{11} can contain w. Thus, matrix terms Δ_{12}, Δ_{13} and Δ_{14} can only contain the three (or $n-1$) remaining coordinates x, y, z. Looking at the next term Δ_{12}, it multiplies x and so it can contain a linear sum of the other two coordinates y and z, as in $\Delta_{12} = \delta_{12}y + \delta_{14}z$. When multiplying-out the vector component, i.e. $\Delta_{12}x = \delta_{12}yx + \delta_{14}zx$, it gives two coordinate pairings yx and zx. These will have to cancel with a pairing formed from the other two terms, Δ_{13} and Δ_{14}, to give a total zero, invariant effect on X. Indeed, $\Delta_{13} = -\delta_{12}x + \delta_{13}z$, has the component $-\delta_{12}x$ and, since Δ_{13} multiplies coordinate y, this component will cancel the $+\delta_{12}y$ component in Δ_{12}, which multiples x. Likewise, for all other components in Δ_{13}, Δ_{14}, which all cancel with each other and Δ_{12} such that $\Delta_{12}x + \Delta_{13}y + \Delta_{14}z = 0$. Thus, finally, the number of variational coefficients is given by the number of such coordinate pairings amongst the three (or $n-1$) coordinates in each row. For the first row, the pairings are xy, xz, yz, which excludes w; for the second row, the pairs are wy, wz, yz, where coordinate x is now excluded. Each row thus having $^3c_2$ (or $^{n-1}c_2$) pairings, and with three (or n) rows in Δ, the total number of variational parameters is given by $^3c_2$ (or $^{n-1}c_2 n$) (8.31).

Note that only first order terms in the coordinates are considered in Δ (8.16), and all variational terms are linear sums of w, x, y, z, there are no second degree or higher terms such as xy, x^2 or xyz etc. This is only ruled-out on the basis of keeping things simple, i.e. linear, with thoughts of physical associations always driving the development. It is certainly possible to use non-linear terms in Δ and, indeed, may provide some interesting physics. However, the field of local and global variations in URMT is enormous and a full exposition has to be deferred to a separate publication because of this.

The combinatoric explanation for the 'excess' formula (8.31) illustrates how the full variational matrix Δ works by cancellation in pairs. An alternative explanation is given earlier of how the URM4 Δ matrix can be thought of in terms of orthogonal URM3 vectors, each of which also only actually has two non-zero elements. The commonality of two non-zero elements in the vectors, and 'cancellation in pairs', gets to the heart of the variational matrix form in general URMT, which is really an embedding of 2-element, URM2 vectors $\overline{\mathbf{X}}_2$ and $\mathbf{X}_2$ into URMn, where $\overline{\mathbf{X}}_2$ is orthogonal to $\mathbf{X}_2$. For example, in URM3, the variational matrix is the simpler form,

$$(8.33) \quad \Delta_3 = \begin{pmatrix} 0 & +\eta z & -\eta y \\ -\delta z & 0 & +\delta x \\ +\varepsilon y & -\varepsilon x & 0 \end{pmatrix}.$$

Concentrating on the top row, and defining 2-element vectors $\overline{\mathbf{X}}_2$ and $\mathbf{X}_2$ as follows:

$$(8.34a) \quad \overline{\mathbf{X}}_2 = \begin{pmatrix} +z & -y \end{pmatrix}$$

(8.34b) $\mathbf{X}_2 = \begin{pmatrix} y \\ z \end{pmatrix}$,

then, for arbitrary integer parameter η, the vectors $\overline{\mathbf{X}}_2$ and $\mathbf{X}_2$ are, indeed, orthogonal as in

(8.35) $\eta\overline{\mathbf{X}}_2\mathbf{X}_2 = 0$.

But, the vector $\eta\overline{\mathbf{X}}_2$ is nothing other than the two non-zero top row elements of $\mathbf{\Delta}_3$, and $\mathbf{X}_2$ comprises the two non-zero elements y and z of the URM3 coordinate vector $\mathbf{X}_3$, with a zero first element in place of coordinate x (A3), i.e. an embedding of a URM2 coordinate vector $\mathbf{X}_2$ in URM3, as in

(8.36) $\mathbf{\Delta}_3 = \begin{pmatrix} 0 & \eta\overline{\mathbf{X}}_2 \\ - & - \end{pmatrix}$, $\mathbf{X}_3 = \begin{pmatrix} 0 \\ \mathbf{X}_2 \end{pmatrix}$.

The middle row of $\mathbf{\Delta}_3$ is slightly trickier in that it is still comprises only two non-zero elements, $-\delta z$ and $+\delta x$, but they are split with a zero in between making it not so obvious how it is a URM2 embedding. Nevertheless, it is still a cancellation in pairs of terms, as in the following orthogonal product (8.39) of the 2-element vectors $\overline{\mathbf{X}}_2$ and $\mathbf{X}_2$, integer parameter δ, i.e.

(8.37) $\overline{\mathbf{X}}_2 = \begin{pmatrix} -z & +x \end{pmatrix}$

(8.38) $\mathbf{X}_2 = \begin{pmatrix} x \\ z \end{pmatrix}$,

with the orthogonal, annihilator relation satisfied

(8.39) $\delta\overline{\mathbf{X}}_2\mathbf{X}_2 = 0$.

This completes the discussion on the form of the variational matrix and how it works. The section now proceeds to usage of $\mathbf{\Delta}$ in URMT.

(8-2) Using $\mathbf{\Delta}$ in URMT

Applying variations to the DCE, individually or in two or more combinations, gives linear, quadratic or higher order perturbation terms in the variational parameters. What to vary is largely a matter of choice, but amongst all possible combinations are special combinations, primarily those that retain a symmetry in the transformed **A** matrix (8.2). The classic example in URM3 being the case when all the variational parameters are equal in magnitude, i.e. equal to within a sign; such a combination is termed a 'global' variation since it affects the whole matrix. In URM4 and beyond, a global variation doesn't have to be all parameters equal, but rather a subset, e.g. those affecting one row or column, as will be seen shortly below. Usually the URM3 global variation is given the symbol δ or m in [1] and m, s and $\bar{s}$ for URM4 in Part I of this book, and t_3 in Part II, where they are also termed evolutionary parameters because they can generally be associated with time under the standard physical interpretation in Appendix (J).

Given the standard URM4 unity root matrix **A**

$$(1.3a) \quad \mathbf{A} = \begin{pmatrix} 0 & S & T & U \\ \bar{S} & 0 & R & \bar{Q} \\ \bar{T} & \bar{R} & 0 & P \\ \bar{U} & Q & \bar{P} & 0 \end{pmatrix},$$

then, by comparison with variational matrix $\mathbf{\Delta}$ (8.16), the dynamical variables transform under the local variations as follows

(8.40)

(8.40a) $S \rightarrow S + \delta_{12} y + \delta_{14} z$

(8.40b) $\bar{S} \rightarrow \bar{S} + \delta_{21} y + \delta_{24} z$

(8.40c) $T \rightarrow T - \delta_{12} x + \delta_{13} z$

(8.40d) $\bar{T} \rightarrow \bar{T} + \delta_{31} x + \delta_{34} z$

(8.40e) $U \rightarrow U - \delta_{14} x - \delta_{13} y$

(8.40f) $\bar{U} \rightarrow \bar{U} + \delta_{41} x + \delta_{43} y$

(8.40g) $Q \rightarrow Q - \delta_{41} w + \delta_{42} y$

(8.40h) $\bar{Q} \rightarrow \bar{Q} - \delta_{24} w - \delta_{23} y$

(8.40i) $R \rightarrow R - \delta_{21} w + \delta_{23} z$

(8.40j) $\bar{R} \rightarrow \bar{R} - \delta_{31}w + \delta_{32}z$

(8.40k) $P \rightarrow P - \delta_{34}w - \delta_{32}x$

(8.40l) $\bar{P} \rightarrow \bar{P} - \delta_{43}w - \delta_{42}x$.

Following URM3 methodology, the standard practice to obtain the dynamical equations and their solutions from first principles, i.e. starting with a conservation equation and an invariance principle, is to apply one or more of the variations to the DCE and equate each unique variational term to zero.

From Section (1), the full DCE is the characteristic equation (1.9)

(1.9) $\lambda^4 + a_2\lambda^2 + a_1\lambda + a_0 = 0$,

with coefficients a_2, a_1, a_0 (1.10).

Inserting eigenvalue $\lambda = C$ as the only known eigenvalue, and that for which the eigenvector $\mathbf{X}$ (1.5) is invariant, makes the full DCE

(8.41) $C^4 + a_2C^2 + a_1C + a_0 = 0$

It is helpful, for forthcoming algebraic manipulation, to completely expand this form of DCE as follows:

(8.42)

$$C^4 = -P\bar{P}C^2 - Q\bar{Q}C^2 - R\bar{R}C^2 - S\bar{S}C^2 - T\bar{T}C^2 - U\bar{U}C^2 +$$
$$- PQRC - \bar{P}\bar{Q}\bar{R}C +$$
$$- RS\bar{T}C - \bar{R}\bar{S}TC +$$
$$- Q\bar{S}UC - \bar{Q}S\bar{U}C +$$
$$- \bar{P}\bar{T}UC - PT\bar{U}C +$$
$$+ P\bar{P}S\bar{S} + Q\bar{Q}T\bar{T} + R\bar{R}U\bar{U} +$$
$$- PRS\bar{U} - \bar{P}\bar{R}\bar{S}U +$$
$$- QR\bar{T}U - \bar{Q}\bar{R}T\bar{U} +$$
$$- \bar{P}\bar{Q}S\bar{T} - PQ\bar{S}T$$

Keep in mind that absolutely no simplifications or assumptions have been made in this full, URM4 DCE form. For example, in Section (4), the coefficient a_1 is zero under URM4 Pythagoras conditions, which reduces the characteristic equation to a quadratic (1.12) in the square of the eigenvalue. Neither has the DCE been partitioned here into a kinetic and Potential components.

(8-3) Dynamical Equations

To obtain a dynamical equation linking coordinates x and w

(8-31) Method 1

With the goal of obtaining some URM4 equations using the variational method, attention is now focussed on a simple example, namely obtaining an equation that links coordinate x with coordinate w in eigenvector **X** (1.5). To do this, look at the third row of Δ (8.16). This is the first row that contains variations in both the coordinates x and w, plus a term in z, i.e.

(8.50a) $\delta_{31}x + \delta_{34}z$

(8.50b) $-\delta_{31}w + \delta_{32}z$

(8.50c) $-\delta_{34}w - \delta_{32}x$.

Each of these three terms transform the dynamical variables as follows, according to (8.40)

(8.51a) $\overline{T} \rightarrow \overline{T} + \delta_{31}x + \delta_{34}z$

(8.51b) $\overline{R} \rightarrow \overline{R} - \delta_{31}w + \delta_{32}z$

(8.51c) $P \rightarrow P - \delta_{34}w - \delta_{32}x$.

The unwanted z term can be very easily removed by choosing coefficients δ_{32} and δ_{34} to be zero,

(8.52) $\delta_{32} = 0$, $\delta_{34} = 0$,

to give variations in just two of the dynamical variables, $\overline{R}$ and $\overline{T}$, with a single variational parameter δ_{31}

(8.53a) $\overline{T} \rightarrow \overline{T} + \delta_{31}x$

(8.53b) $\overline{R} \rightarrow \overline{R} - \delta_{31}w$.

That is, the variations $\delta \overline{T}$ and $\delta \overline{R}$ are

(8.54a) $\delta \overline{T} = \delta_{31} x$

(8.54b) $\delta \overline{R} = -\delta_{31} w$,

with the third dynamical variable P (8.51c) having no variation with δ_{31}.

This single δ_{31} variation is as simple as it can get but, nevertheless, is ideal to extract an equation in x and w.

To proceed, the two variations in $\overline{R}$ and $\overline{T}$ are applied to the full DCE (8.42). However, given only terms in $\overline{R}$ and $\overline{T}$ produce any variation, it is useful to first reduce the DCE to terms only in $\overline{R}$ and $\overline{T}$ giving

(8.55)

$$0 = -R\overline{R}C^2 - T\overline{T}C^2 +$$
$$- \overline{PQ}RC +$$
$$- RS\overline{T}C - \overline{RS}TC +$$
$$- P\overline{T}UC +$$
$$+ Q\overline{Q}T\overline{T} + R\overline{R}U\overline{U} +$$
$$- \overline{PRS}U +$$
$$- QR\overline{T}U - \overline{QR}T\overline{U} +$$
$$- \overline{PQ}S\overline{T}$$

There are twelve terms in all, six in $\overline{R}$ and six in $\overline{T}$; equal numbers being expected by symmetry arguments alone. Loosely speaking, $\overline{R}$ and $\overline{T}$ are on equal terms, one is no more special than the other.

It is now a simple matter here to substitute for the variations in $\overline{R}$ and $\overline{T}$ using $\delta \overline{T} = \delta_{31} x$ (8.54a) and $\delta \overline{R} = -\delta_{31} w$ (8.54b), which means replacing $\overline{T}$ with x and $\overline{R}$ with $-w$, where the variational integer δ_{31} becomes a coefficient to every term, i.e.

(8.56)

$$0 = \delta_{31}(RwC^2 - TxC^2 +$$
$$+ \overline{PQ}wC +$$
$$- RSxC + w\overline{S}TC +$$
$$- \overline{P}xUC +$$
$$+ Q\overline{Q}Tx - RwU\overline{U} +$$
$$+ \overline{P}w\overline{S}U +$$
$$- QRxU + \overline{Q}wT\overline{U} +$$
$$- \overline{PQ}Sx)$$

giving six variational terms in x (for $\overline{T}$) and six in w (for $\overline{R}$). Grouping terms in x and w, and equating to zero for arbitrary variations in δ_{31}, as per URM3 variational methodology, gives the expression

$$(8.57) \quad \begin{aligned} w(RC^2 + \overline{PQ}C + \overline{S}TC - RU\overline{U} + \overline{PS}U + \overline{Q}T\overline{U}) = \\ x(TC^2 + RSC + \overline{P}UC - Q\overline{Q}T + QRU + \overline{PQ}S) \end{aligned} .$$

Note that, as an aside check, all bracketed terms in this expression (8.57) are of the same degree, which is cubic in the dynamical variables when the eigenvalue has the same units as the dynamical variables; see Appendix (J). Each side in (8.57) is also linear in the coordinates.

Whilst (8.57) is a simple linear relation in x and w (as expected), the bracketed coefficients are obviously a lot more involved, i.e. there are a lot more terms than the simple URM3 equivalent solutions, e.g. (8.15), and it is not obvious where to apply simplifications. The usual simplifications will be made further below, i.e. application of Pythagoras conditions but, as would be expected, algebraic manipulation in URM4 becomes a lot more involved than in URM3, and the most general form has no obvious simplification. In this case, URM4 uses the hindsight of Pythagoras conditions, introduced by analogy with URM3, rather than seeing them as immediately obvious from a variational expression, as is the case for the URM3 'global delta variation' in [1],#1.

Having obtained one form of solution (8.57) for coordinate x in terms of w, or vice-versa, the next stage would be to find y and z, also in terms of w, to give a solution for eigenvector $\mathbf{X}$, with each element x, y and z parameterised by w (the first element of $\mathbf{X}$). Thus w acts a scale factor on the elements of $\mathbf{X}$, noting that eigenvectors are only unique to within a scale factor. Normally in an eigenvalue problem this would be the end. However, URMT goes one step further and effects a quantisation of the theory by restricting the coordinates to integers and, additionally, placing a co-primality criterion on all coordinates as in $\gcd(x, y, w, z) = 0$ (1.6).

Note that because w is used a parameter, the single gcd criterion (1.6) is actually three criteria:

(8.58) $\gcd(x, w) = 1$, $\gcd(y, w) = 1$ and $\gcd(z, w) = 1$.

Up to this stage there has been no explicit requirement for any variable to be an integer. However, applying the specific gcd criterion $\gcd(x, w) = 1$ means that the bracketed terms in (8.57) satisfy, for some integer j, the following relations

(8.59)

(8.59a) $jx = (RC^2 + \overline{PQ}C + \overline{S}TC - RU\overline{U} + \overline{PS}U + \overline{Q}T\overline{U})$, $j \in \mathbb{Z}$

(8.59b) $jw = (TC^2 + \overline{RS}C + \overline{P}UC - Q\overline{Q}T + QRU + \overline{PQ}S)$.

This completes the integer solution for x in terms of w using a third row variation, and termed here 'method 1' because it is noticed that there is also a second method whereby the fourth row can also be used to obtain x in terms of w, see Section (8-32) next. Furthermore, a complete solution also requires solutions for y and z in terms of w before a solution for the eigenvector $\mathbf{X}$ (1.5) can be obtained. Each solution, for all three coordinates x, y and z in terms of w, can be obtained by two unique methods, with x in terms of w already solved using 'method 1' (above), the remaining five possible solutions are given following. However, the provision of these general solutions is primarily for completeness because, due to their complexity, they are not used further in this book. The reader may well wish to skip to Section (8-5) which considers variations under Pythagoras conditions, and is much more relevant to the remainder of the book.

(8-32) Method 2

The fourth row of Δ (8.16) also contains variations in both the x and w coordinates, plus a term in y, i.e.

(8.40e) $\overline{U} \to \overline{U} + \delta_{41}x + \delta_{43}y$

(8.40g) $Q \to Q - \delta_{41}w + \delta_{42}y$

(8.40l) $\overline{P} \to \overline{P} - \delta_{43}w - \delta_{42}x$.

Removing the unwanted y term by choosing coefficients δ_{42} and δ_{43} to be zero,

(8.60) $\delta_{42} = 0, \delta_{43} = 0$,

gives variations in just two of the dynamical variables, $\overline{U}$ and Q, with a single variational parameter δ_{41}

(8.61)

(8.61a) $\overline{U} \to \overline{U} + \delta_{41}x$

(8.61b) $Q \to Q - \delta_{41}w$.

That is, the variations $\delta\overline{U}$ and δQ are

(8.62a) $\delta\overline{U} = \delta_{41}x$

(8.62b) $\delta Q = -\delta_{41}w$,

with the third dynamical variable $\overline{P}$ (8.40l) having no variation with δ_{41}.

The DCE (8.42) reduces to the following terms in $\overline{U}$ and Q

(8.63)

$$0 = -Q\overline{Q}C^2 - U\overline{U}C^2 +$$
$$- PQRC +$$
$$- Q\overline{S}UC - \overline{Q}S\overline{U}C +$$
$$- PT\overline{U}C +$$
$$+ Q\overline{Q}T\overline{T} + R\overline{R}U\overline{U} +$$
$$- PRS\overline{U} +$$
$$- QR\overline{T}U - \overline{Q}RT\overline{U} +$$
$$- PQ\overline{S}T$$

Replacing $\overline{U}$ with x, and Q with $-w$, using (8.61), then (8.63) becomes

(8.64)

$$0 = \delta_{41}(w\overline{Q}C^2 - UxC^2 +$$
$$+ PwRC +$$
$$+ w\overline{S}UC - \overline{Q}SxC +$$
$$- PTxC +$$
$$- w\overline{Q}T\overline{T} + R\overline{R}Ux +$$
$$- PRSx +$$
$$+ wR\overline{T}U - \overline{Q}RTx +$$
$$+ Pw\overline{S}T)$$

Grouping terms in x and w gives

$$(8.65) \quad \begin{aligned} w(\overline{Q}C^2 + PRC + \overline{S}UC - \overline{Q}T\overline{T} + R\overline{T}U + P\overline{S}T) = \\ x(UC^2 + \overline{Q}SC + PTC - R\overline{R}U + PRS + \overline{Q}RT) \end{aligned}$$

Applying the gcd criterion $\gcd(x, w) = 1$ (8.58) means that the solution (8.65) satisfies, for some integer k,

(8.66)

(8.66a) $kx = (\overline{Q}C^2 + PRC + \overline{S}UC - \overline{Q}T\overline{T} + R\overline{T}U + P\overline{S}T), \; k \in \mathbb{Z}$

(8.66b) $kw = (UC^2 + \overline{Q}SC + PTC - R\overline{R}U + PRS + \overline{Q}RT).$

This completes method 2 to obtain x in terms of w.

To obtain a dynamical equation linking coordinates y and w

(8-33) Method 1

The second row of $\mathbf{\Delta}$ (8.16) is the first row that contain variations in both the y and w coordinates, plus a term in z, i.e.

(8.40b) $\overline{S} \rightarrow \overline{S} + \delta_{21}y + \delta_{24}z$

(8.40h) $\overline{Q} \rightarrow \overline{Q} - \delta_{24}w - \delta_{23}y$

(8.40i) $R \rightarrow R - \delta_{21}w + \delta_{23}z$

Removing the unwanted z term by choosing coefficients δ_{23} and δ_{24} to be zero,

(8.70) $\delta_{23} = 0$, $\delta_{24} = 0$,

gives variations in just two of the dynamical variables, $\overline{S}$ and R, with a single variational parameter δ_{21}

(8.71)

(8.71a) $\overline{S} \rightarrow \overline{S} + \delta_{21}y$

(8.71b) $R \rightarrow R - \delta_{21}w$

That is, the variations $\delta\overline{S}$ and δR are

(8.72a) $\delta\overline{S} = \delta_{21}y$

(8.72b) $\delta R = -\delta_{21}w$,

with the third dynamical variable $\overline{Q}$ (8.40h) having no variation with δ_{21}.

The DCE (8.42) reduces to the following terms in $\overline{S}$ and R

(8.73)

$$0 = -R\overline{R}C^2 - S\overline{S}C^2 +$$
$$- PQRC +$$
$$- RS\overline{T}C - \overline{R}STC +$$
$$- Q\overline{S}UC +$$
$$+ P\overline{P}S\overline{S} + R\overline{R}U\overline{U} +$$
$$- PRS\overline{U} - \overline{P}R\overline{S}U +$$
$$- QR\overline{T}U +$$
$$- PQ\overline{S}T$$

Replacing $\overline{S}$ with y, and R with $-w$, using (8.71), then (8.73) becomes

(8.74)

$$0 = \delta_{21}(w\overline{R}C^2 - SyC^2 +$$
$$+ PQwC +$$
$$+ wS\overline{T}C - \overline{R}yTC +$$
$$- QyUC +$$
$$+ P\overline{P}Sy - w\overline{R}U\overline{U} +$$
$$+ PwS\overline{U} - \overline{P}RyU +$$
$$+ Qw\overline{T}U +$$
$$- PQyT)$$

Grouping terms in y and w gives

(8.75) $\quad \begin{aligned} w(\overline{R}C^2 + PQC + S\overline{T}C - \overline{R}U\overline{U} + PS\overline{U} + Q\overline{T}U) = \\ y(SC^2 + \overline{R}TC + QUC - P\overline{P}S + \overline{P}RU + PQT) \end{aligned}$

Applying the gcd criterion $\gcd(y, w) = 1$ (8.58) means that the solution (8.75) satisfies, for some integer i,

(8.76)

(8.76a) $iy = (\overline{R}C^2 + PQC + S\overline{T}C - \overline{R}U\overline{U} + PS\overline{U} + Q\overline{T}U)$, $i \in \mathbb{Z}$

(8.76b) $iw = (SC^2 + \overline{R}TC + QUC - P\overline{P}S + \overline{P}RU + PQT)$

This completes method 1 to obtain y in terms of w.

(8-34) Method 2

The fourth row of $\mathbf{\Delta}$ (8.16) also contains variations in both the y and w coordinates, plus terms in x, i.e.

(8.40f) $\overline{U} \rightarrow \overline{U} + \delta_{41}x + \delta_{43}y$

(8.40g) $Q \rightarrow Q - \delta_{41}w + \delta_{42}y$

(8.40l) $\overline{P} \rightarrow \overline{P} - \delta_{43}w - \delta_{42}x$

Removing the unwanted x term by choosing coefficients δ_{41} and δ_{42} to be zero,

(8.80) $\delta_{41} = 0$, $\delta_{42} = 0$,

gives variations in just two of the dynamical variables, $\overline{U}$ and $\overline{P}$, with a single variational parameter δ_{43}

(8.81)

(8.81a) $\overline{U} \rightarrow \overline{U} + \delta_{43}y$

(8.81b) $\overline{P} \rightarrow \overline{P} - \delta_{43}w$

That is, the variations $\delta\overline{U}$ and $\delta\overline{P}$ are

(8.82)

(8.82a) $\delta\overline{U} = \delta_{43}y$

(8.82a) $\delta\overline{P} = -\delta_{43}w$,

with the third dynamical variable Q (8.40g) having no variation with δ_{43}.

The DCE (8.42) reduces to the following terms in $\overline{U}$ and $\overline{P}$

(8.84)

$$0 = -P\overline{P}C^2 - U\overline{U}C^2 +$$
$$-\overline{PQ}\,\overline{R}C +$$
$$-\overline{Q}S\overline{U}C +$$
$$-\overline{P}\overline{T}UC - PT\overline{U}C +$$
$$+P\overline{P}S\overline{S} + R\overline{R}\,U\overline{U} +$$
$$-PRS\overline{U} - \overline{P}\,\overline{R}\,\overline{S}U +$$
$$-\overline{Q}\,\overline{R}T\overline{U} +$$
$$-\overline{P}\,\overline{Q}S\overline{T}$$

Replacing $\overline{U}$ with y, and $\overline{P}$ with $-w$, using (8.81), then (8.84) becomes

(8.85)

$$0 = PwC^2 - UyC^2 +$$
$$+w\overline{Q}\,\overline{R}C +$$
$$-\overline{Q}SyC +$$
$$+w\overline{T}UC - PTyC +$$
$$-PwS\overline{S} + R\overline{R}\,Uy +$$
$$-PRSy + w\overline{R}\,\overline{S}U +$$
$$-\overline{Q}\,\overline{R}Ty +$$
$$+w\overline{Q}S\overline{T}$$

Grouping terms in y and w gives

(8.86)
$$w(PC^2 + \overline{Q}\,\overline{R}C + \overline{T}UC - PS\overline{S} + \overline{R}\,\overline{S}U + \overline{Q}S\overline{T}) =$$
$$y(UC^2 + \overline{Q}SC + PTC - R\overline{R}U + PRS + \overline{Q}\,\overline{R}T)$$

Comparing this with (8.65) implies that, for the same integer k as used in (8.66), the solution (8.86) satisfies,

(8.87) $ky = (PC^2 + \overline{Q}\,\overline{R}C + \overline{T}UC - PS\overline{S} + \overline{R}\,\overline{S}U + \overline{Q}S\overline{T})$.

This completes method 2 to obtain y in terms of w.

To obtain a dynamical equation linking coordinates z and w (8-35) Method 1

The second row of Δ (8.16) is the first row that contain variations in both the z and w coordinates, plus a term in y, i.e.

(8.40b) $\bar{S} \rightarrow \bar{S} + \delta_{21} y + \delta_{24} z$

(8.40h) $\bar{Q} \rightarrow \bar{Q} - \delta_{24} w - \delta_{23} y$

(8.40i) $R \rightarrow R - \delta_{21} w + \delta_{23} z$

Removing the unwanted y term by choosing coefficients δ_{21} and δ_{23} to be zero,

(8.90) $\delta_{21} = 0$, $\delta_{23} = 0$,

gives variations in just two of the dynamical variables, $\bar{S}$ and $\bar{Q}$, with a single variational parameter δ_{24}

(8.91)

(8.91a) $\bar{S} \rightarrow \bar{S} + \delta_{24} z$

(8.91b) $\bar{Q} \rightarrow \bar{Q} - \delta_{24} w$

That is, the variations $\delta \bar{S}$ and $\delta \bar{Q}$ are

(8.92a) $\delta \bar{S} = \delta_{24} z$

(8.92b) $\delta \bar{Q} = -\delta_{24} w$

with the third dynamical variable R (8.40i) having no variation with δ_{24}.

The DCE (8.42) reduces to the following terms in $\bar{S}$ and $\bar{Q}$

(8.93)

$$0 = -Q\overline{Q}C^2 - S\overline{S}C^2 +$$
$$-\overline{PQ}RC +$$
$$-\overline{R}\overline{S}TC +$$
$$-Q\overline{S}UC - \overline{Q}S\overline{U}C +$$
$$+P\overline{P}S\overline{S} + Q\overline{Q}T\overline{T} +$$
$$-\overline{PR}\overline{S}U +$$
$$-\overline{Q}R T\overline{U} +$$
$$-\overline{PQ}S\overline{T} - PQ\overline{S}T$$

Replacing $\overline{S}$ with z, and $\overline{Q}$ with $-w$, using (8.91), then (8.93) becomes

(8.94)

$$0 = QwC^2 - SzC^2 +$$
$$+\overline{P}w\overline{R}C +$$
$$-\overline{R}zTC +$$
$$-QzUC + wS\overline{U}C +$$
$$+P\overline{P}Sz - QwT\overline{T} +$$
$$-\overline{PR}zU +$$
$$+w\overline{R}T\overline{U} +$$
$$+\overline{P}wS\overline{T} - PQzT$$

Grouping terms in z and w gives

$$(8.95) \quad \begin{aligned} w(QC^2 + \overline{PR}C + S\overline{U}C - QT\overline{T} + \overline{R}T\overline{U} + \overline{P}S\overline{T}) = \\ z(SC^2 + \overline{R}TC + QUC - P\overline{P}S + \overline{PR}U + PQT) \end{aligned}.$$

Comparing this with (8.75) implies that, for the same integer i as used in (8.76), the solution (8.95) satisfies,

$$(8.96) \quad iz = (QC^2 + \overline{PR}C + S\overline{U}C - QT\overline{T} + \overline{R}T\overline{U} + \overline{P}S\overline{T}).$$

This completes method 1 to obtain z in terms of w.

(8-36) Method 2

The third row of Δ (8.16) also contains variations in both the z and w coordinates, plus a term in x, i.e.

(8.40d) $\overline{T} \to \overline{T} + \delta_{31}x + \delta_{34}z$

(8.40j) $\overline{R} \to \overline{R} - \delta_{31}w + \delta_{32}z$

(8.40k) $P \to P - \delta_{34}w - \delta_{32}x$.

The unwanted x term can be very easily removed by choosing coefficients δ_{31} and δ_{32} to be zero,

(8.100) $\delta_{31} = 0$, $\delta_{32} = 0$,

to give variations in just two of the dynamical variables, $\overline{T}$ and P, with a single variational parameter δ_{34}

(8.101)

(8.101a) $\overline{T} \to \overline{T} + \delta_{34}z$

(8.101b) $P \to P - \delta_{34}w$.

That is, the variations $\delta\overline{T}$ and δP are

(8.102a) $\delta\overline{T} = \delta_{34}z$

(8.102b) $\delta P = -\delta_{34}w$.

with the third dynamical variable $\overline{R}$ (8.40j) having no variation with δ_{31}.

The DCE (8.42) reduces to the following terms in $\overline{T}$ and P

(8.103)

$$0 = -P\bar{P}C^2 - T\bar{T}C^2 +$$
$$- PQRC +$$
$$- RS\bar{T}C +$$
$$- \bar{P}\bar{T}UC - PT\bar{U}C +$$
$$+ P\bar{P}S\bar{S} + Q\bar{Q}T\bar{T} +$$
$$- PRS\bar{U} +$$
$$- QR\bar{T}U +$$
$$- \bar{P}\bar{Q}S\bar{T} - PQ\bar{S}T$$

Replacing $\bar{T}$ with z, and P with $-w$, using (8.101), then (8.103) becomes

(8.104)

$$0 = w\bar{P}C^2 - TzC^2 +$$
$$+ wQRC +$$
$$- RSzC +$$
$$- \bar{P}zUC + wT\bar{U}C +$$
$$- w\bar{P}S\bar{S} + Q\bar{Q}Tz +$$
$$+ wRS\bar{U} +$$
$$- QRzU +$$
$$- \bar{P}\bar{Q}Sz + wQ\bar{S}T$$

Grouping terms in z and w gives

(8.105)
$$w(\bar{P}C^2 + QRC + T\bar{U}C - \bar{P}S\bar{S} + RS\bar{U} + Q\bar{S}T) = z(TC^2 + RSC + \bar{P}UC - Q\bar{Q}T + QRU + \bar{P}\bar{Q}S)$$.

Comparing this with (8.57) implies, for the same integer j as used in (8.59), the solution (8.105) satisfies,

(8.106) $jz = (\bar{P}C^2 + QRC + T\bar{U}C - \bar{P}S\bar{S} + RS\bar{U} + Q\bar{S}T)$

This completes method 2 to obtain z in terms of w.

(8-37) Summarising the results

By applying a single variation δ_{ij} to Δ (8.16), a solution relating two coordinates, e.g. using δ_{31} for x in terms of w in Section (8-31), can be obtained in terms of the dynamical variables of $\mathbf{A}$ (1.3a). This enables a complete eigenvector solution for $\mathbf{X}$ (1.5) to be obtained by obtaining, for example, all coordinates x, y, z in terms of w. At the same time, by applying gcd conditions (8.58), three extra integers i (8.76), j (8.59) and k (8.66) are introduced. These integers are actually the URM4 equivalents of the URM3 scale factors $\alpha_3, \beta_3, \gamma_3$, which are also related to the URM4 scale factors α, β, γ (3.56), albeit no further detail is provided herein for the following reason:

The use of only a single variation in the preceding workings just scratches the surface of the full power of URMT's variational method. As for URM3 in [1],#1, two or more variations can be applied to obtain numerous more interesting results. However, given the number of possible variations (8.30), the field is too large to explore here and, as mentioned earlier in this Section (8), deferred to a separate publication

To finish this summary, the complete URM4 coordinate solution, for x, y, z in terms of w, is reproduced below. Note that these solutions have no simplifying preconditions, such as Pythagoras conditions (4.1), which are now considered following the summary.

$$(8.57) \quad \begin{aligned} w(RC^2 + \overline{P}\overline{Q}C + \overline{S}TC - RU\overline{U} + \overline{P}\overline{S}U + \overline{Q}T\overline{U}) = \\ x(TC^2 + RSC + \overline{P}UC - Q\overline{Q}T + QRU + \overline{P}\overline{Q}S) \end{aligned}$$

$$(8.59a) \quad jx = (RC^2 + \overline{P}\overline{Q}C + \overline{S}TC - RU\overline{U} + \overline{P}\overline{S}U + \overline{Q}T\overline{U})$$

$$(8.59b) \quad jw = (TC^2 + \overline{R}\overline{S}C + \overline{P}UC - Q\overline{Q}T + QRU + \overline{P}\overline{Q}S).$$

$$(8.65) \quad \begin{aligned} w(\overline{Q}C^2 + PRC + \overline{S}UC - \overline{Q}T\overline{T} + RTU + P\overline{S}T) = \\ x(UC^2 + \overline{Q}SC + PTC - R\overline{R}U + PRS + \overline{Q}RT) \end{aligned}, $$

$$(8.66a) \quad kx = (\overline{Q}C^2 + PRC + \overline{S}UC - \overline{Q}T\overline{T} + RTU + P\overline{S}T)$$

$$(8.66b) \quad kw = (UC^2 + \overline{Q}SC + PTC - R\overline{R}U + PRS + \overline{Q}RT).$$

$$(8.75) \quad \begin{aligned} w(\overline{R}C^2 + PQC + S\overline{T}C - \overline{R}U\overline{U} + PS\overline{U} + Q\overline{T}U) = \\ y(SC^2 + \overline{R}TC + QUC - P\overline{P}S + \overline{P}RU + PQT) \end{aligned}$$

148

(8.76a) $iy = (\overline{R}C^2 + PQC + S\overline{T}C - \overline{R}U\overline{U} + PS\overline{U} + Q\overline{T}U)$

(8.76b) $iw = (SC^2 + \overline{R}TC + QUC - P\overline{P}S + \overline{P}\overline{R}U + PQT)$

(8.86) $\begin{aligned} w(PC^2 + \overline{Q}\overline{R}C + \overline{T}UC - PS\overline{S} + \overline{R}SU + \overline{Q}S\overline{T}) = \\ y(UC^2 + \overline{Q}SC + PTC - R\overline{R}U + PRS + \overline{Q}\overline{R}T) \end{aligned}$

(8.87) $ky = (PC^2 + \overline{Q}\overline{R}C + \overline{T}UC - PS\overline{S} + \overline{R}SU + \overline{Q}S\overline{T})$

(8.95) $\begin{aligned} w(QC^2 + \overline{P}\overline{R}C + S\overline{U}C - QT\overline{T} + \overline{R}T\overline{U} + \overline{P}S\overline{T}) = \\ z(SC^2 + \overline{R}TC + QUC - P\overline{P}S + \overline{P}\overline{R}U + PQT) \end{aligned}$

(8.96) $iz = (QC^2 + \overline{P}\overline{R}C + S\overline{U}C - QT\overline{T} + \overline{R}T\overline{U} + \overline{P}S\overline{T})$

(8.105) $\begin{aligned} w(\overline{P}C^2 + QRC + T\overline{U}C - \overline{P}S\overline{S} + RS\overline{U} + Q\overline{S}T) = \\ z(TC^2 + RSC + \overline{P}UC - Q\overline{Q}T + QRU + \overline{P}\overline{Q}S) \end{aligned}$

(8.106) $jz = (\overline{P}C^2 + QRC + T\overline{U}C - \overline{P}S\overline{S} + RS\overline{U} + Q\overline{S}T)$

(8-5) Pythagoras Variations

Under URM4 Pythagoras conditions (4.1), the URM4 unity root matrix becomes

(4.2) $\mathbf{A} = \begin{pmatrix} 0 & S & T & U \\ -S & 0 & R & Q \\ -T & -R & 0 & P \\ U & Q & P & 0 \end{pmatrix}.$

The local variations δ_{ij} in (8.16) must now satisfy the six constraints (4.1), which, as shown following, only leaves four independent global variations; contrast this with a single, URM3 δ variation.

Equating S and $\overline{S}$ such that $\overline{S} = -S$ implies

(8.110) $\delta_{21} = -\delta_{12}, \ \delta_{24} = -\delta_{14},$

and modifies the second row as follows:

(8.111)

$$\Delta = \begin{pmatrix} 0 & \delta_{12}y + \delta_{14}z & -\delta_{12}x + \delta_{13}z & -\delta_{14}x - \delta_{13}y \\ -\delta_{12}y - \delta_{14}z & 0 & \delta_{12}w + \delta_{23}z & -\delta_{14}w - \delta_{23}y \\ \delta_{31}x + \delta_{34}z & -\delta_{31}w + \delta_{32}z & 0 & -\delta_{34}w - \delta_{32}x \\ \delta_{41}x + \delta_{43}y & -\delta_{41}w + \delta_{42}y & -\delta_{43}w - \delta_{42}x & 0 \end{pmatrix}.$$

Equating T and $\overline{T}$ such that $\overline{T} = -T$, implies

(8.112) $\delta_{31} = \delta_{12}$, $\delta_{34} = -\delta_{13}$,

and equating R and $\overline{R}$ such that $\overline{R} = -R$ additionally implies

(8.113) $\delta_{32} = -\delta_{23}$,

and modifies the third row as follows:

(8.114)

$$\Delta = \begin{pmatrix} 0 & \delta_{12}y + \delta_{14}z & -\delta_{12}x + \delta_{13}z & -\delta_{14}x - \delta_{13}y \\ -\delta_{12}y - \delta_{14}z & 0 & \delta_{12}w + \delta_{23}z & \delta_{14}w - \delta_{23}y \\ \delta_{12}x - \delta_{13}z & -\delta_{12}w - \delta_{23}z & 0 & \delta_{13}w + \delta_{23}x \\ \delta_{41}x + \delta_{43}y & -\delta_{41}w + \delta_{42}y & -\delta_{43}w \doteq \delta_{42}x & 0 \end{pmatrix}.$$

Lastly, equating U and $\overline{U}$ such that $\overline{U} = U$ implies

(8.115) $\delta_{41} = -\delta_{14}$, $\delta_{43} = -\delta_{13}$,

and equating Q and $\overline{Q}$ such that $\overline{Q} = Q$ additionally implies

(8.116) $\delta_{42} = -\delta_{23}$,

and modifies the fourth row as follows, to give the complete, global Pythagoras delta variation Δ^P, as follows:

(8.117)

$$\Delta^P = \begin{pmatrix} 0 & \delta_{12}y + \delta_{14}z & -\delta_{12}x + \delta_{13}z & -\delta_{14}x - \delta_{13}y \\ -\delta_{12}y - \delta_{14}z & 0 & \delta_{12}w + \delta_{23}z & \delta_{14}w - \delta_{23}y \\ \delta_{12}x - \delta_{13}z & -\delta_{12}w - \delta_{23}z & 0 & \delta_{13}w + \delta_{23}x \\ -\delta_{14}x - \delta_{13}y & \delta_{14}w - \delta_{23}y & \delta_{13}w + \delta_{23}x & 0 \end{pmatrix}$$

Having applied five Pythagoras conditions $\overline{S} = -S$, $\overline{T} = -T$, $\overline{U} = U$, $\overline{R} = -R$ and $\overline{Q} = Q$, the variational matrix Δ^P (8.117) now satisfies all six Pythagoras conditions (4.1), without further application of the last condition $\overline{P} = P$. It is seen that there are four independent parameters remaining, $\delta_{12}, \delta_{13}, \delta_{14}, \delta_{23}$, unlike URM3's Δ^P in [1],#3, which only had one global variation δ.

By the 'symmetry' of Δ^P, which is identical to that of the URM4 matrix $\mathbf{A}$ under Pythagoras conditions, i.e. (4.2), the transformed matrix $\mathbf{A} = \mathbf{A}' + \Delta$ (8.2) has the same symmetry about the diagonal as the original, untransformed matrix $\mathbf{A}$. Because of this, the transformed dynamical variables, (8.118) below, still satisfy the URM4 Pythagoras conditions (4.1), and therefore the eigenvectors, for the non-zero eigenvalues, are Pythagorean quadruples. However, the URM4 Potential V (4.3c) is not invariant under Pythagoras conditions, neither is it generally zero. This is unlike URM3, which retains an invariant, zero Potential under a global, Pythagoras delta transformation Δ^P. This fact means a lot more effort in URM4 (and URMn in general) has to be expended to obtain the highly desirable, URMT goal of an invariant, zero Potential solution, as will be examined in subsequent sections of this book; see also Section (9) for the reasons behind this desire.

By comparison of $\mathbf{A}$ (4.2), under Pythagoras conditions, with variational matrix Δ^P (8.117), the dynamical variables now transform under a global transformation as follows:

(8.118)

(8.118a) $S \rightarrow S + \delta_{12} y + \delta_{14} z$, $(\overline{S} = -S)$

(8.118b) $T \rightarrow T - \delta_{12} x + \delta_{13} z$, $(\overline{T} = -T)$

(8.118c) $U \rightarrow U - \delta_{14} x - \delta_{13} y$, $(\overline{U} = U)$

(8.118d) $P \rightarrow P + \delta_{13} w + \delta_{23} x$, $(\overline{P} = P)$

(8.118e) $Q \rightarrow Q + \delta_{14} w - \delta_{23} y$, $(\overline{Q} = Q)$

(8.118f) $R \rightarrow R + \delta_{12} w + \delta_{23} z$, $(\overline{R} = -R)$.

Making all four parameters equal, i.e. $\delta_{12} = \delta_{13} = \delta_{14} = \delta_{23} = \delta$, then makes the variation analogous to the single, URM3 global δ variation, albeit this is not a reduction to the identical URM3 case.

These four parameters, $\delta_{12}, \delta_{13}, \delta_{14}, \delta_{23}$ are used throughout the remainder of the variational work herein, and are simply relabelled $\delta_a, \delta_b, \delta_c, \delta_d$

(8.119)

(8.119a) $\delta_a = \delta_{12}$

(8.119b) $\delta_b = \delta_{13}$

(8.119c) $\delta_c = \delta_{14}$

(8.119d) $\delta_d = \delta_{23}$.

Using these relabelled variational parameters, the dynamical variables transform as follows:

(8.120)

(8.120a) $S \rightarrow S + \delta_a y + \delta_c z$, $(\overline{S} = -S)$

(8.120b) $T \rightarrow T - \delta_a x + \delta_b z$, $(\overline{T} = -T)$

(8.120c) $U \rightarrow U - \delta_c x - \delta_b y$, $(\overline{U} = U)$

(8.120d) $P \rightarrow P + \delta_b w + \delta_d x$, $(\overline{P} = P)$

(8.120e) $Q \rightarrow Q + \delta_c w - \delta_d y$, $(\overline{Q} = Q)$

(8.120f) $R \rightarrow R + \delta_a w + \delta_d z$, $(\overline{R} = -R)$,

and the global, variational matrix $\mathbf{\Delta}^P$ (8.117) becomes,

(8.121)

$$\mathbf{\Delta}^P = \begin{pmatrix} 0 & \delta_a y + \delta_c z & -\delta_a x + \delta_b z & -\delta_c x - \delta_b y \\ -\delta_a y - \delta_c z & 0 & \delta_a w + \delta_d z & \delta_c w - \delta_d y \\ \delta_a x - \delta_b z & -\delta_a w - \delta_d z & 0 & \delta_b w + \delta_d x \\ -\delta_c x - \delta_b y & \delta_c w - \delta_d y & \delta_b w + \delta_d x & 0 \end{pmatrix}.$$

(8-6) Decomposition of Δ^P

The Pythagorean variational matrix Δ^P can be decomposed into four smaller forms, each essentially a 3x3 embedding, with features almost identical to URM3 variational matrices. The advantage of doing this is for ease of analysis and algebraic manipulation, and hopefully it all helps in understanding the variational method.

The matrix Δ^P is split into the following four separate, 4x4 variational matrices Δ_a^P, Δ_b^P, Δ_b^P and Δ_d^P, one for each variational parameter $\delta_a, \delta_b, \delta_c, \delta_d$, where the parameter now becomes a coefficient to the individual matrices, i.e.

$$(8.130)\quad \Delta^P = \delta_a \Delta_a^P + \delta_b \Delta_b^P + \delta_c \Delta_b^P + \delta_d \Delta_d^P$$

$$(8.131a)\quad \Delta_a^P = \begin{pmatrix} 0 & +y & -x & 0 \\ -y & 0 & +w & 0 \\ +x & -w & 0 & 0 \\ 0 & 0 & 0 & 0 \end{pmatrix}$$

$$(8.131b)\quad \Delta_b^P = \begin{pmatrix} 0 & 0 & +z & -y \\ 0 & 0 & 0 & 0 \\ -z & 0 & 0 & +w \\ -y & 0 & +w & 0 \end{pmatrix}$$

$$(8.131c)\quad \Delta_c^P = \begin{pmatrix} 0 & +z & 0 & -x \\ -z & 0 & 0 & +w \\ 0 & 0 & 0 & 0 \\ -x & +w & 0 & 0 \end{pmatrix}$$

$$(8.131d)\quad \Delta_d^P = \begin{pmatrix} 0 & 0 & 0 & 0 \\ 0 & 0 & +z & -y \\ 0 & -z & 0 & +x \\ 0 & -y & +x & 0 \end{pmatrix}.$$

Since the coordinates w, x, y, z are invariant to a general transformation Δ (8.16), including Δ^P (8.121), each of these four matrices is a constant matrix with the variational parameter acting as a coefficient.

It can be seen that the three matrices $\mathbf{\Delta}_b^P$, $\mathbf{\Delta}_c^P$ and $\mathbf{\Delta}_d^P$ each have the same sign structure which, discarding the zero rows and columns, is that of the URM3 variational matrix sign structure as in [1] for $\mathbf{\Delta}_3^P$, i.e.

$$(8.132) \quad \begin{pmatrix} 0 & + & - \\ - & 0 & + \\ - & + & 0 \end{pmatrix} \text{ or } \begin{pmatrix} 0 & - & + \\ + & 0 & - \\ + & - & 0 \end{pmatrix}, \text{ for } \mathbf{\Delta}_b^P, \mathbf{\Delta}_c^P, \mathbf{\Delta}_d^P,$$

where the URM3 matrix $\mathbf{\Delta}_3^P$ is as follows, also denoted by symbol $\mathbf{A}_+$ in [1],#5

$$(8.133) \quad \mathbf{\Delta}_3^P = \begin{pmatrix} 0 & +z & -y \\ -z & 0 & +x \\ -y & +x & 0 \end{pmatrix} \sim \mathbf{A}_+ \text{ in } [1],\#5.$$

However, note that the first matrix $\mathbf{\Delta}_a^P$ (8.131a) is skew-symmetric, and not of the same structure as the other three $\mathbf{\Delta}_b^P$, $\mathbf{\Delta}_c^P$ and $\mathbf{\Delta}_d^P$. The sign structure for a skew-symmetric 3x3 matrix is, by definition,

$$(8.134) \quad \begin{pmatrix} 0 & + & - \\ - & 0 & + \\ + & - & 0 \end{pmatrix} \text{ or } \begin{pmatrix} 0 & - & + \\ + & 0 & - \\ - & + & 0 \end{pmatrix},$$

which is actually just a sign reversal of the bottom row of the 3x3 matrices in (8.132). With this in mind, a URM3 compatible $\mathbf{\Delta}^P$ matrix is obtained with the following settings

$$(8.135) \quad \delta_a = 0, \ \delta_b = 0, \ \delta_c = 0, \ \delta_d = \delta,$$

to give

$$(8.136) \quad \mathbf{\Delta}^P = \delta \begin{pmatrix} 0 & 0 & 0 & 0 \\ 0 & 0 & +z & -y \\ 0 & -z & 0 & +x \\ 0 & -y & +x & 0 \end{pmatrix}.$$

As an aside note, under the unifying concepts in [1],#5, the URM3 matrix $\mathbf{\Delta}_3^P$ (see also (8.133) above) is given by the $\mathbf{T}$ operator transform (Appendix (E)) of the standard, global skew variation, i.e.

$\mathbf{\Delta}_3^P = \mathbf{T}\mathbf{\Delta}^s$, which simply reverses the sign in the bottom row, as mentioned above. The matrices $\mathbf{T}$ and $\mathbf{\Delta}^S$ are reproduced from [1],#1 as follows:

$$(8.137) \quad \mathbf{T} = \mathbf{T}^T = \mathbf{T}^{-1} = \begin{pmatrix} 1 & 0 & 0 \\ 0 & 1 & 0 \\ 0 & 0 & -1 \end{pmatrix},$$

$$(8.138) \quad \mathbf{\Delta}^s = \begin{pmatrix} 0 & +z & -y \\ -z & 0 & +x \\ +y & -x & 0 \end{pmatrix}.$$

(8-7) A Single Global Variation Matrix

By setting all four parameters $\delta_a, \delta_b, \delta_c, \delta_d$ equal to the single variational parameter δ as in,

(8.140)

$(8.140\mathrm{e}) \quad \delta = \delta_a = \delta_{12}$

$(8.140\mathrm{b}) \quad \delta = \delta_b = \delta_{13}$

$(8.140\mathrm{c}) \quad \delta = \delta_c = \delta_{14}$

$(8.140\mathrm{d}) \quad \delta = \delta_d = \delta_{23}$,

then a global variational matrix $\mathbf{\Delta}^P$ is obtained with one variation δ acting as a coefficient of $\mathbf{\Delta}^P$.

For URM3 this was the most used form since it preserved the URM3 Potential V_3 when under Pythagoras conditions. Whilst this preservation of a zero Potential will not always be seen to be the case in most URM4 example solutions, it is, nevertheless, useful because of its simplicity; see, for example, Sections (9) and (10).

In URM3 the evolution parameter m was physically associated with time, i.e. a temporal parameter defined simply in terms of δ as in

$(8.141) \quad m = -\delta$.

From here onward, the evolutionary parameter m will generally be used in place of δ, and explicitly removed from the definition of matrix $\mathbf{\Delta}^P$ such that $\mathbf{\Delta}^P$ is merely a constant function of the coordinates w, x, y, z, with no embedded factor i.e.

(8.142) $\mathbf{\Delta}^P \rightarrow -m\mathbf{\Delta}^P \sim \mathbf{\Delta}^P \rightarrow \delta\mathbf{\Delta}^P$

$$(8.143)\ \mathbf{\Delta}^P = \begin{pmatrix} 0 & +y+z & -x+z & -x-y \\ -y-z & 0 & +w+z & +w-y \\ +x-z & -w-z & 0 & +w+x \\ -x-y & +w-y & +w+x & 0 \end{pmatrix}$$

This evolutionary parameter m is used in preference to δ when varying URM4 solutions under Pythagorean conditions, and where the study of the variation in the eigenvectors solution becomes a study of their time evolution, as in URM3. Note, however, this association of m with time is still tentative, i.e. not set in stone in URMT. Nevertheless it is a very consistent and compelling association when considering physical aspects. Furthermore, in Part II of this book, m is replaced again by the equivalent time parameter t_3, since each dimension, three and higher, is given its own evolutionary parameter t_n, $n \geq 3$. However, for Part I, m is retained for URM3 compatibility.

Finally, as will be used in the next few sections when applying variational methods to some specific solutions, the unity root matrix $\mathbf{A}$, when transformed by a global, delta transformation $\mathbf{\Delta}^P$ at evolutionary time m, is denoted by $\mathbf{A}_m$, and algebraically specified in much the same way as (8.2) by

(8.144) $\mathbf{A}_m = \mathbf{A}' - m\mathbf{\Delta}^P$, cf. (8.2),

where the primed matrix $\mathbf{A}'$ is its initial value for $m = 0$, i.e. subject to no variation

(8.145) $\mathbf{A}' = \mathbf{A}(m = 0)$.

(9) An Invariant Zero Potential

This section illustrates the variational methods as presented in Section (8-5), when under Pythagoras conditions, and applied to the 'PS+RU', zero Potential solution given in Section (5). A brief summary of the PS+RU solution is provided below before moving on to applying variational methods to the solution. Note that the singular, invariant, zero Potential solution determined in this section is a one-off, and not the more general solution obtained via lifting in Section (7), and subject to variational methods in Section (11). It is this latter solution that is carried forward to Part II where it is shown to exhibit 'compactification' behaviour. Nevertheless, it is illustrative that such an invariant, zero Potential solution can be obtained within URM4, as for URM3.

(9.0) Before proceeding, it is worth noting exactly why such an invariant, zero Potential solution is important in URMT. Firstly, an invariant Potential, not specifically zero, is very important in URMT physics because, as in any energy conservation equation, it means the kinetic energy is a constant in a closed system. In this case, the DCE (1.14) has a constant kinetic term K equal to the total energy E, i.e. $E = C^2 = K$ (1.14b). It also means there is no kinetic/Potential energy interchange and, hence, no force, or at least no forces that do any work. Given there is no apparent mass in the energy conservation equation (1.14b), an invariant Potential solution is thought to represent the path of a massless particle, e.g. a photon or graviton, moving in a force-free environment. That the Potential is also zero, and not just constant (invariant), means the particle has its maximum kinetic energy, i.e. speed, given the Potential in URM4 (4.3c) is always greater than or equal to zero. In brief, a massless particle, moving under zero force, in a zero Potential, is as basic and fundamental as it can physically get. If URMT can accommodate this physical aspect, then it has some solid foundations.

(9-1) Solution Summary

The zero Potential solution, Section (5), is given under URM4 Pythagoras conditions (4.1), with the additional condition (4.16) on the dynamical variables to enforce a zero Potential (4.3c), i.e.

(4.16) $V = 0 \Rightarrow QT = PS + RU$.

This condition can be satisfied with the following conditions on the dynamical variables

(5.1)

(5.1a) $Q = 0$

(5.1b) $T = 0$

(5.1c) $PS + RU = 0$.

That both $Q = 0$ and $T = 0$ is not strictly necessary to meet $V = 0$ (4.16) but, in fact, $T = 0$ only holds for zero variation, and it becomes non-zero for arbitrary variations, as shall be seen later. The two conditions, (4.1) and (4.16), give four real eigenvalues

(4.14a) $\lambda = \pm C, 0, 0$.

The solution for the non-zero dynamical variables P, R, S, U , such that $PS + RU = 0$ (5.6c), is parameterised in terms of the coordinates w, x, y, z as follows:

(5.12)

(5.12a) $P = yz$

(5.12b) $Q = 0$

(5.12c) $R = xy$

(5.12d) $S = -wx$

(5.12e) $T = 0$

(5.12f) $U = wz$

(1.5) $\gcd(w, x, y, z) = 1$.

With conditions (5.1), and a zero Potential (4.16), the DCE (1.14) is just the Kinetic energy K

(5.2a) $K = C^2 = (P^2 + U^2) - (R^2 + S^2)$, the DCE.

and, using the parameterisations (5.12), this becomes

(5.13) $C^2 = (z^2 - x^2)(w^2 + y^2)$.

The two, bracketed factors on the right are each chosen to equal the eigenvalue, i.e.

(5.14)

(5.14a) $C = (z^2 - x^2)$

(5.14b) $C = (w^2 + y^2)$.

The choice is made such that C is positive and greater than zero by URMT convention.

(5.15) $C > 0 \Rightarrow |z| > |x|$.

By equating the two expressions (5.14a) and (5.14b), the quadruple (w, x, y, z) naturally satisfies the 4D Pythagoras equation:

(5.16) $z^2 = w^2 + x^2 + y^2$.

The four parameters w, x, y, z are re-parameterised in terms of four new integers i, j, s, t (5.24) to generate all primitive, Pythagorean quadruples with a positive value for z. However, this parameterisation is not explicitly required here, as all working is in w, x, y, z, so the reader is referred back to equations (5.24) for the solution.

The eigenvectors $\mathbf{X}_+$, $\mathbf{X}_-$, $\mathbf{X}_{0A}$, $\mathbf{X}_{0B}$ and their reciprocals $\mathbf{X}^+$, $\mathbf{X}^-$, $\mathbf{X}^{0A}$ and $\mathbf{X}^{0B}$ are

$$(5.55)\ \mathbf{X}_+ = \begin{pmatrix} w \\ x \\ y \\ z \end{pmatrix},\ \mathbf{X}_{0A} = \begin{pmatrix} -y \\ 0 \\ +w \\ 0 \end{pmatrix},\ \mathbf{X}_{0B} = \begin{pmatrix} 0 \\ z \\ 0 \\ x \end{pmatrix},\ \mathbf{X}_- = \begin{pmatrix} +Cw \\ -Cx \\ +Cy \\ -Cz \end{pmatrix}$$

(5.56)

$$\mathbf{X}^+ = \begin{pmatrix} +Cw & -Cx & +Cy & +Cz \end{pmatrix}$$

$$\mathbf{X}^{0A} = \begin{pmatrix} -Cy & 0 & +Cw & 0 \end{pmatrix}$$

$$\mathbf{X}^{0B} = \begin{pmatrix} 0 & +Cz & 0 & -Cx \end{pmatrix}$$

$$\mathbf{X}^{\neg} = \begin{pmatrix} +w & +x & +y & -z \end{pmatrix},$$

and the $\mathbf{A}$ matrix is

$$(5.53) \ \mathbf{A} = \begin{pmatrix} 0 & -wx & 0 & +wz \\ +wx & 0 & +xy & 0 \\ 0 & -xy & 0 & +yz \\ +wz & 0 & +yz & 0 \end{pmatrix}.$$

This completes the summary of the PS+RU solution in Section (5).

(9-2) Applying the variational method

Given the zero Potential solution is under URM4 Pythagoras conditions, the Pythagoras variational matrix $\mathbf{\Delta}^P$ (8.121), reproduced below, is used instead of the full form $\mathbf{\Delta}$ (8.16). This means there are a maximum of four variational parameters $\delta_a, \delta_b, \delta_c, \delta_d$ available

(8.121)

$$\mathbf{\Delta}^P = \begin{pmatrix} 0 & \delta_a y + \delta_c z & -\delta_a x + \delta_b z & -\delta_c x - \delta_b y \\ -\delta_a y - \delta_c z & 0 & \delta_a w + \delta_d z & \delta_c w - \delta_d y \\ \delta_a x - \delta_b z & -\delta_a w - \delta_d z & 0 & \delta_b w + \delta_d x \\ -\delta_c x - \delta_b y & \delta_c w - \delta_d y & \delta_b w + \delta_d x & 0 \end{pmatrix}.$$

Comparing $\mathbf{\Delta}^P$ with $\mathbf{A}$ (below), when subject to the conditions (5.12) on the dynamical variables,

$$(8.526) \ \mathbf{A} = \begin{pmatrix} 0 & S & 0 & U \\ -S & 0 & R & 0 \\ 0 & -R & 0 & P \\ U & 0 & P & 0 \end{pmatrix}, \ Q = 0, \ T = 0,$$

shows that the dynamical variables transform as per (8.120), also reproduced following:

(8.120)

$$(8.120a) \ S \rightarrow S + \delta_a y + \delta_c z, \ (\bar{S} = -S)$$

(8.120b) $T \to 0 - \delta_a x + \delta_b z$, ($\overline{T} = -T$)

(8.120c) $U \to U - \delta_c x - \delta_b y$, ($\overline{U} = U$)

(8.120c) $P \to P + \delta_b w + \delta_d x$, ($\overline{P} = P$)

(8.120d) $Q \to 0 + \delta_c w - \delta_d y$, ($\overline{Q} = Q$)

(8.120e) $R \to R + \delta_a w + \delta_d z$, ($\overline{R} = -R$).

The purpose of the variational method, as used in this section, is to retain the zero Potential constraint (4.16) under arbitrary variations of the dynamical variables (8.120). This will be seen to involve several more simplifications and actually resulting in all four parameters $\delta_a, \delta_b, \delta_c, \delta_d$ set equal.

The first simplification is to constrain the dynamical variable Q (8.120d) to zero

(9.1) $Q = 0 \ \forall \ \delta_c, \delta_d \in \mathbb{Z}$.

This is simply so that the product QT remains zero regardless of the value of dynamical variable T, which will become non-zero when variations are applied,

(9.2) $QT = 0 \ \forall \ T \in \mathbb{Z}$.

Constraining Q to a constant, zero value means that it must have no variation, hence looking at the variational term in Q (8.120d) this means

(9.3) $\delta_c w - \delta_d y = 0$.

The coordinates w and y must remain invariant to arbitrary variations δ_c, δ_d since they are elements of eigenvector solution $\mathbf{X}_+$ (5.55), which is invariant by definition. However, the above constraint (9.3) cannot generally satisfy this invariance requirement on w and y for arbitrary δ_c and δ_d without some simplification. Therefore, the next step is to set the two arbitrary variations δ_c and δ_d equal, i.e.

(9.4) $\delta_d = \delta_c$,

so that the variational expression (9.3) now becomes

(9.5) $\delta_c(w-y)=0$.

and, for arbitrary variations δ_c, this constrains the w and y coordinate solutions to be equal

(9.6) $y = w$.

Note that setting w and y zero is not generally an option for arbitrary δ_c and δ_d in (9.3), primarily because it gives a trivial solution; see (9.19) further below. Setting δ_c and δ_d equal does reduce the number of possible variations from four to three but is the only choice. In fact, shortly, all variations will be set equal.

Using equality (9.6), the eigenvalue C, as given by (5.14b), becomes

(9.7) $C = 2w^2$,

and the Pythagoras equation (5.16) becomes

(9.8) $z^2 = 2w^2 + x^2$.

Thus so far, the four arbitrary variations $\delta_a, \delta_b, \delta_c, \delta_d$ in Δ^P have been reduced to three $\delta_a, \delta_b, \delta_c$ by (9.4), the dynamical variable Q has been constrained to zero by (9.1), and the coordinate solution for $\mathbf{X}_+$ (5.55), in terms of w, x, y, z, is such that w and y are now equal, by (9.6). As usual, w is used as the parameter, and hence substitutes for y in $\mathbf{X}_+$ to become

(9.9) $\mathbf{X}_+ = \begin{pmatrix} w \\ x \\ w \\ z \end{pmatrix}$.

With $y = w$, the solution (5.12) in the dynamical variables now simplifies such that

(9.10)

(9.10a) $U = P = wz$

(9.10b) $S = -R = xw$.

Using (9.10), dynamical variables U and S can be eliminated from the DCE in favour of P and R respectively, to give

(9.11) $C^2 = 2(P^2 - R^2)$.

Note that this means that for C to be a perfect square, either P and R are both even or both odd, but not mixed odd and even.

With Q constrained to zero, the next important step is to constrain the remaining non-zero term in the Potential (4.3c) to zero under arbitrary variations, i.e.

(9.12) $PS + RU = 0 \;\; \forall \; \delta_a, \delta_b, \delta_c \in \mathbb{Z}$

Using $U = P$, $S = -R$ (9.10), $\delta_d = \delta_c$ (9.4) and $y = w$ (9.6), the variations in the dynamical variables P, S, R, U can now be considerably simplified to the following terms, eliminating the initial value solutions (no variation) for U, S, δ_d and y in favour of P, R, δ_c and w respectively

(9.13)

(9.13a) $S \rightarrow -R + \delta_a w + \delta_c z$

(9.13b) $P \rightarrow P + \delta_b w + \delta_c x$

(9.13c) $R \rightarrow R + \delta_a w + \delta_c z$

(9.13d) $U \rightarrow P - \delta_c x - \delta_b w$.

Denoting the change in S by Δ_S, P by Δ_P, etc., the above variations (9.13) are written as

(9.14)

(9.14a) $\Delta_S = \delta_a w + \delta_c z$

(9.14b) $\Delta_P = \delta_b w + \delta_c x$

(9.14c) $\Delta_R = \delta_a w + \delta_c z$

(9.14d) $\Delta_U = -\delta_b w - \delta_c x$.

Note that these changes in the dynamical variables (9.14) are actually finite differences, and hence the usage of symbol Δ. Furthermore, note that whilst $U = P$ (9.10a) and $S = -R$ (9.10b), the difference relations, (9.52) below, are of the opposite sign. It is stressed that $U = P$ and $S = -R$ are initial conditions only, they do not hold for arbitrary variations, and only the overall, zero Potential condition (4.16) remains true.

From (9.14) it is seen that the changes in the dynamical variables are related as follows:

(9.15)

(9.15a) $\Delta_S = \Delta_R$

(9.15b) $\Delta_U = -\Delta_P$.

Taking variations (differences) in the Potential term $PS + RU = 0$ (9.12) gives

(9.16) $\Delta_P S + P\Delta_S + \Delta_R U + R\Delta_U = 0$,

and using $\Delta_S = \Delta_R$, $\Delta_U = -\Delta_P$ (9.15), $U = P$ and $S = -R$ (9.10), this simplifies to

(9.17) $-\Delta_P R + \Delta_R P = 0$.

Substituting for Δ_P (9.14b) and Δ_R (9.14c) into (9.17), and using solutions $P = wz$ and $R = xw$ (9.10), the entire expression is now written in terms of the coordinates as

(9.18) $-(\delta_b w^2 x + \delta_c x^2 w) + (\delta_a w^2 z + \delta_c wz^2) = 0$.

Firstly, if $w = 0$ then the expression (9.18) is identically zero for all variations. Given $w = y$ (9.6), then $y = 0$ and, for a Pythagorean solution, this implies $z = \pm x$ and the eigenvector solution $\mathbf{X}_+$ is thus

$$(9.19) \quad \mathbf{X}_+ = \begin{pmatrix} 0 \\ x \\ 0 \\ \pm x \end{pmatrix}.$$

Whilst this is legitimate, it won't currently be considered further as too trivial.

So, assuming $w \neq 0$, then dividing throughout by w, and using $z^2 - x^2 = 2w^2$ (9.8), the variational expression (9.18) can be re-written as the following simple linear equation in $\delta_a, \delta_b, \delta_c$ and w, x, z:

(9.20) $\delta_c 2w - \delta_b x + \delta_a z = 0$.

Since it is required that (9.20) remains completely invariant to arbitrary variations in $\delta_a, \delta_b, \delta_c$, this means there is only one choice, other than w, x, z all zero, and that is to make all three variations equal, now denoted by symbol δ, i.e.

(9.21) $\delta = \delta_a = \delta_b = \delta_c$, note also $\delta_d = \delta_c$ (9.4),

which then enables the variation (9.20) to factor as

(9.22) $\delta(2w - x + z) = 0$.

Thus, for arbitrary variations δ, this gives the solution

(9.23) $2w - x + z = 0 \, \forall \, \delta \in \mathbb{Z}$.

Equating this constraint with the Pythagoras equation (9.8) to eliminate z implies that

(9.24) $w = 2x$,

and since $y = w$ (9.6) then

(9.25) $y = 2x$.

Substituting $w = 2x$ (9.24) into the solution (9.23) then gives z in terms of x as

(9.26) $z = -3x$,

Therefore the complete coordinate solution, expressed in terms of x as parameter, is

(9.27) $\mathbf{X}_+ = x \begin{pmatrix} +2 \\ +1 \\ +2 \\ -3 \end{pmatrix}$,

which makes the primitive solution for $x = 1$ as follows:

$$(9.28) \quad \mathbf{X}_+ = \begin{pmatrix} +2 \\ +1 \\ +2 \\ -3 \end{pmatrix}, \quad w = 2 \ x = 1, \ y = 2, \ z = -3.$$

This seems a remarkable result, to put it mildly! In essence, it shows that a solution under Pythagoras conditions, that leaves the Potential invariant (and zero) for arbitrary variations, can only be obtained with a global variation, (9.4) and (9.21), and then the eigenvector solution $\mathbf{X}_+$ is the simplest possible Pythagorean quadruple $(2,1,2,-3)$, with no other primitive solution possible. There is not even any flexibility on the sign of the coordinates, it has to be that given above. Whilst the Pythagoras equation (5.16) would not be affected by a sign change in any of the elements, the Potential certainly would - it has to be the sign combination $(+2,+1,+2,-3)$, and nothing else will do.

Note that it is not just an invariant Potential that is required, but that it is also identically zero, because a non-zero URM4 Potential leads to complex eigenvalues; see (4.14b).

Using the solution in the coordinates, the dynamical variables now simplify to

(9.29)

$(9.29a) \ P = -6x^2$

$(9.29b) \ Q = 0$

$(9.29c) \ R = 2x^2$

$(9.29d) \ S = -2x^2$

$(9.29e) \ T = 0$

$(9.29f) \ U = -6x^2.$

Using $C = 2w^2$ (9.7), the eigenvalue is calculated in terms of x (9.24) as

$(9.30) \ C = 8x^2,$

which, for the primitive solution with $x = 1$, takes its smallest value of eight,

(9.31)

(9.31a) $C = 8$ for $x = 1$,

and the dynamical variables (9.29) are

(9.31b) $P = -6$, $Q = 0$, $R = 2$

(9.31f) $S = -2$, $T = 0$, $U = -6$.

Given the co-primality constraint (1.6), then x must be restricted to unity. However, if the parameter x is retained as arbitrary, it simply scales all dynamical variables and coordinates, hence it also scales the **A** matrix and eigenvalue by x^2 as in $C = 8x^2$ (9.30). This scaling will not be used and x will be constrained to unity to keep $\mathbf{X}_+$ (9.27) primitive.

The coordinate solution for $\mathbf{X}_+$ (9.28) has not been derived from the parameters i, j, s, t as in (5.27). However, for completeness, the four parameters i, j, s, t that give this solution are as follows, using the parameterisation (5.27) for $z < 0$,

(9.32) $i = -1$, $j = +1$, $s = +1$, $t = 0$.

To complete the example, and for easy reference, the entire solution is repeated below for all the standard URM quantities. Note that the eigenvectors (9.33e) are obtained from the w, x, y, z solution (9.33d) substituted into (5.55) and (5.56).

(9.33)

$$(9.33a)\ \mathbf{A} = \begin{pmatrix} 0 & -2 & 0 & -6 \\ +2 & 0 & +2 & 0 \\ 0 & -2 & 0 & -6 \\ -6 & 0 & -6 & 0 \end{pmatrix}$$

(9.33b)

$$P = \overline{P} = -6,\ Q = \overline{Q} = 0,\ R = -\overline{R} = 2,$$

$$S = -\overline{S} = -2,\ T = -\overline{T} = 0,\ U = \overline{U} = -6,$$

(9.33c) $C = 8$, $K = C^2 = 64$, $V = 0$

(9.33d) $w = 2$, $x = 1$, $y = 2$, $z = -3$

(9.33e)

$$\mathbf{X}_+ = \begin{pmatrix} +2 \\ +1 \\ +2 \\ -3 \end{pmatrix}, \; \mathbf{X}_{0A} = \begin{pmatrix} -2 \\ 0 \\ +2 \\ 0 \end{pmatrix}, \; \mathbf{X}_{0B} = \begin{pmatrix} 0 \\ -3 \\ 0 \\ +1 \end{pmatrix}, \; \mathbf{X}_- = \begin{pmatrix} +16 \\ -8 \\ +16 \\ +24 \end{pmatrix}$$

(9.33f)

$$\mathbf{X}^+ = (+16 \quad -8 \quad +16 \quad -24)$$

$$\mathbf{X}^{0A} = (-16 \quad 0 \quad +16 \quad 0)$$

$$\mathbf{X}^{0B} = (0 \quad -24 \quad 0 \quad -8)$$

$$\mathbf{X}^- = (+2 \quad +1 \quad +2 \quad +3).$$

The inner products between the eigenvectors are as follows, derived from (5.57), and confirm the same scalar invariants as per URM3 given in Appendix (F)

(9.34)

$$\mathbf{X}^+\mathbf{X}_{0A} = 0, \; \mathbf{X}^-\mathbf{X}_{0A} = 0$$

$$\mathbf{X}^+\mathbf{X}_{0B} = 0, \; \mathbf{X}^-\mathbf{X}_{0B} = 0$$

$$\mathbf{X}^+\mathbf{X}_+ = 128, \; \mathbf{X}^-\mathbf{X}_- = 128 = +2C^2, \; C = 8$$

$$\mathbf{X}^+\mathbf{X}_- = 0, \; \mathbf{X}^-\mathbf{X}_+ = 0$$

$$\mathbf{X}^{0A}\mathbf{X}_{0A} = 64, \; \mathbf{X}^{0B}\mathbf{X}_{0B} = 64 = +C^2, \; C = 8$$

$$\mathbf{X}^{0A}\mathbf{X}_{0B} = 0, \; \mathbf{X}^{0B}\mathbf{X}_{0A}.$$

(9-3) Variational Methods

Having obtained an invariant, zero Potential solution in the previous section, the next step, as in all URMT, is to apply variational methods to see how the complete solution changes, i.e. how all the eigenvectors evolve under variational transformations. Obviously the Potential is invariant and zero by design, as is $\mathbf{X}_+$ and the eigenvalues $\lambda = \pm C, 0, 0$. The full list of quantities that remain invariant is reproduced below.

(9.40)

$$V = 0,\ C = 8,\ K = 64,$$

$$\lambda = \pm 8, 0, 0$$

$$w = 2,\ x = 1,\ y = 2,\ z = -3$$

$$\mathbf{X}_+ = \begin{pmatrix} +2 \\ +1 \\ +2 \\ -3 \end{pmatrix}$$

$$\mathbf{X}^- = \begin{pmatrix} 2 & 1 & 2 & 3 \end{pmatrix}$$

Note that using $\mathbf{X}^- = (\mathbf{TX}_+)^T$ (E5), and since $\mathbf{T}$ (E3) is a constant and therefore invariant, then $\mathbf{X}^-$ is also invariant given $\mathbf{X}_+$ is invariant. All other quantities, e.g. dynamical variables and eigenvectors such as $\mathbf{X}_-$, $\mathbf{X}_{0A}$, $\mathbf{X}_{0B}$, are not invariant and will be seen to evolve as a function of the variational parameter m.

Since the initial solution is under Pythagoras conditions (4.1), these conditions must be retained, and so the variational matrix to be used is $\mathbf{\Delta}^P$ (8.121). Indeed, it was part of the process in obtaining an invariant solution that all four variations in $\mathbf{\Delta}^P$ are set equal to the single δ parameter (9.21). As per URM3, and detailed at the end of Section (8), this parameter δ is replaced by the evolutionary parameter m, i.e.

(8.141) $\delta = -m$, note $\delta = \delta_a = \delta_b = \delta_c = \delta_d$ (9.21).

The simplest possible variational transform $\mathbf{\Delta}^P$ is given by (8.143), reproduced below

$$(8.143)\ \mathbf{\Delta}^P = \begin{pmatrix} 0 & +y+z & -x+z & -x-y \\ -y-z & 0 & +w+z & +w-y \\ +x-z & -w-z & 0 & +w+x \\ -x-y & +w-y & +w+x & 0 \end{pmatrix}$$

Given that there is only one invariant solution $w = 2$, $x = 1$, $y = 2$, $z = -3$ (9.33d), then $\mathbf{\Delta}^P$ becomes, upon substitution for w, x, y, z,

$$(9.41) \quad \mathbf{\Delta}^P = \begin{pmatrix} 0 & -1 & -4 & -3 \\ +1 & 0 & -1 & 0 \\ +4 & +1 & 0 & +3 \\ -3 & 0 & +3 & 0 \end{pmatrix}.$$

Using this $\mathbf{\Delta}^P$, and the solution for $\mathbf{A}$ (9.33a), then the full unity root matrix, denoted by $\mathbf{A}_m$, where $\mathbf{A}_m = \mathbf{A}' - m\mathbf{\Delta}^P$ (8.144), expands to

(9.42)

$$\mathbf{A}_m = \begin{pmatrix} 0 & -2 & 0 & -6 \\ +2 & 0 & +2 & 0 \\ 0 & -2 & 0 & -6 \\ -6 & 0 & -6 & 0 \end{pmatrix} - m \begin{pmatrix} 0 & -1 & -4 & -3 \\ +1 & 0 & -1 & 0 \\ +4 & +1 & 0 & +3 \\ -3 & 0 & +3 & 0 \end{pmatrix}.$$

Comparing $\mathbf{A}_m$ with the standard form for $\mathbf{A}$ (1.3a), the dynamical variables thus transform as

(9.44)

$(9.44a)$ $S = -\overline{S} = -2 + m$,

$(9.44b)$ $T = -\overline{T} = 4m$

$(9.44c)$ $U = \overline{U} = -6 + 3m$

$(9.44d)$ $P = \overline{P} = -6 - 3m$

$(9.44e)$ $Q = \overline{Q} = 0$

$(9.44f)$ $R = -\overline{R} = 2 + m$

The dynamical variable Q has no variation and remains invariant at zero, so too $\overline{Q}$, since $\overline{Q} = Q$ under Pythagoras conditions (4.1).

The relations $U = P$ and $S = -R$ (9.10) are no longer maintained for arbitrary variations m but, regardless, the Potential is verified as invariant and zero by substituting for the dynamical variables from (9.44) into (4.3c).

(9-4) The process to calculate the eigenvectors

The following process is used to calculate the eigenvectors as they vary with parameter m.

Calculate residual matrix $\mathbf{E}_+$ from the quadratic polynomial (5.31a)

(5.31a) $\mathbf{E}_+(m) = \mathbf{A}_m(\mathbf{A}_m + C\mathbf{I}) = \mathbf{A}_m^2 + C\mathbf{A}_m$.

Use $\mathbf{E}_+(m)$ and $\mathbf{X}_+$ (9.28) to calculate $\mathbf{X}^+(m)$ from

(5.30a) $\mathbf{E}_+(m) = \mathbf{X}_+\mathbf{X}^+(m)$.

Use $\mathbf{X}_+$ (9.28) and $\mathbf{T}$ (E3) to calculate $\mathbf{X}^-$

(E5) $\mathbf{X}^- = (\mathbf{TX}_+)^T$

Calculate the residual matrix $\mathbf{E}_-(m)$ from the quadratic polynomial (5.31c)

(5.31c) $\mathbf{E}_-(m) = \mathbf{A}_m(\mathbf{A}_m - C\mathbf{I}) = \mathbf{A}_m^2 - C\mathbf{A}_m$.

Use $\mathbf{X}^-$ and $\mathbf{E}_-(m)$ to calculate $\mathbf{X}_-(m)$ from

(5.30c) $\mathbf{E}_-(m) = \mathbf{X}_-(m)\mathbf{X}^-$

Use $\mathbf{X}_-(m)$ and $\mathbf{T}$ (E3) to verify

(E7) $\mathbf{X}^+(m) = (\mathbf{TX}_-(m))^T$

End of process

Notes

The quadratic expression for $\mathbf{E}_+$ (5.31a) is valid in accordance with Section (3-2).

Calculation of $\mathbf{X}_{0A}$ and $\mathbf{X}_{0B}$ is best done analytically for arbitrary m ($m \neq 0$) because the single residual matrix $\mathbf{E}_0$ is not found to split nicely into separate $\mathbf{E}_{0A}$ and $\mathbf{E}_{0B}$ matrices, which makes it hard to decipher $\mathbf{X}_{0A}$ and $\mathbf{X}_{0B}$ from $\mathbf{E}_0$ in the residual matrix solution, Section (3). Furthermore, when obtaining an analytic solution, the orthogonality between $\mathbf{X}_{0A}$ and $\mathbf{X}_{0B}$ in the unvaried, initial solution ($m = 0$) is not seen in the solution for arbitrary m, $m \neq 0$. Ultimately, because this PS+RU solution and the '2a2p1' solution, Section (6), are not pursued further beyond Part I of this book, the $\mathbf{X}_{0A}$ and $\mathbf{X}_{0B}$ solutions are quoted but not examined further.

(9-5) Implementing the process

Using $\mathbf{A}_m = \mathbf{A}' - m\Delta^P$ (8.144) then $\mathbf{E}_+(m) = \mathbf{A}_m(\mathbf{A}_m + C\mathbf{I})$ (5.31a) expands to

$$(9.50) \quad \begin{aligned} \mathbf{E}_+(m) &= \mathbf{A}_m^2 + C\mathbf{A}_m = \\ &(\mathbf{A}^2 + C\mathbf{A}) - m(\Delta^P\mathbf{A} + \mathbf{A}\Delta^P + C\Delta^P) + m^2(\Delta^P)^2 \end{aligned}$$

Denoting the initial eigenvector solutions for $m = 0$ with a superscript prime, and since eigenvectors $\mathbf{X}_+$ and $\mathbf{X}^-$ are static and invariant to arbitrary variations, i.e. the same for all m, then

$$(9.51) \quad \mathbf{X}'_+ = \mathbf{X}_+(m) = \mathbf{X}_+ = \begin{pmatrix} +2 \\ +1 \\ +2 \\ -3 \end{pmatrix} \quad \forall \ m \in \mathbb{Z}$$

$$(9.52) \quad \mathbf{X}'^- = \mathbf{X}^-(m) = \mathbf{X}^- = \begin{pmatrix} 2 & 1 & 2 & 3 \end{pmatrix} \ \forall \ m \in \mathbb{Z}.$$

The initial values for the other, non-invariant eigenvectors are, by (9.33e) and (9.33f),

$$(9.53) \quad \mathbf{X}'_{0A} = \begin{pmatrix} -2 \\ 0 \\ +2 \\ 0 \end{pmatrix}, \quad \mathbf{X}'_{0B} = \begin{pmatrix} 0 \\ -3 \\ 0 \\ +1 \end{pmatrix}, \quad \mathbf{X}'_- = \begin{pmatrix} +16 \\ -8 \\ +16 \\ +24 \end{pmatrix}$$

$(9.54a) \quad \mathbf{X}'^+ = \begin{pmatrix} +16 & -8 & +16 & -24 \end{pmatrix}$

$(9.54b) \quad \mathbf{X}'^{0A} = \begin{pmatrix} -16 & 0 & +16 & 0 \end{pmatrix}$

$(9.54c) \quad \mathbf{X}'^{0B} = \begin{pmatrix} 0 & -24 & 0 & -8 \end{pmatrix}$

Returning to $\mathbf{E}_+(m)$ (9.50), to determine $\mathbf{X}^+(m)$ from $\mathbf{X}_+$, then the first term $(\mathbf{A}_m^2 + C\mathbf{A}_m)$ is simply the unvaried residual matrix $\mathbf{E}_+(0)$, for $m = 0$, i.e.

(9.55) $\mathbf{E}_+(0) = (\mathbf{A}^2 + C\mathbf{A}) = \mathbf{X}'_+\mathbf{X}'^+$, $m = 0$

Using $\mathbf{A}$ (9.33a) and $C = 8$ (9.33c) gives $\mathbf{A}^2$, $C\mathbf{A}$ and $\mathbf{E}_+(0)$ as

$$(9.56)\ \mathbf{A}^2 = \begin{pmatrix} +32 & 0 & +32 & 0 \\ 0 & -8 & 0 & -24 \\ +32 & 0 & +32 & 0 \\ 0 & +24 & 0 & +72 \end{pmatrix}$$

$$(9.57)\ C\mathbf{A} = \begin{pmatrix} 0 & -16 & 0 & -48 \\ +16 & 0 & +16 & 0 \\ 0 & -16 & 0 & -48 \\ -48 & 0 & -48 & 0 \end{pmatrix}$$

$$(9.58)\ \mathbf{E}_+(0) = \mathbf{A}^2 + C\mathbf{A} = \begin{pmatrix} +32 & -16 & +32 & -48 \\ +16 & -8 & +16 & -24 \\ +32 & -16 & +32 & -48 \\ -48 & +24 & -48 & +72 \end{pmatrix}$$

Resolving this matrix $\mathbf{E}_+(0)$ in terms of an outer product of $\mathbf{X}_+$ and a 1x4 row vector gives

$$(9.59)\ \mathbf{E}_+(0) = \begin{pmatrix} +2 \\ +1 \\ +2 \\ -3 \end{pmatrix} (+16\ \ -8\ \ +16\ \ -24),$$

which is, indeed, the static term $\mathbf{E}_+(0) = \mathbf{X}'_+\mathbf{X}'^+$ using $\mathbf{X}'_+ = \mathbf{X}_+$ (9.51) and $\mathbf{X}'^+$ (9.54a).

Given $\mathbf{X}_+$ does not evolve with m, then the residual matrix $\mathbf{E}_+(m)$ will always be an outer product of a static $\mathbf{X}_+$ with an evolutionary vector $\mathbf{X}^+(m)$, i.e. $\mathbf{E}_+ = \mathbf{X}_+\mathbf{X}^+(m)$. This means that each matrix term in the expression (9.50) for $\mathbf{E}_+$ will also be an outer product of a row-vector with the column eigenvector $\mathbf{X}_+$. For example, the last m^2 term in (9.50), containing $\boldsymbol{\Delta}^P$, evaluates as follows, using the solution (9.41) for $\boldsymbol{\Delta}^P$:

$$(9.60) \quad \left(\boldsymbol{\Delta}^P\right)^2 = \begin{pmatrix} -8 & -4 & -8 & -12 \\ -4 & -2 & -4 & -6 \\ -8 & -4 & -8 & -12 \\ +12 & +6 & +12 & +18 \end{pmatrix}.$$

Resolving this $\left(\boldsymbol{\Delta}^P\right)^2$ expression in terms of an outer product of $\mathbf{X}_+$ and a 1x4 row vector gives

$$(9.61) \quad \left(\boldsymbol{\Delta}^P\right)^2 = \begin{pmatrix} +2 \\ +1 \\ +2 \\ -3 \end{pmatrix} \begin{pmatrix} -4 & -2 & -4 & -6 \end{pmatrix}.$$

To evaluate the linear expression $\boldsymbol{\Delta}^P\mathbf{A} + \mathbf{A}\boldsymbol{\Delta}^P + C\boldsymbol{\Delta}^P$ in (9.50), then using $\mathbf{A}$ (9.33a), $\boldsymbol{\Delta}^P$ (9.41), and $C = 8$ (9.40) gives

$$(9.62) \quad \boldsymbol{\Delta}^P\mathbf{A} = \begin{pmatrix} +16 & +8 & +16 & +24 \\ 0 & 0 & 0 & 0 \\ -16 & -8 & -16 & -24 \\ 0 & 0 & 0 & 0 \end{pmatrix},$$

$$(9.63) \quad \mathbf{A}\boldsymbol{\Delta}^P = \begin{pmatrix} +16 & 0 & -16 & 0 \\ +8 & 0 & -8 & 0 \\ +16 & 0 & -16 & 0 \\ -24 & 0 & +24 & 0 \end{pmatrix},$$

$$(9.64) \quad C\boldsymbol{\Delta}^P = \begin{pmatrix} 0 & -8 & -32 & -24 \\ +8 & 0 & -8 & 0 \\ +32 & +8 & 0 & +24 \\ -24 & 0 & +24 & 0 \end{pmatrix},$$

and summing terms

$$(9.65) \quad \boldsymbol{\Delta}^P \mathbf{A} + \mathbf{A}\boldsymbol{\Delta}^P + C\boldsymbol{\Delta}^P = \begin{pmatrix} +32 & 0 & -32 & 0 \\ +16 & 0 & -16 & 0 \\ +32 & 0 & -32 & 0 \\ -48 & 0 & +48 & 0 \end{pmatrix}.$$

Resolving this matrix in terms of an outer product of $\mathbf{X}_+$ and a 1x4 row vector gives

$$(9.66) \quad \boldsymbol{\Delta}^P \mathbf{A} + \mathbf{A}\boldsymbol{\Delta}^P + C\boldsymbol{\Delta}^P = \begin{pmatrix} +2 \\ +1 \\ +2 \\ -3 \end{pmatrix} \begin{pmatrix} +16 & 0 & -16 & 0 \end{pmatrix}.$$

Finally, substituting terms (9.59), (9.61) and (9.66) into (9.50) for $\mathbf{E}_+(m)$, gives the evolution of $\mathbf{X}^+$ as a function of m, denoted by $\mathbf{X}^+(m)$,

$$(9.67) \quad \begin{aligned} \mathbf{X}^+(m) &= \begin{pmatrix} +16 & -8 & +16 & -24 \end{pmatrix} + \\ &- m \begin{pmatrix} +16 & 0 & -16 & 0 \end{pmatrix} + \\ &+ m^2 \begin{pmatrix} -4 & -2 & -4 & -6 \end{pmatrix} \end{aligned}$$

Some example values of $\mathbf{X}^+(m)$ for $m = 0..3$ are

$$(9.68a) \quad \mathbf{X}^+(0) = \begin{pmatrix} +16 & -8 & +16 & -24 \end{pmatrix}, \; m = 0$$

$$(9.68b) \quad \mathbf{X}^+(1) = \begin{pmatrix} -4 & -10 & +28 & -30 \end{pmatrix}, \; m = 1$$

$$(9.68c) \quad \mathbf{X}^+(2) = \begin{pmatrix} -32 & -16 & +32 & -48 \end{pmatrix}, \; m = 2$$

$$(9.68d) \quad \mathbf{X}^+(3) = \begin{pmatrix} -68 & -26 & +28 & -78 \end{pmatrix}, \; m = 3.$$

Each one of these solutions is a Pythagorean triple, as expected since the solution is under Pythagoras conditions, which are invariant with respect to m by virtue of the particular solution for $\mathbf{X}_+$ (9.51).

Now turning the attention to the determination of the evolving $\mathbf{X}_-(m)$ eigenvector. Using $\mathbf{A}_m = \mathbf{A}' - m\boldsymbol{\Delta}^P$ (8.144), then $\mathbf{E}_-$ (5.31c) expands to

(9.69)
$$\mathbf{E}_-(m) = \mathbf{A}_m^2 - C\mathbf{A}_m =$$
$$(\mathbf{A}^2 - C\mathbf{A}) - m(\mathbf{\Delta}^P\mathbf{A} + \mathbf{A}\mathbf{\Delta}^P - C\mathbf{\Delta}^P) + m^2(\mathbf{\Delta}^P)^2.$$

The first term $(\mathbf{A}^2 - C\mathbf{A})$ is simply the unvaried residual matrix $\mathbf{E}_-(0)$, $m = 0$, i.e.

(9.70) $\mathbf{E}_-(0) = (\mathbf{A}^2 - C\mathbf{A}) = \mathbf{X}'_-\mathbf{X}'^-$, $m = 0$.

Using $\mathbf{A}^2$ (9.56) and $C\mathbf{A}$ (9.57) gives

(9.71) $\mathbf{A}^2 - C\mathbf{A} = \begin{pmatrix} +32 & +16 & +32 & +48 \\ -16 & -8 & -16 & -24 \\ +32 & +16 & +32 & +48 \\ +48 & +24 & +48 & +72 \end{pmatrix}.$

Resolving this matrix in terms of an outer product of a 1x4 column vector and $\mathbf{X}'^-$ (9.52) gives

(9.72) $\mathbf{A}^2 - C\mathbf{A} = \begin{pmatrix} +16 \\ -8 \\ +16 \\ +24 \end{pmatrix}(+2 \quad +1 \quad +2 \quad +3)$

This is, indeed, the static term $\mathbf{E}_-(0) = \mathbf{X}'_-\mathbf{X}'^-$ using $\mathbf{X}'^- = \mathbf{X}^-$ (9.52) and $\mathbf{X}'_-$ (9.53).

To evaluate the linear expression $\mathbf{\Delta}^P\mathbf{A} + \mathbf{A}\mathbf{\Delta}^P - C\mathbf{\Delta}^P$ (9.69), then using $\mathbf{\Delta}^P\mathbf{A}$ (9.62), $\mathbf{A}\mathbf{\Delta}^P$ (9.63) and $C\mathbf{\Delta}^P$ (9.64) gives

(9.73) $\mathbf{\Delta}^P\mathbf{A} + \mathbf{A}\mathbf{\Delta}^P - C\mathbf{\Delta}^P = \begin{pmatrix} +32 & +16 & +32 & +48 \\ 0 & 0 & 0 & 0 \\ -32 & -16 & -32 & -48 \\ 0 & 0 & 0 & 0 \end{pmatrix}.$

Resolving this matrix in terms of an outer product of a 1x4 column vector and $\mathbf{X}'^-$ (9.52) gives

$$(9.74) \quad \mathbf{\Delta}^P \mathbf{A} + \mathbf{A}\mathbf{\Delta}^P - C\mathbf{\Delta}^P = \begin{pmatrix} +16 \\ 0 \\ -16 \\ 0 \end{pmatrix} (+2 \quad +1 \quad +2 \quad +3)$$

The matrix $(\mathbf{\Delta}^P)^2$ is as per (9.60) and can be written as the outer product of a 1x4 column vector and $\mathbf{X}'^-$ (9.52) as follows

$$(9.75) \quad (\mathbf{\Delta}^P)^2 = \begin{pmatrix} -4 \\ -2 \\ -4 \\ +6 \end{pmatrix} (+2 \quad +1 \quad +2 \quad +3)$$

Finally, substituting terms (9.72), (9.74) and (9.75) into (9.69) for $\mathbf{E}_-(m)$, gives the evolution of $\mathbf{X}_-$ as a function of m, and denoted by $\mathbf{X}_-(m)$, as

$$(9.76) \quad \mathbf{X}_-(m) = \begin{pmatrix} +16 \\ -8 \\ +16 \\ +24 \end{pmatrix} - m \begin{pmatrix} +16 \\ 0 \\ -16 \\ 0 \end{pmatrix} + m^2 \begin{pmatrix} -4 \\ -2 \\ -4 \\ +6 \end{pmatrix}.$$

Comparing this expression for $\mathbf{X}_-(m)$ with that for $\mathbf{X}^+(m)$ (9.67), the relation $\mathbf{X}^+(m) = (\mathbf{T}\mathbf{X}_-(m))^T$ (E7) is seen to be verified.

The three inner products $\mathbf{X}^+\mathbf{X}_+$, $\mathbf{X}^-\mathbf{X}_-$ and $\mathbf{X}^+\mathbf{X}_-$, which are superficially functions of m given that the individual vectors are functions of m, can also be verified as invariant and independent of m, noting that the fourth inner product $\mathbf{X}^-\mathbf{X}_+ = 0$ is always invariant for arbitrary m since $\mathbf{X}_+$ and $\mathbf{X}^-$ are both invariant.

$(9.77a) \quad \mathbf{X}^+(m)\mathbf{X}_+ = 128 = +2C^2$, (F9)

$(9.77b) \quad \mathbf{X}^-\mathbf{X}_-(m) = 128 = +2C^2$, (F9)

$(9.77c) \quad \mathbf{X}^+(m)\mathbf{X}_-(m) = 0$, (F8)

$(9.77d) \quad \mathbf{X}^-\mathbf{X}_+ = 0$, (F7).

Some example values of $\mathbf{X}_-(m)$ for $m = 0..3$ are as follows, noting that they all satisfy the relation $\mathbf{X}_-(m) = (\mathbf{TX}^+(m))^T$ (E7), as expected, see (9.68) for the $\mathbf{X}^+(m)$ example values.

$$(9.78) \quad \mathbf{X}_-(0) = \begin{pmatrix} +16 \\ -8 \\ +16 \\ +24 \end{pmatrix}, \quad \mathbf{X}_-(1) = \begin{pmatrix} -4 \\ -10 \\ +28 \\ +30 \end{pmatrix},$$

$$\mathbf{X}_-(2) = \begin{pmatrix} -32 \\ -16 \\ +32 \\ +48 \end{pmatrix}, \quad \mathbf{X}_-(3) = \begin{pmatrix} -68 \\ -26 \\ +28 \\ +78 \end{pmatrix}.$$

(9-6) The zero eigenvectors $\mathbf{X}_{0A}$, $\mathbf{X}_{0B}$

As mentioned earlier in Section (9-4), the computation of $\mathbf{X}_{0A}$, $\mathbf{X}_{0B}$ and their reciprocals, $\mathbf{X}^{0A}$ and $\mathbf{X}^{0B}$, is best done analytically. However, very briefly, if the residual method were to be used it would proceed as normal, calculating the residual matrix $\mathbf{E}_0(m)$ as

$$(9.80) \quad \mathbf{E}_0(m) = (\mathbf{A}_m^2 - C^2 \mathbf{I})$$

which, using $\mathbf{A}_m = \mathbf{A}' - m\Delta^P$ (8.144), expands to

$$(9.81) \quad \mathbf{E}_0(m) = (\mathbf{A}^2 - C^2 \mathbf{I}) - m(\Delta^P \mathbf{A} + \mathbf{A}\Delta^P) + m^2(\Delta^P)^2.$$

The first term $(\mathbf{A}^2 - C^2 \mathbf{I})$ is simply the initial $m = 0$ solution, i.e. $\mathbf{E}_0(m) = (\mathbf{A}^2 - C^2 \mathbf{I})$, where $\mathbf{E}_0 = \mathbf{X}_{0A}\mathbf{X}^{0A} + \mathbf{X}_{0B}\mathbf{X}^{0B}$ for $m = 0$, with $\mathbf{X}_{0A}$, $\mathbf{X}_{0B}$, $\mathbf{X}^{0A}$ and $\mathbf{X}^{0B}$ given by (9.33e) and (9.33f). The remaining terms for $m \neq 0$ do not, however, split into outer products of the initial eigenvector solution for $m = 0$. This is confirmed when calculating the zero eigenvectors using the standard algebraic method.

When the zero eigenvalue equation $\mathbf{AX}_0 = 0$ is solved algebraically then, since there are two zero eigenvalues, a general solution for arbitrary integers a and b is given in terms of the dynamical variables as follows:

$$(9.82) \quad \mathbf{X}_0 = a \begin{pmatrix} RP \\ 0 \\ SP \\ TR \end{pmatrix} + b \begin{pmatrix} 0 \\ RP \\ 0 \\ RR \end{pmatrix}.$$

Removing scale factors a and b splits the vector nicely into two, linearly independent vectors $\mathbf{X}_{0A}$ and $\mathbf{X}_{0B}$ as follows:

$$(9.83) \quad \mathbf{X}_{0A} = \begin{pmatrix} RP \\ 0 \\ SP \\ TR \end{pmatrix}, \quad \mathbf{X}_{0B} = \begin{pmatrix} 0 \\ RP \\ 0 \\ RR \end{pmatrix}.$$

It is noted that $\mathbf{X}_{0B}$ has a redundant factor R that could be removed. Comparing P (9.44d) and R (9.44f), then the dynamical variable P is related to R, for arbitrary m, by

$$(9.84) \quad P = -3R.$$

Substituting in this relation (9.84), and removing the common factor R, simplifies both vectors to the following forms, which are now first order in the dynamical variables, unlike (9.83), which are quadratic:

$$(9.85) \quad \mathbf{X}_{0A} = \begin{pmatrix} -3R \\ 0 \\ -3S \\ +T \end{pmatrix}, \quad \mathbf{X}_{0B} = \begin{pmatrix} 0 \\ -3R \\ 0 \\ +R \end{pmatrix}.$$

Finally, substituting the expressions (9.44) for the dynamical variables, and tidying, gives for $\mathbf{X}_{0A}$

$$(9.86) \quad \mathbf{X}_{0A} = 3 \begin{pmatrix} -2 \\ 0 \\ +2 \\ 0 \end{pmatrix} - m \begin{pmatrix} +3 \\ 0 \\ +3 \\ -4 \end{pmatrix}, \quad \mathbf{X}_{0B} = 2 \begin{pmatrix} 0 \\ -3 \\ 0 \\ +1 \end{pmatrix} + m \begin{pmatrix} 0 \\ -3 \\ 0 \\ +1 \end{pmatrix}.$$

For $m = 0$ it can be seen that, disregarding the factors of 3 and 2, then $\mathbf{X}_{0A}$ and $\mathbf{X}_{0B}$ reduce to their initial values, $\mathbf{X}'_{0A}$ and $\mathbf{X}'_{0B}$ respectively, as per (9.33e)

$$(9.87) \quad \mathbf{X}'_{0A} = \begin{pmatrix} -2 \\ 0 \\ +2 \\ 0 \end{pmatrix}, \ \mathbf{X}'_{0B} = \begin{pmatrix} 0 \\ -3 \\ 0 \\ +1 \end{pmatrix}, \ m = 0.$$

This completes the evaluation of the PS+RU eigenvectors for arbitrary variations.

(10) Variational Methods 2a2p1 Solution

This section examines the behaviour of the URM4 '2a2p1' solution, Section (6), under invariance transformations, Section (8).

(10-1) Solution Summary

To recap, this 2a2p1 solution is essentially a variant of the PS+RU zero solution in Section (5), but possesses a unity eigenvalue, i.e.

(6.1a) $C = 1$.

The Potential is initially zero as for the PS+RU solution but, in this case, instead of constraints (5.1), where $Q = 0$, $T = 0$ and $(PS + RU) = 0$, only the single dynamical variable U is initially zero (6.2a), and the other condition is that $QT - PS = 0$ (6.2b), i.e.

(6.2a) $U = 0$

(6.2b) $QT - PS = 0$,

With these conditions, the Potential V is zero, i.e.

(6.3) $V = \dfrac{[QT - (PS + RU)]^2}{C^2} = 0$.

Because the Potential is zero and the eigenvalue unity, the DCE (6.4) is simply the unity kinetic term

(6.4) $K = 1$, the DCE.

To satisfy conditions (6.2), the dynamical variables are assigned in terms of a single parameter a and derived parameter b as in

(6.11) $b = 2a^2 + 1$

(6.7) $P = ab$, $Q = -b$, $R = P = ab$, $S = -2a$, $T = 2a^2$, $U = 0$.

For a zero Potential and unity eigenvalue, the full set of four eigenvalues is, by (4.14a),

(6.16) $\lambda = \pm 1, 0, 0$.

Substituting for dynamical variables P, Q, R, S, T, U (6.7) into $\mathbf{A}$ (4.2) gives the unity root matrix as

$$(6.17) \quad \mathbf{A} = \begin{pmatrix} 0 & -2a & +2a^2 & 0 \\ +2a & 0 & +ab & -b \\ -2a^2 & -ab & 0 & +ab \\ 0 & -b & +ab & 0 \end{pmatrix}.$$

(10.1) The complete eigenvector solution is reproduced below from Section (6) :

(6.18a) $w = 2a$

(6.18b) $x = b - 2$

(6.18c) $y = 2a$

(6.18d) $z = b$.

$$(6.19) \quad \mathbf{X}_+ = \begin{pmatrix} 2a \\ b-2 \\ 2a \\ b \end{pmatrix}$$

(6.21) $\mathbf{X}^+ = \left(-a(b+1) \quad -(a^2b+1) \quad -a(b-1) \quad +b(a^2+1) \right)$

(6.24) $\mathbf{X}^- = \left(2a \quad b-2 \quad 2a \quad -b \right)$

$$(6.25) \quad \mathbf{X}_- = \begin{pmatrix} -a(b+1) \\ -(a^2b+1) \\ -a(b-1) \\ -b(a^2+1) \end{pmatrix}$$

(6.24a) $\mathbf{X}_{0A} = \begin{pmatrix} b \\ 0 \\ 0 \\ 2a \end{pmatrix}$,

(6.24b) $\mathbf{X}_{0B} = \begin{pmatrix} 0 \\ a \\ 1 \\ a \end{pmatrix}$

(6.27) $\mathbf{X}^{0A} = \begin{pmatrix} b & 0 & 0 & -2a \end{pmatrix}$, $\mathbf{X}^{0B} = \begin{pmatrix} 0 & a & 1 & -a \end{pmatrix}$

(10-2) Variational Methods

Using the unity eigenvalue, zero Potential solution (10.1), the next step, as in all URMT, is to apply variational methods to see how the solution, in particular the Potential, changes under variational transforms. This will be done using the first example in Section (6-2), which uses zero for parameter a, and consequently the derived parameter b (6.11) is unity, i.e.

(10.2) $a = 0 \Rightarrow b = 1$.

The treatment of the 2a2p1 example in Section (6-2) illustrates the emergence of complex eigenvalues under a variation in the dynamical variables, whereby the URM4 Pythagoras conditions are maintained, although this does not guarantee an invariant (and zero) Potential in URM4, unlike URM3, see [1],#1. The variation applied herein triggers such a non-zero Potential $V > 0$, and generates two real and two complex eigenvalues in accordance with $\lambda = \pm C, \pm i\sqrt{V}$ (4.14b), for $C = 1$ (6.1a). This is a change from the initial, four eigenvalues $\lambda = \pm 1, 0, 0$ (6.16) for the unvaried solution.

Applying $a = 0$, $b = 1$ to (10.1) gives the initial solution

(10.3)

$w = 0$, $x = -1$, $y = 0$, $z = +1$

$$\mathbf{X}_+ = \begin{pmatrix} 0 \\ -1 \\ 0 \\ +1 \end{pmatrix}, \ \mathbf{X}_- = \begin{pmatrix} 0 \\ -1 \\ 0 \\ -1 \end{pmatrix}$$

$$\mathbf{X}^+ = \begin{pmatrix} 0 & -1 & 0 & +1 \end{pmatrix},$$

$$\mathbf{X}^- = \begin{pmatrix} 0 & -1 & 0 & -1 \end{pmatrix}$$

$$\mathbf{X}^+\mathbf{X}_+ = \mathbf{X}^-\mathbf{X}_- = 2C^2 = 2$$

$$\mathbf{X}^+\mathbf{X}_- = \mathbf{X}^-\mathbf{X}_+ = 0$$

$$P = 0, \ Q = -1, \ R = P = 0, \ S = 0, \ T = 0, \ U = 0$$

$C = 1$, unity eigenvalue

$$\mathbf{A} = \begin{pmatrix} 0 & 0 & 0 & 0 \\ 0 & 0 & 0 & -1 \\ 0 & 0 & 0 & 0 \\ 0 & -1 & 0 & 0 \end{pmatrix}$$

The sparseness of $\mathbf{A}$ will be very useful later when evaluating residual matrices with many terms, and it is a key reason such a simple solution is chosen.

There is a downside to using a unity eigenvalue $C = 1$, and that is that all expressions no longer appear dimensionally correct when every equation has the eigenvalue symbol (C) removed, effectively replaced by an invisible factor '1'. Nevertheless, because it has been removed and replaced by unity, the dimensional correctness of all algebraic expressions here will have to be accepted as is.

Since the initial solution is under Pythagoras conditions (4.1), these conditions must be retained, and so the variational matrix to be used is $\mathbf{\Delta}^P$ (8.121). Exactly as in the PS+RU case, Section (9-3), all four variations in $\mathbf{\Delta}^P$ are set equal to the single δ parameter (9.21) and δ is replaced by the evolutionary parameter m, i.e.

(9.21) $\delta = \delta_a = \delta_b = \delta_c = \delta_d$.

(8.141) $\delta = -m$.

The simplest possible variational transform Δ^P is given by (8.143), reproduced below

$$(8.143) \quad \Delta^P = \begin{pmatrix} 0 & +y+z & -x+z & -x-y \\ -y-z & 0 & +w+z & +w-y \\ +x-z & -w-z & 0 & +w+x \\ -x-y & +w-y & +w+x & 0 \end{pmatrix}$$

Given the invariant solution $w=0$, $x=-1$, $y=0$ and $z=+1$ (10.3), then Δ^P (8.143) becomes, upon substitution for w, x, y, z,

$$(10.4) \quad \Delta^P = \begin{pmatrix} 0 & +1 & +2 & +1 \\ -1 & 0 & +1 & 0 \\ -2 & -1 & 0 & -1 \\ +1 & 0 & -1 & 0 \end{pmatrix}.$$

Using this Δ^P (10.4) and the solution for $\mathbf{A}$ (10.3), then the full unity root matrix $\mathbf{A}_m$ (8.144) expands to

$$(10.5) \quad \mathbf{A}_m = \begin{pmatrix} 0 & 0 & 0 & 0 \\ 0 & 0 & 0 & -1 \\ 0 & 0 & 0 & 0 \\ 0 & -1 & 0 & 0 \end{pmatrix} - m \begin{pmatrix} 0 & +1 & +2 & +1 \\ -1 & 0 & +1 & 0 \\ -2 & -1 & 0 & -1 \\ +1 & 0 & -1 & 0 \end{pmatrix}.$$

Reading off the dynamical variables from $\mathbf{A}_m$ for non-zero evolutionary parameter, $m \neq 0$, they evolve according to

(10.6a) $P = m$

(10.6b) $Q = -1$, no variation

(10.6c) $R = -m$

(10.6d) $S = -m$

(10.6e) $T = -2m$

(10.6f) $U = -m$.

With these values, the kinetic term K (6.8) becomes

(10.7) $K = 1 - (2m)^2$,

185

and the Potential V (4.3c)

(10.8) $V = (2m)^2$.

The general DCE $C^2 = K + V$ (1.14) is thus verified for unity eigenvalue $C = 1$.

In accordance with (4.14b), the eigenvalues are

(10.9) $\lambda = \pm 1, \pm 2mi$,

i.e. there are two real (integer) eigenvalues, $\lambda = \pm 1$, and two pure imaginary eigenvalues $\lambda = \pm 2mi$. That they are pure imaginary for URM4, under Pythagoras conditions, simplifies the eigenvector solution considerably.

Denoting the initial eigenvector solutions for $m = 0$ with a superscript prime, and since eigenvectors $\mathbf{X}_+$ and $\mathbf{X}^-$ are invariant to variations, i.e. the same for all m, then as per (10.3)

(10.10a) $\mathbf{X}'_+ = \mathbf{X}_+(m) = \mathbf{X}_+ = \begin{pmatrix} 0 \\ -1 \\ 0 \\ +1 \end{pmatrix} \quad \forall \ m \in \mathbb{Z}$

(10.10b) $\mathbf{X}'^- = \mathbf{X}^-(m) = \mathbf{X}^- = \begin{pmatrix} 0 & -1 & 0 & -1 \end{pmatrix} \ \forall \ m \in \mathbb{Z}$.

The initial values for the other, non-invariant eigenvectors are, from (10.3),

(10.11a) $\mathbf{X}'_- = \begin{pmatrix} 0 \\ -1 \\ 0 \\ -1 \end{pmatrix}$, $m = 0$

(10.11b) $\mathbf{X}'^+ = \begin{pmatrix} 0 & -1 & 0 & +1 \end{pmatrix}$, $m = 0$.

Since for non-zero m there are no zero eigenvalues, there are also, therefore, no associated zero eigenvectors, having been replaced by their complex counterparts, and denoted as follows:

(10.12a) $\mathbf{X}_{i+}$, $\mathbf{X}^{i+}$, $\forall m \neq 0$, $\lambda = +2mi$

(10.12b) $\mathbf{X}_{i-}$, $\mathbf{X}^{i-}$, $\forall m \neq 0$, $\lambda = -2mi$.

As usual, the residual matrix method will be used to determine the eigenvectors, see Section (3). This time, however, the Cayley-Hamilton polynomial is now factored according to the eigenvalues $\lambda = \pm 1, \pm 2mi$ (10.9) as follows, using the evolved matrix $\mathbf{A}_m$ (8.144)

(10.13) $0 = (\mathbf{A}_m - \mathbf{I})(\mathbf{A}_m + \mathbf{I})(\mathbf{A}_m - 2mi\mathbf{I})(\mathbf{A}_m + 2mi\mathbf{I})$.

With the quadratic factor $(\mathbf{A}_m - 2mi\mathbf{I})(\mathbf{A}_m + 2mi\mathbf{I})$ evaluated as the real expression $(\mathbf{A}_m^2 + 4m^2\mathbf{I})$, the residual matrices for $\lambda = \pm 1$ as functions of m, denoted by $\mathbf{E}_+(m)$ and $\mathbf{E}_-(m)$, are given by the following polynomials, see Section (3) for the factoring method,

(10.14) $\mathbf{E}_{\pm}(m) = (\mathbf{A}_m \pm \mathbf{I})(\mathbf{A}_m^2 + 4m^2\mathbf{I})$, $\lambda = \pm 1$

Denoting the following matrix expressions in $\mathbf{\Delta}^P$ and $\mathbf{A}$ by $\mathbf{B}$, $\mathbf{D}$ and $\mathbf{F}$ (because capital letters C and E are already used for the eigenvalue and energy respectively)

(10.15) $\mathbf{B}_{\pm} = \mathbf{A} \pm \mathbf{I}$

(10.16) $\mathbf{D} = \mathbf{\Delta}^P \mathbf{A} + \mathbf{A} \mathbf{\Delta}^P$

(10.17) $\mathbf{F} = \left(\mathbf{\Delta}^P\right)^2 + 4\mathbf{I}$,

and substituting for $\mathbf{A}_m$ from (8.144) into (10.14), gives $\mathbf{E}_{\pm}(m)$ expanded in full as

(10.18)
$$\begin{aligned}
\mathbf{E}_{\pm}(m) = {} & \mathbf{B}_{\pm}\mathbf{A}^2 + \\
& - m(\mathbf{B}_{\pm}\mathbf{D} + \mathbf{\Delta}^P\mathbf{A}^2) + \\
& + m^2(\mathbf{B}_{\pm}\mathbf{F} + \mathbf{\Delta}^P\mathbf{D}) + \\
& - m^3\mathbf{\Delta}^P\mathbf{F}
\end{aligned}$$

This is seen to be a cubic polynomial in m, quite unlike the quadratic expression for real eigenvalues when under Pythagoras conditions with a zero Potential.

Using $\mathbf{A}$ (10.3) and $\mathbf{\Delta}^P$ (10.4), the individual terms for $\mathbf{B}$ (10.15), $\mathbf{D}$ (10.16) and $\mathbf{F}$ (10.17) evaluate as

$$(10.19)\ \mathbf{B}_\pm = \begin{pmatrix} \pm1 & 0 & 0 & 0 \\ 0 & \pm1 & 0 & -1 \\ 0 & 0 & \pm1 & 0 \\ 0 & -1 & 0 & \pm1 \end{pmatrix}$$

$$(10.20)\ \mathbf{D} = \begin{pmatrix} 0 & -1 & 0 & -1 \\ -1 & 0 & +1 & 0 \\ 0 & +1 & 0 & +1 \\ +1 & 0 & -1 & 0 \end{pmatrix}$$

$$(10.21)\ \mathbf{F} = \begin{pmatrix} 0 & -2 & 0 & -2 \\ -2 & +2 & -2 & -2 \\ 0 & -2 & 0 & -2 \\ +2 & +2 & +2 & +6 \end{pmatrix}$$

Firstly, to calculate $\mathbf{E}_\pm(m)$ (10.18), calculate the static, initial, invariant terms $\mathbf{B}_\pm \mathbf{A}$ and $\mathbf{B}_+ \mathbf{A}^2$

$$(10.22)\ \mathbf{B}_\pm \mathbf{A} = \begin{pmatrix} 0 & 0 & 0 & 0 \\ 0 & +1 & 0 & \mp1 \\ 0 & 0 & 0 & 0 \\ 0 & \mp1 & 0 & +1 \end{pmatrix}$$

$$(10.23)\ \mathbf{B}_\pm \mathbf{A}^2 = \begin{pmatrix} 0 & 0 & 0 & 0 \\ 0 & \pm1 & 0 & -1 \\ 0 & 0 & 0 & 0 \\ 0 & -1 & 0 & \pm1 \end{pmatrix}$$

Resolving this matrix in terms of an outer product of $\mathbf{X}_+$ and a 1x4 row vector gives

$$(10.24) \quad \mathbf{B}_+\mathbf{A}^2 = \begin{pmatrix} 0 \\ -1 \\ 0 \\ +1 \end{pmatrix} \begin{pmatrix} 0 & -1 & 0 & +1 \end{pmatrix}.$$

This is, indeed, the static term $\mathbf{E}_+(0) = \mathbf{X}'_+\mathbf{X}'^+$, where $\mathbf{X}'_+ = \mathbf{X}_+$ (10.10a) and $\mathbf{X}'^+$ is given by (10.11b).

Noting that $\mathbf{E}_+$, for $m = 0$, can be expressed in two forms:

$(10.25) \quad \mathbf{E}_+ = \mathbf{B}_+\mathbf{A}^2$ or $\mathbf{E}_+ = \mathbf{B}_+\mathbf{A}$, for $m = 0$.

This means that for the initial $m = 0$ case only the quadratic polynomial expression $\mathbf{B}_+\mathbf{A} = (\mathbf{A}+\mathbf{I})\mathbf{A}$ is required for $\mathbf{E}_+$, as per Section (3-2). This is also true for $\mathbf{E}_-(m = 0)$, to within a sign, since $\mathbf{B}_-\mathbf{A}^2 = -\mathbf{B}_-\mathbf{A}$ and $\mathbf{B}_-\mathbf{A} = (\mathbf{A}-\mathbf{I})\mathbf{A}$, which is the quadratic form for $\mathbf{E}_-(m = 0)$.

Also note that since $\mathbf{B}_+\mathbf{A}^2 = (\mathbf{A}+\mathbf{I})\mathbf{A}^2$, and is therefore only an expression in $\mathbf{A}$, then $\mathbf{A}^2\mathbf{B}_+ = \mathbf{B}_+\mathbf{A}^2$ as $\mathbf{A}$ naturally commutes with itself.

Resolving the product matrix $\mathbf{B}_-\mathbf{A}^2$ (10.23) in terms of an outer product of the initial vector $\mathbf{X}_-$ and a 1x4 row vector gives

$$(10.26) \quad \mathbf{B}_-\mathbf{A}^2 = \begin{pmatrix} 0 \\ -1 \\ 0 \\ -1 \end{pmatrix} \begin{pmatrix} 0 & +1 & 0 & +1 \end{pmatrix} = -\mathbf{X}'_-\mathbf{X}'^-.$$

$(10.27) \quad \mathbf{X}'^- = \begin{pmatrix} 0 & +1 & 0 & +1 \end{pmatrix}$, the opposite sign to (10.10b)

(10.28) Ordinarily this would be expected to be the same as the static term $\mathbf{E}_- = \mathbf{X}'_-\mathbf{X}'^-$ for $m = 0$ but, in fact, the sign of $\mathbf{X}'^-$ (10.26) is opposite to that in (10.10b) where $\mathbf{X}^- = \begin{pmatrix} 0 & -1 & 0 & -1 \end{pmatrix}$. Because it is only a sign difference, i.e. a scaling of -1, then this is perfectly valid for an eigenvector. Nevertheless, the commonly used $\mathbf{T}$ operator relation $\mathbf{X}^- = (\mathbf{TX}_+)^T$ (A34e) will also give the sign form

$\mathbf{X}^- = \begin{pmatrix} 0 & -1 & 0 & -1 \end{pmatrix}$ (10.10b) instead of $\mathbf{X}^- = \begin{pmatrix} 0 & +1 & 0 & +1 \end{pmatrix}$ in (10.26).

The reason for this sign discrepancy comes from the difference in the original derivation of $\mathbf{X}^-$ (6.24), which derives $\mathbf{X}^-$ from $\mathbf{X}_+$ using the $\mathbf{T}$ operator relation $\mathbf{X}^- = (\mathbf{TX}_+)^T$ (E7), instead of from first principles using the full cubic $\mathbf{B\_A}^2$ derivation (10.26). Indeed, $\mathbf{X}_+$ itself was originally calculated using only the quadratic expression $\mathbf{B}_+\mathbf{A}$ as opposed to $\mathbf{B\_A}^2$ (10.24), albeit for $\mathbf{X}_+$ with a positive, unity eigenvalue $+1$ this is immaterial. However, for $\mathbf{X}^-$, using the quadratic expression $\mathbf{B\_A}$ (10.22) for $\mathbf{E}_-$ $(= \mathbf{X\_X}^-)$ gives the opposite sign to the cubic form $\mathbf{B\_A}^2$ because this includes an extra multiplication by $\mathbf{A}$, which introduces the sign change by virtue of the eigenvalue for $\mathbf{X}_-$ (and $\mathbf{X}^-$) being -1, i.e. $\mathbf{B\_A}^2 = -\mathbf{B\_A}$.

Since the quadratic polynomial usage for the determination of $\mathbf{X}_-$ and $\mathbf{X}^-$ from $\mathbf{E}_-$ gives a different sign to the cubic derivation then, from the remark above about deriving $\mathbf{X}^-$ from $\mathbf{X}_+$ using the $\mathbf{T}$ operator relation $\mathbf{X}^- = (\mathbf{TX}_+)^T$ (E7), this implies that the current usage of the $\mathbf{T}$ operator is in conflict with a full, cubic derivation of the minus eigenvectors (I9). Indeed this is so, and consistent usage of the $\mathbf{T}$ operator appears to only apply to quadratic derivations of the residual matrices and associated eigenvectors. Noting too that the $\mathbf{T}$ operator only enters when discussing a complete set of eigenvectors obtained under Pythagoras conditions. In the most general form of URMT there is only the single eigenvector $\mathbf{X}$ $(\sim \mathbf{X}_+)$ and its reciprocal $\mathbf{X}^+$, which is because there is only a single, mandated eigenvalue equation $\mathbf{AX} = C\mathbf{X}$ and its reciprocal equivalent $\mathbf{X}^+\mathbf{A} = C\mathbf{X}^+$. The reciprocal $\mathbf{X}^+$ is not related to $\mathbf{X}_+$ by any $\mathbf{T}$ operator relation, which, strictly speaking, relates conjugate vectors, e.g. $\mathbf{X}^-$, with their standard forms, $\mathbf{X}_+$ in this example.

Summarising the above: using the cubic polynomial forms to determine $\mathbf{E}_+$ and $\mathbf{E}_-$, and hence $\mathbf{X}_+$, $\mathbf{X}^+$ and $\mathbf{X}_-$, $\mathbf{X}^-$ respectively, the $\mathbf{T}$ operator relations linking the minus (I9) with the plus eigenvectors (I11) are

(10.29a) $\mathbf{X}^- = -\left(\mathbf{TX}_+\right)^T$, a sign change from (E7)

(10.29b) $\mathbf{X}_- = \left(\mathbf{X}^+\mathbf{T}\right)^T$, as per (E7).

Of course, a simple sign change in an eigenvector is immaterial from a purely mathematical perspective since eigenvectors are only unique to within a scale factor, which includes -1. Physically, this sign change could be very important, and is something to be wary of in the future. Prior to this second book, URMT has only been studied for the three-dimensional URM3 case in [1], and this sign issue has not arisen given the residual matrices, under Pythagoras conditions, are quadratic expressions anyhow (the full cubic form is the unfactored, Cayley-Hamilton polynomial, which equates to zero by definition).

Having discussed this sign issue at length, then as regards further work, the full cubic form for $\mathbf{E}_\pm(m)$ (10.18) has to be used for $m \neq 0$.

However, one last reason to use the quadratic sign-form for $\mathbf{X}^- = \begin{pmatrix} 0 & -1 & 0 & -1 \end{pmatrix}$ (10.10b) is that it means the inner product $\mathbf{X}^-\mathbf{X}_- = +1$ is positive, rather than negative when using the cubic sign form $\mathbf{X}^- = \begin{pmatrix} 0 & +1 & 0 & +1 \end{pmatrix}$. Noting that for $\mathbf{X}_-$ the sign is the same in both the quadratic and cubic forms. This could be reversed such that $\mathbf{X}_-$ in the cubic expression (10.26) now has the opposite sign to that of $\mathbf{X}'_-$ (10.11a) and, instead, $\mathbf{X}'^-$ (10.10b) and $\mathbf{X}^-$ (10.26) are chosen to agree. It is currently all a matter of convention, and whether physical reasons enforce a particular sign-convention is yet to be decided in URMT.

Proceeding on to the eigenvector derivations for general $m \neq 0$.

The linear coefficient m, matrix term $\mathbf{B}_\pm\mathbf{D}+\Delta^P\mathbf{A}^2$ in (10.18), is

$$(10.30) \quad \mathbf{B}_-\mathbf{D}+\Delta^P\mathbf{A}^2 = \begin{pmatrix} 0 & +2 & 0 & +2 \\ 0 & 0 & 0 & 0 \\ 0 & -2 & 0 & -2 \\ 0 & 0 & 0 & 0 \end{pmatrix}$$

the quadratic coefficient m^2, matrix term $\mathbf{B}_\pm\mathbf{F}+\Delta^P\mathbf{D}$ in (10.18), is

$$(10.31) \quad \mathbf{B}_-\mathbf{F}+\Delta^P\mathbf{D} = \begin{pmatrix} 0 & +4 & 0 & +4 \\ 0 & -2 & 0 & -2 \\ 0 & +4 & 0 & +4 \\ 0 & -6 & 0 & -6 \end{pmatrix}$$

and the cubic coefficient m^3, matrix term $\Delta^P\mathbf{F}$ in (10.18), is

$$(10.32) \quad \Delta^P\mathbf{F} = \begin{pmatrix} 0 & 0 & 0 & 0 \\ 0 & 0 & 0 & 0 \\ 0 & 0 & 0 & 0 \\ 0 & 0 & 0 & 0 \end{pmatrix}.$$

That $\Delta^P\mathbf{F}$ is zero is not surprising because this cancelling of the cubic variation (here m^3) is also noted in URM3, Appendix (A) of [1],#1, where only linear and quadratic terms appear, and the cubic term '$\eta\delta\varepsilon$' (8.4) in the variational product '$\eta\delta\varepsilon xyz$' cancels. If this did not cancel it would mean that, for arbitrary, local variations in η,δ,ε, one of the coordinates x,y,z would have to be unconditionally zero, reducing URM3 to two-dimensional. Furthermore if, for example, coordinate x were chosen to be zero, then this would force $z=\pm y$ making URM3 trivially URM2, Appendix (H).

The full cubic form for the evolving residual matrix $\mathbf{E}_+(m)$ is, by (10.18),

$$(10.33) \quad \mathbf{E}_+(m) = \mathbf{B}_+\mathbf{A}^2 - m(\mathbf{B}_+\mathbf{D}+\Delta^P\mathbf{A}^2) + m^2(\mathbf{B}_+\mathbf{F}+\Delta^P\mathbf{D}),$$

and combining the terms calculated above gives

$$(10.34) \quad \mathbf{E}_+(m) = \begin{pmatrix} 0 & 0 & 0 & 0 \\ 0 & +1 & 0 & -1 \\ 0 & 0 & 0 & 0 \\ 0 & -1 & 0 & +1 \end{pmatrix} +$$

$$-m \begin{pmatrix} 0 & 0 & 0 & 0 \\ -2 & 0 & +2 & 0 \\ 0 & 0 & 0 & 0 \\ +2 & 0 & -2 & 0 \end{pmatrix} +$$

$$+m^2 \begin{pmatrix} 0 & 0 & 0 & 0 \\ -4 & +2 & -4 & -6 \\ 0 & 0 & 0 & 0 \\ +4 & -2 & +4 & +6 \end{pmatrix}.$$

Given $\mathbf{X}_+$ does not evolve with m, then the evolving residual matrix $\mathbf{E}_+(m)$ will always be an outer product of a static $\mathbf{X}_+$ with an evolutionary vector $\mathbf{X}^+(m)$, i.e. $\mathbf{E}_+(m) = \mathbf{X}_+\mathbf{X}^+(m)$. This means that each matrix term in the expression (10.34) for $\mathbf{E}_+$ will also be an outer product of an evolving row-vector with the static, column eigenvector $\mathbf{X}_+$. Therefore, re-writing $\mathbf{E}_+(m)$ (10.34) in outer product form gives

$$(10.35) \quad \mathbf{E}_+(m) = \begin{pmatrix} 0 \\ -1 \\ 0 \\ +1 \end{pmatrix} (0 \quad -1 \quad 0 \quad +1) +$$

$$-m \begin{pmatrix} 0 \\ -1 \\ 0 \\ +1 \end{pmatrix} (+2 \quad 0 \quad -2 \quad 0) +$$

$$+ m^2 \begin{pmatrix} 0 \\ -1 \\ 0 \\ +1 \end{pmatrix} (+4 \quad -2 \quad +4 \quad +6),$$

and it is clear that the vector $\mathbf{X}^+(m)$ evolves with m according to

$$\begin{aligned} \mathbf{X}^+(m) &= (0 \quad -1 \quad 0 \quad +1) + \\ (10.36) \quad &- m(+2 \quad 0 \quad -2 \quad 0) + \\ &m^2(+4 \quad -2 \quad +4 \quad +6) \end{aligned}$$

The first few $\mathbf{X}^+(m)$ for $m = 0..3$ are

(10.36a) $\mathbf{X}^+(0) = (0 \quad -1 \quad 0 \quad +1)$

(10.36b) $\mathbf{X}^+(1) = (+2 \quad -3 \quad +6 \quad +7)$

(10.36c) $\mathbf{X}^+(2) = (+12 \quad -9 \quad +20 \quad +25)$

(10.36d) $\mathbf{X}^+(3) = (+30 \quad -19 \quad +42 \quad +55)$

Each one of these solutions is a Pythagorean quadruple, as expected, since the variation $\mathbf{\Delta}^P$ (8.143) is such that the dynamical variables (10.6) maintain the Pythagoras conditions (4.1), even though the Potential (10.3) is no longer zero for non-zero m.

For the $\mathbf{X}_-$ and $\mathbf{X}^-$ eigenvectors, the full, cubic form for the evolving residual matrix $\mathbf{E}_-(m)$ (10.18) is

(10.37) $\mathbf{E}_-(m) = \mathbf{B}_-\mathbf{A}^2 - m(\mathbf{B}_-\mathbf{D} + \mathbf{\Delta}^P\mathbf{A}^2) + m^2(\mathbf{B}_-\mathbf{F} + \mathbf{\Delta}^P\mathbf{D})$

which evaluates as follows, using the above calculations for the individual terms,

$$(10.38) \quad \mathbf{E}_-(m) = \begin{pmatrix} 0 & 0 & 0 & 0 \\ 0 & -1 & 0 & -1 \\ 0 & 0 & 0 & 0 \\ 0 & -1 & 0 & -1 \end{pmatrix} +$$

$$-m\begin{pmatrix} 0 & +2 & 0 & +2 \\ 0 & 0 & 0 & 0 \\ 0 & -2 & 0 & -2 \\ 0 & 0 & 0 & 0 \end{pmatrix} +$$

$$m^2\begin{pmatrix} 0 & +4 & 0 & +4 \\ 0 & -2 & 0 & -2 \\ 0 & +4 & 0 & +4 \\ 0 & -6 & 0 & -6 \end{pmatrix}.$$

Using similar arguments to $\mathbf{E}_+(m)$ and $\mathbf{X}_+$, given $\mathbf{X}^- = (\mathbf{T}\mathbf{X}_+)^T$ (E7), at least to within a sign/scale-factor (10.28), then since $\mathbf{X}_+$ does not evolve with m, neither does $\mathbf{X}^-$, and the evolving residual matrix $\mathbf{E}_-(m)$ will always be an outer product of a static $\mathbf{X}^-$ with an evolving eigenvector $\mathbf{X}_-(m)$, i.e. $\mathbf{E}_-(m) = \mathbf{X}_-(m)\mathbf{X}^-$. This means that each matrix term in the expression for $\mathbf{E}_-(m)$ (10.38), will also be an outer product of a column vector with the row eigenvector $\mathbf{X}^-$ (10.27). Therefore, re-writing $\mathbf{E}_-(m)$ in outer product form gives

$$(10.39) \quad \mathbf{E}_-(m) = \begin{pmatrix} 0 \\ -1 \\ 0 \\ -1 \end{pmatrix}(0 \quad +1 \quad 0 \quad +1)+$$

$$-m\begin{pmatrix} +2 \\ 0 \\ -2 \\ 0 \end{pmatrix}(0 \quad +1 \quad 0 \quad +1)+$$

$$+m^2\begin{pmatrix} +4 \\ -2 \\ +4 \\ -6 \end{pmatrix}(0 \quad +1 \quad 0 \quad +1),$$

and it is clear that the vector $\mathbf{X}_-(m)$ evolves with m according to

$$(10.40) \quad \mathbf{X}_{-}(m) = \begin{pmatrix} 0 \\ -1 \\ 0 \\ -1 \end{pmatrix} - m \begin{pmatrix} +2 \\ 0 \\ -2 \\ 0 \end{pmatrix} + m^2 \begin{pmatrix} +4 \\ -2 \\ +4 \\ -6 \end{pmatrix}.$$

Note that the row eigenvector $\mathbf{X}^-$ (10.27) in the outer product $\mathbf{E}_{-}(m)$ (10.39), is of the opposite sign to that in (10.10b), as explained in Point (10.28).

The first few $\mathbf{X}_{-}(m)$ for $m = 0..3$ are

(10.41)

$$\mathbf{X}_{-}(0) = \begin{pmatrix} 0 \\ -1 \\ 0 \\ -1 \end{pmatrix}, \quad \mathbf{X}_{-}(1) = \begin{pmatrix} +2 \\ -3 \\ +6 \\ -7 \end{pmatrix},$$

$$\mathbf{X}_{-}(2) = \begin{pmatrix} +12 \\ -9 \\ +20 \\ -25 \end{pmatrix}, \quad \mathbf{X}_{-}(3) = \begin{pmatrix} +30 \\ -19 \\ +42 \\ -55 \end{pmatrix}.$$

Evaluation of the complex eigenvectors $\mathbf{X}_{i+}$, $\mathbf{X}^{i+}$ and $\mathbf{X}_{i-}$, $\mathbf{X}^{i-}$, for complex eigenvalues $\lambda = +2mi$ and $\lambda = -2mi$ respectively, is much the same as evaluating the zero eigenvectors for arbitrary m, $m > 0$, which is usually better done by direct algebraic manipulation of the dynamical equations rather than determination from residual matrices. Because the area of complex URMT is well beyond the scope of this book, this shall not be done here. Suffice to conclude this section with an outline of the residual method, in the context of complex eigenvectors, as it has some theoretical value.

The Cayley-Hamilton polynomial for the eigenvalues $\lambda = \pm 1, \pm 2mi$ (10.9) is

$$(10.42) \quad 0 = (\mathbf{A}_m - \mathbf{I})(\mathbf{A}_m + \mathbf{I})(\mathbf{A}_m - 2mi\mathbf{I})(\mathbf{A}_m + 2mi\mathbf{I}).$$

With the quadratic factor $(\mathbf{A}_m - \mathbf{I})(\mathbf{A}_m + \mathbf{I})$ evaluated as the real expression $(\mathbf{A}_m^2 - \mathbf{I})$, the residual matrix $\mathbf{E}_{i+} = \mathbf{X}_{i+}\mathbf{X}^{i+}$ is given by the following polynomial

(10.43) $\mathbf{E}_{i+} = (\mathbf{A}_m^2 - \mathbf{I})(\mathbf{A}_m + 2mi\mathbf{I})$.

This factors nicely into real and imaginary components

(10.44) $\mathbf{E}_{i+} = (\mathbf{A}_m^2 - \mathbf{I})\mathbf{A}_m + 2m(\mathbf{A}_m^2 - \mathbf{I})i$.

Doing the same for $\mathbf{E}_{i-} = \mathbf{X}_{i-}\mathbf{X}^{i-}$ gives

(10.45) $\mathbf{E}_{i-} = (\mathbf{A}_m^2 - \mathbf{I})\mathbf{A}_m - 2m(\mathbf{A}_m^2 - \mathbf{I})i$.

It is clear that $\mathbf{E}_{i+}$ and $\mathbf{E}_{i-}$ are complex conjugates of each other, i.e.

(10.46a) $\mathrm{Re}(\mathbf{E}_{i+}) = \mathrm{Re}(\mathbf{E}_{i-})$

(10.46b) $\mathrm{Im}(\mathbf{E}_{i+}) = -\mathrm{Im}(\mathbf{E}_{i-})$

(10.46c) $\mathbf{E}_{i+}^* = \mathbf{E}_{i-}$, $\mathbf{E}_{i-}^* = \mathbf{E}_{i+}$.

It is expected therefore that $\mathbf{X}_{i+}$ and $\mathbf{X}_{i-}$ are complex conjugates of each other, as are $\mathbf{X}^{i+}$ and $\mathbf{X}^{i-}$, i.e.

(10.47a) $\mathbf{X}_{i+}^* = \mathbf{X}_{i-}$, $\mathbf{X}_{i-}^* = \mathbf{X}_{i+}$

(10.47b) $\mathbf{X}^{i+*} = \mathbf{X}^{i-}$, $\mathbf{X}^{i-*} = \mathbf{X}^{i+}$.

Note that for no variation, i.e. $m = 0$, the residual matrices $\mathbf{E}_{i+}$ and $\mathbf{E}_{i-}$ are one and the same, i.e.

(10.48) $\mathbf{E}_{i+} = \mathbf{E}_{i-} = (\mathbf{A}_m^2 - \mathbf{I})\mathbf{A}_m$, $m = 0$,

and comparing with the Cayley-Hamilton polynomial for $\mathbf{E}_0$, eigenvectors $\mathbf{X}_0$ and $\mathbf{X}^0$,

(10.49) $\mathbf{E}_0 = \mathbf{A}(\mathbf{A}^2 - \mathbf{I}) = \mathbf{X}_0\mathbf{X}^0$,

it is seen that $\mathbf{E}_{i+}$, $\mathbf{E}_{i-}$ and $\mathbf{E}_0$ are one and the same, i.e.

(10.50) $\mathbf{E}_{i+} = \mathbf{E}_{i-} = \mathbf{E}_0$, $m = 0$.

This should come as no surprise since, for $m = 0$, there are no complex eigenvalues and, hence, any complex components of the residual matrices should reduce to zero, i.e. comprise real components only.

This concludes the section on variational methods applied to the 2a2p1 solution.

(11) Variational Methods on Lifting

This section applies variational methods to the URM4 Pythagoras solutions, as first given in Section (7) as examples of 'lifting' solutions.

(11-1) Solution Summary

The unity root matrix $\mathbf{A}_{40}$ (7.5) is reproduced below from Section (7), where the embedded URM3 matrix $\mathbf{A}_3$ is explicitly shown as a function of the variational (evolutionary) parameter m, denoted by $\mathbf{A}_{30}(m)$, with the extra '0' in the subscript denoting, as per URM3 in [1], that it is a unity root matrix under URM3 Pythagoras conditions (A18)

$$(7.5) \quad \mathbf{A}_{40} = \begin{pmatrix} 0 & -s\mathbf{X}^{3-} \\ s\mathbf{X}_3 & \mathbf{A}_{30}(m) \end{pmatrix}.$$

Note that, as in Section (7), the entire matrix $\mathbf{A}_{40}$ satisfies the URM4 Pythagoras conditions (4.1).

From [1], the URM3 matrix $\mathbf{A}_{30}(m)$ is also split into a sum of its initial value $\mathbf{A}'_{30}$, for $m=0$, and a variational matrix $\mathbf{\Delta}_3^P$ ($\sim \mathbf{\Delta}^P$ (8.136)), with $m=-\delta$,

$$(11.1) \quad \mathbf{A}_{30}(m) = \mathbf{A}'_{30} - m\mathbf{\Delta}_3^P$$

$$(11.2) \quad \mathbf{\Delta}_3^P = \begin{pmatrix} 0 & +z & -y \\ -z & 0 & +x \\ -y & +x & 0 \end{pmatrix}.$$

Substituting the decomposition of $\mathbf{A}_{30}(m)$ (11.1) into matrix $\mathbf{A}_{40}$ (7.5) gives

$$(11.3) \quad \mathbf{A}_{40} = \begin{pmatrix} 0 & 0 \\ 0 & \mathbf{A}'_{30} \end{pmatrix} + \begin{pmatrix} 0 & -s\mathbf{X}^{3-} \\ s\mathbf{X}_{3+} & -m\Delta_3^P \end{pmatrix},$$

Denoting the two separate matrix components in (11.3) by an initial matrix $\mathbf{A}'_{40}$ for both $m = 0$ and $s = 0$, and a variational, parametric matrix Δ_4, defined as follows:

$$(11.4) \quad \mathbf{A}'_{40} = \begin{pmatrix} 0 & 0 \\ 0 & \mathbf{A}'_{30} \end{pmatrix},$$

$$(11.5) \quad \Delta_4 = \begin{pmatrix} 0 & -s\mathbf{X}^{3-} \\ s\mathbf{X}_{3+} & -m\Delta_3^P \end{pmatrix},$$

then $\mathbf{A}_{40}$ (11.3) is written as

$$(11.6) \quad \mathbf{A}_{40} = \mathbf{A}'_{40} + \Delta_4.$$

Using $\mathbf{X}_{4+}$ $(\sim \mathbf{X})$ (1.5), then it can be verified that Δ_4 satisfies the annihilator property (8.1) required of all variational matrices, i.e.

$$(8.1) \quad \Delta_4 \mathbf{X}_{4+} = 0.$$

The variational term Δ_4 (11.5) is really just a special case of the full, twelve-parameter, variational matrix Δ (8.16). The relation between $-s$ in Δ_4 (11.5) and the δ_{1j} (first row variational parameters) in Δ (8.16) is not a simple one but, nevertheless, exists and makes Δ_4 a special case of Δ. Thus, under URM4 Pythagoras conditions, Δ_4 is parameterised by two arbitrary, integer parameters m and s.

For arbitrary variations m and s, $\mathbf{A}_{40}$ (7.5) is expanded in full as follows, and is seen to adhere to URM4 Pythagoras conditions (4.1) by inspection of its elemental, dynamical variables, (11.8) further below

$$(11.7) \quad \mathbf{A}_{40} = \begin{pmatrix} 0 & -sx & -sy & sz \\ sx & 0 & R-mz & Q+my \\ sy & -R+mz & 0 & P-mx \\ sz & Q+my & P-mx & 0 \end{pmatrix}.$$

Comparing (11.7) with the general form of $\mathbf{A}_{40}$ (4.2), then the URM4-specific dynamical variables S, T, U are now

(11.8)

(11.8a) $\bar{S} = -S = sx$

(11.8b) $\bar{T} = -T = sy$

(11.8c) $\bar{U} = U = sz$,

and the URM3 dynamical variables transform under variations in m as

(11.8d) $Q \rightarrow Q + my$

(11.8e) $P \rightarrow P - mx$

(11.8f) $R \rightarrow R - mz$

Furthermore, and most importantly, with these forms for the dynamical variables, the URM4 Potential V remains zero for arbitrary variations in m and s, i.e.

(11.9) $V = \dfrac{[QT - (PS + RU)]^2}{C^2} = 0$, $\forall m, s \in \mathbb{Z}$.

Remember, a zero URM4 Potential is not one of the URM4 Pythagoras conditions, nor a consequence of them, whereas, under URM3 Pythagoras conditions, a zero URM3 Potential naturally follows.

That the Potential remains zero and invariant, for arbitrary variations in m and s, can be shown by substituting for the dynamical variables P, Q, R and S, T, U from (11.8) into (11.9) to give

(11.10) $QT - (PS + RU) = s(Px - Qy - Rz) - ms(x^2 + y^2 - z^2)$.

Since the triple (x, y, z), forming both $\mathbf{X}^{3-}$ (A35c) and $\mathbf{X}_{3+}$ (A33a), is a Pythagorean triple, then the bracketed term $x^2 + y^2 - z^2$ on the right is zero. The bracketed term $(Px - Qy - Rz)$ is also zero by the 'delta equation' (F5), and so the entire right of (11.10) is zero, and therefore $QT - (PS + RU)$ is also zero. Thus, the URM4 Potential (11.9) is invariant and zero for arbitrary variations in m and s:

(11.11) $V = 0$, $\forall s, m \in \mathbb{Z}$.

That the URM4 Potential is invariant and zero, for arbitrary variations in s and m, is not coincidental in so far as this particular combination and selection of two variational parameters, out of a maximum of twelve, is known to give the desired result. The outcome is extremely sensitive to usage of more parameters, generally never giving an invariant, zero Potential, even when Pythagoras conditions are satisfied. More often than not, the Potential can be non-zero giving complex eigenvalues and eigenvectors as per (4.14b).

(11-2) Eigenvector solution

Because the Potential is zero, the eigenvalues are always $\lambda = \pm C, 0, 0$ (4.14a) for arbitrary variations in m and s. Additionally, the top row of $\mathbf{A}_{40}$ is the vector $\mathbf{X}^{3-}$ (scaled by $-s$) so that, for the same reasons given in Section (3-2), the quadratic polynomial form $\mathbf{E}_{4+} = \mathbf{A}_{40}^2 + C\mathbf{A}_{40}$ (7.49) can be used to compute the residual matrix $\mathbf{E}_{4+}$. Thus, with $\mathbf{X}_{4+}$ defined by (2.4), and invariant (by definition) to any variations (independent of parameters m and s), the row eigenvector $\mathbf{X}^{4+}$ is obtained from the residual matrix $\mathbf{E}_{4+} = \mathbf{X}_{4+}\mathbf{X}^{4+}$ by calculating $\mathbf{E}_{4+}$ using $\mathbf{A}_{40}$ (7.5). The work to calculate $\mathbf{E}_{4+}$ ($= \mathbf{XX}^+ = \mathbf{A}^2 + C\mathbf{A}$) has already been done in Section (3), and gives the result

$$(3.22) \quad \mathbf{E}_{4+} = \begin{pmatrix} 0 & 0 \\ 2sC\mathbf{X}_{3+} & -s^2\mathbf{X}_{3+}\mathbf{X}^{3-} + \mathbf{X}_{3+}\mathbf{X}^{3+}(m) \end{pmatrix}.$$

Using (2.3) for $\mathbf{X}_{4+}$, the row eigenvector $\mathbf{X}^{4+}$ is thus

$$(11.20) \quad \mathbf{X}^{4+} = \left(2sC \quad -s^2\mathbf{X}^{3-} + \mathbf{X}^{3+}(m) \right).$$

What has not been done so far is the explicit decomposition of $\mathbf{X}^{3+}(m)$ as a function of URM3 variational parameter m. From Appendix (B), the vector $\mathbf{X}^{3+}(m)$ evolves with m according to the following:

$$(11.21) \quad \mathbf{X}^{3+}(m) = -m^2\mathbf{X}^{3-} + 2m\mathbf{X}'^{30} + \mathbf{X}'^{3+}, \text{ see (B2f),}$$

where the vectors $\mathbf{X}^{3-}$, $\mathbf{X}'^{30}$ and $\mathbf{X}'^{3+}$ are their initial $m = 0$ forms (B1).

Note that $\mathbf{X}_{3+}$ and $\mathbf{X}^{3-}$ are static vectors in URM3 as $\mathbf{X}_{3+}$ is invariant to all arbitrary variations, by definition, and $\mathbf{X}^{3-}$ is simply related to $\mathbf{X}_{3+}$ by the usual $\mathbf{T}$ operator relation $\mathbf{X}^- = \left(\mathbf{TX}_+\right)^T$ (A34e). Consequently, they have no variation with any parameter and are always equal to their initial value, so the primed superscript notation is not generally used in their case.

Substituting for $\mathbf{X}^{3+}(m)$ from (11.21) into (11.20) gives $\mathbf{X}^{4+}$ as

(11.22) $\mathbf{X}^{4+} = \left(2sC \quad -(s^2 + m^2)\mathbf{X}^{3-} + 2m\mathbf{X}'^{30} + \mathbf{X}'^{3+}\right)$

This last expression makes clear the parametric dependence of $\mathbf{X}^{4+}$ on both m and s. The initial forms of the vectors $\mathbf{X}'^{3+}$, $\mathbf{X}'^{30}$ and $\mathbf{X}^{3-}$, as used in the above, are reproduced below, where the URM3 scale factors α_3, β_3, γ_3 and dynamical variables P, Q, R are also their initial, $m = 0$ forms

(A35a) $\mathbf{X}'^{3+} = \left(\alpha_3 \quad \beta_3 \quad \gamma_3\right)$, $m = 0$

(A35b) $\mathbf{X}'^{30} = \left(P \quad -Q \quad -R\right)$, $m = 0$

(A35c) $\mathbf{X}^{3-} = \left(x \quad y \quad -z\right)$, $m = 0$.

The four-element, row eigenvector $\mathbf{X}^{4+}$ is written, as follows, in terms of the URM4 scale factors $\alpha_4, \beta_4, \gamma_4$, also known as just α, β, γ (3.56) :

(11.23) $\mathbf{X}^{4+} = \left(2sC \quad \alpha_4 \quad \beta_4 \quad \gamma_4\right)$,

Comparing $\mathbf{X}^{4+}$ with (11.22), then the URM4 scale factors are given by

(11.24)

(11.24a) $\alpha_4 = -(s^2 + m^2)x + 2mP + \alpha_3$

(11.24b) $\beta_4 = -(s^2 + m^2)y - 2mQ + \beta_3$

(11.24c) $\gamma_4 = +(s^2 + m^2)z - 2mR + \gamma_3$.

Since this solution is under Pythagoras conditions, the $\mathbf{X}^{4-}$ and $\mathbf{X}_{4-}$ vectors can be obtained from $\mathbf{X}_{4+}$ and $\mathbf{X}^{4+}$ respectively, using the standard, $\mathbf{T}$ operator relations, Appendix (E), to give

(11.25)

(11.25a) $\mathbf{X}^{4-} = \begin{pmatrix} 0 & x & y & -z \end{pmatrix}$

(11.25b) $\mathbf{X}_{4-} = \begin{pmatrix} 2sC \\ +\alpha_4 \\ +\beta_4 \\ -\gamma_4 \end{pmatrix}$.

It is noted that $\mathbf{X}_{4-}$ is a function of parameters m and s by virtue of its dependence on the scale factors $\alpha_4, \beta_4, \gamma_4$ (11.24).

(11-3) Inner Products

By the rules of matrix algebra (F25), all inner products between a row eigenvector and a column eigenvector will be zero if the two vectors have distinct eigenvalues. In the case of a repeated zero eigenvalue, as is the case in this example, it has to be remembered that inner products between the zero eigenvectors are not necessarily zero. Whatever the case, all inner products are algebraically evaluated in this section as confirmation that the vectors are specified correctly. Since this example forms the backdrop against which URM4 Physical comparisons are made with URM3, it is reassuring to verify all expressions are correct.

Using $\mathbf{X}^{4+}$ (11.22) and $\mathbf{X}_{4+}$ (2.4), the URM4 Potential equation $\mathbf{X}^{4+}\mathbf{X}_{4+}$ is

(11.30) $\mathbf{X}^{4+}\mathbf{X}_{4+} = \begin{pmatrix} 2sC & -(s^2 + m^2)\mathbf{X}^{3-} + 2m\mathbf{X}'^{30} + \mathbf{X}'^{3+} \end{pmatrix} \begin{pmatrix} 0 \\ \mathbf{X}_{3+} \end{pmatrix}$.

Given $\mathbf{X}^{3-}\mathbf{X}_{3+} = 0$ (F1) and $\mathbf{X}'^{30}\mathbf{X}_{3+} = 0$ (F5), then $\mathbf{X}^{4+}\mathbf{X}_{4+}$ reduces to the URM3 Potential equation (F4), i.e.

(11.31) $\mathbf{X}^{4+}\mathbf{X}_{4+} = \mathbf{X}^{3+}\mathbf{X}_{3+} = +2C^2$.

In fact, this should really be $+2C^3$, according to (3.112) for $V_3 = 0$, and is quadratic, not cubic, for the same quadratic/cubic reasons given in Section (3-2); see also point (5.32).

Alternatively, using the forms $\mathbf{X}^{4+}$ (11.23) and $\mathbf{X}_{4+}$ (2.3) for $\mathbf{X}^{4+}\mathbf{X}_{4+}$ gives

$$(11.32) \quad \mathbf{X}^{4+}\mathbf{X}_{4+} = \begin{pmatrix} 2sC & \alpha_4 & \beta_4 & \gamma_4 \end{pmatrix} \begin{pmatrix} 0 \\ x \\ y \\ z \end{pmatrix}.$$

Multiplying out and equating with (11.31) implies that

$$(11.33) \quad \mathbf{X}^{4+}\mathbf{X}_{4+} = \alpha_4 x + \beta_4 y + \gamma_4 z = +2C^2.$$

Since this is the same as the URM3 Potential equation for $V_3 = 0$, i.e.

$$(F4) \quad \mathbf{X}^{3+}\mathbf{X}_{3+} = \alpha_3 x + \beta_3 y + \gamma_3 z = +2C^2, \; V_3 = 0,$$

then the following relation is deduced

$$(11.34) \quad \alpha_3 x + \beta_3 y + \gamma_3 z = \alpha_4 x + \beta_4 y + \gamma_4 z, \; V_3 = 0,$$

This does not, of course, mean that the URM3 and URM4 scale factors are identical, as their definition (11.24) clearly shows. However, this relation is nicely illustrated by multiplying the URM4 scale factors α_4, β_4, γ_4 by x, y, z respectively, as follows:

(11.35)

$$(11.35a) \quad \alpha_4 x = -(s^2 + m^2)x^2 + 2mPx + \alpha_3 x$$

$$(11.35b) \quad \beta_4 y = -(s^2 + m^2)y^2 - 2mQy + \beta_3 y$$

$$(11.35c) \quad \gamma_4 z = +(s^2 + m^2)z^2 - 2mRz + \gamma_3 z.$$

and summing these three expressions gives

$$(11.36) \quad \begin{aligned} \alpha_4 x + \beta_4 x + \gamma_4 z = \\ -(s^2 + m^2)(x^2 + y^2 - z^2) \\ + 2m(Px - Qy - Rz) \\ + \alpha_3 x + \beta_3 x + \gamma_3 z \end{aligned}$$

Using Pythagoras (A23) and the delta equation (F5), this expression simplifies to that of (11.34), thus confirming its correctness.

Moving on to the inner product $\mathbf{X}^{4-}\mathbf{X}_{4-}$, then using $\mathbf{X}^{4-}$ (11.25a) and $\mathbf{X}_{4-}$ (11.25b) it is trivially seen identical to $\mathbf{X}^{4+}\mathbf{X}_{4+}$ (11.33), as expected.

Lastly, using the scale factors α_4, β_4, γ_4 (11.35), and with the help of the URM3 evolutionary form of $\mathbf{X}_{3-}$ (B2c), the eigenvector $\mathbf{X}_{4-}$ (11.25b) can be written in another form, used later in Section (14-4),

$$(11.37) \quad \mathbf{X}_{4-} = -s^2 \begin{pmatrix} 0 \\ \mathbf{X}_{3+} \end{pmatrix} + 2s \begin{pmatrix} C \\ \mathbf{0}_3 \end{pmatrix} + \begin{pmatrix} 0 \\ \mathbf{X}_{3-} \end{pmatrix}, \quad s \sim t_4 \text{ in (14.4).}$$

(11-4) Calculation of the Zero Eigenvectors

By the residual method, the matrix $\mathbf{E}_{40}$ is calculated, when under Pythagoras conditions, from the quadratic expression

$$(7.57) \quad \mathbf{E}_{40} = \mathbf{A}^2 - C^2 \mathbf{I}.$$

Using $\mathbf{A}$ (7.24) to determine $\mathbf{A}^2$, and incorporating relations $\mathbf{X}^{3-}\mathbf{X}_{3+} = 0$ (F1) and $\mathbf{A}_{30}\mathbf{X}_{3+} = C\mathbf{X}_{3+}$ (A4), then $\mathbf{E}_{40}$ (7.57) evaluates to

$$(11.40) \quad \mathbf{E}_{40} = \begin{pmatrix} -C^2 & sC\mathbf{X}^{3-} \\ sC\mathbf{X}_{3+} & -s^2\mathbf{X}_{3+}\mathbf{X}^{3-} + (\mathbf{A}_{30}(m))^2 - C^2\mathbf{I}_3 \end{pmatrix}$$

Note that this expression can also be determined using (3.7), with $\bar{s} = -s$ (7.4) and $\mathbf{X}^{3-} = \overline{\mathbf{X}}_{3+}$ (A37a).

Knowing that there are two zero eigenvectors, $\mathbf{X}_{40A}$ and $\mathbf{X}_{40B}$, gives a hint that this matrix $\mathbf{E}_{40}$ can be decomposed into two components, $\mathbf{E}_{40A}$ and $\mathbf{E}_{40B}$, i.e.

$$(11.41) \quad \mathbf{E}_{40} = \mathbf{E}_{40A} + \mathbf{E}_{40B},$$

where, by inspection of (11.40), $\mathbf{E}_{40A}$ comprises the URM3 part,

$$(11.42) \quad \mathbf{E}_{40A} = \begin{pmatrix} 0 & 0 \\ 0 & (\mathbf{A}_3(m))^2 - C^2\mathbf{I}_3 \end{pmatrix},$$

and $\mathbf{E}_{40B}$ comprises the URM4 part

$$(11.43)\ \mathbf{E}_{40B} = \begin{pmatrix} -C^2 & sC\mathbf{X}^{3-} \\ sC\mathbf{X}_{3+} & -s^2\mathbf{X}_{3+}\mathbf{X}^{3-} \end{pmatrix}.$$

From Section (3), specifically equation (3.74), it is also known that a URM4 zero eigenvector is as follows, when using the method of embedding, which will be taken to be $\mathbf{X}_{40B}$ and justified shortly

$$(11.44)\ \mathbf{X}_{40B} = \begin{pmatrix} C \\ -s\mathbf{X}_{3+} \end{pmatrix}, \text{ see } (3.74).$$

This eigenvector $\mathbf{X}_{40B}$ is usefully split into the sum of a URM4, time-dependent (parameter s) form and a constant, time-independent form, as follows, and used later in (14.41d)

$$(11.45)\ \mathbf{X}_{40B} = -s\begin{pmatrix} 0 \\ \mathbf{X}_{3+} \end{pmatrix} + \begin{pmatrix} C \\ \mathbf{0}_3 \end{pmatrix}, \ s \sim t_4 \text{ in } (14.4).$$

Using (11.44), and given $\mathbf{E}_{40B}$ is defined by the outer product

$$(11.46)\ \mathbf{E}_{40B} = -\mathbf{X}_{40B}\mathbf{X}^{40B},$$

then the residual matrix $\mathbf{E}_{40B}$ (11.43) is seen to be the outer product

$$(11.47)\ \mathbf{E}_{40B} = -\begin{pmatrix} C \\ -s\mathbf{X}_{3+} \end{pmatrix}\begin{pmatrix} C & -s\mathbf{X}^{3-} \end{pmatrix}.$$

By comparing this with $\mathbf{E}_{40B}$ (11.46), then $\mathbf{X}^{40B}$ is

$$(11.48)\ \mathbf{X}^{40B} = \begin{pmatrix} C & -s\mathbf{X}^{3-} \end{pmatrix}.$$

Expanding both vectors $\mathbf{X}_{40B}$ and $\mathbf{X}^{40B}$ in full gives

$$(11.49)\ \mathbf{X}_{40B} = \begin{pmatrix} +C \\ -sx \\ -sy \\ -sz \end{pmatrix}$$

$$(11.50)\ \mathbf{X}^{40B} = \begin{pmatrix} +C & -sx & -sy & +sz \end{pmatrix}.$$

From these, it is confirmed that $\mathbf{X}^{40B}$ is related to $\mathbf{X}_{40B}$ by the $\mathbf{T}$ operator relation $\mathbf{X}^{40B} = \left(\mathbf{T}\mathbf{X}_{40B}\right)^{T}$ (E6).

Since $\mathbf{X}_{3+}$ and $\mathbf{X}^{3-}$ are static in URM3, i.e. they do not evolve, then from the forms $\mathbf{X}_{40B}$ (11.44) and $\mathbf{X}^{40B}$ (11.48), it is seen that $\mathbf{X}_{40B}$ and $\mathbf{X}^{40B}$ only evolve with respect to parameter s, i.e. they have a URM4-unique, evolution given s is unique to URM4, and not in URM3.

Using the orthogonality (Pythagoras) relation $\mathbf{X}^{3-}\mathbf{X}_{3+} = 0$ (F1), the inner product $\mathbf{X}^{40B}\mathbf{X}_{40B}$ is seen to be the same as per URM3, which is just the URM3 DCE (F3), i.e.

(11.51) $\mathbf{X}^{40B}\mathbf{X}_{40B} = C^{2}$.

To get the other eigenvectors $\mathbf{X}_{40A}$ and $\mathbf{X}^{40A}$ then, by looking at $\mathbf{E}_{40A}$, the bottom right, non-zero term is simply the evolved URM3 residual matrix $\mathbf{E}_{30}$ as in

(11.52) $\mathbf{E}_{30} = \left(\mathbf{A}_{3}(m)\right)^{2} - C^{2}\mathbf{I}_{3}$, URM3

Of course, this is already a solved problem in URM3, and $\mathbf{E}_{30}$ is given by

(11.53) $\mathbf{E}_{30} = -\mathbf{X}_{30}\mathbf{X}^{30} = -\begin{pmatrix} +P \\ -Q \\ +R \end{pmatrix}\left(+P \quad -Q \quad -R\right)$, note '-' sign.

Given all other elements of $\mathbf{E}_{40A}$ (11.42) are zero, by simply adding an extra zero to the URM3 eigenvectors $\mathbf{X}_{30}$ and $\mathbf{X}^{30}$, then the URM4 eigenvectors, $\mathbf{X}_{40A}$ and $\mathbf{X}^{40A}$, are given in terms of URM3 $\mathbf{X}_{30}$ and $\mathbf{X}^{30}$ as

(11.54a) $\mathbf{X}_{40A} = \begin{pmatrix} 0 \\ \mathbf{X}_{30} \end{pmatrix}$,

(11.54b) $\mathbf{X}^{40A} = \left(0 \quad \mathbf{X}^{30}\right)$.

Expanded in full, using the definitions of $\mathbf{X}_{30}$ (A33b) and $\mathbf{X}^{30}$ (A35b), gives

$$(11.55a) \quad \mathbf{X}_{40A} = \begin{pmatrix} 0 \\ +P \\ -Q \\ +R \end{pmatrix},$$

$$(11.55b) \quad \mathbf{X}^{40A} = \begin{pmatrix} 0 & +P & -Q & -R \end{pmatrix}$$

$$(11.55c) \quad \mathbf{X}^{40A} = \left(\mathbf{T}\mathbf{X}_{40A}\right)^{T}.$$

Notes

As regards the minus sign in (11.53), it was noted in equation (3.70) that URM3 actually changed the sign of $\mathbf{X}^{30}$ to the negative of what it should actually be when computed via the URM3 residual matrix $\mathbf{E}_{0}$. This is a harmless scaling because $\mathbf{X}^{0}$ is an eigenvector, and it was done so that a consistent relation between $\mathbf{X}_{30}$ and $\mathbf{X}^{30}$ is obtained using the $\mathbf{T}$ operator, i.e. $\mathbf{X}^{30} = \left(\mathbf{T}\mathbf{X}_{30}\right)^{T}$ and not $\mathbf{X}^{30} = -\left(\mathbf{T}\mathbf{X}_{30}\right)^{T}$. It also makes the invariant inner product $\mathbf{X}^{30}\mathbf{X}_{30}$ positive, i.e. $\mathbf{X}^{30}\mathbf{X}_{30} = +C^{2}$ instead of $\mathbf{X}^{30}\mathbf{X}_{30} = -C^{2}$. This convention has been retained for URM4.

Since the URM3 vectors $\mathbf{X}_{30}$ and $\mathbf{X}^{30}$ are not static, but evolve with respect to parameter m, then both $\mathbf{X}_{40A}$ and $\mathbf{X}^{40A}$ also evolve with respect to parameter m.

From Appendix (B), both $\mathbf{X}_{30}$ and $\mathbf{X}^{30}$ evolve with m according to the following relations:

$$(B2b) \quad \mathbf{X}_{30}(m) = -m\mathbf{X}_{3+} + \mathbf{X}'_{30}$$

$$(B2e) \quad \mathbf{X}^{30}(m) = -m\mathbf{X}^{3-} + \mathbf{X}'^{30},$$

where the vectors $\mathbf{X}'_{30}$ and $\mathbf{X}'^{30}$ are their initial, $m = 0$, forms (B1).

When considering any products with $\mathbf{X}_{40A}$ and $\mathbf{X}^{40A}$, the evolving forms of $\mathbf{X}_{30}(m)$ (B2b) and $\mathbf{X}^{30}(m)$ (B2e), must be used, i.e. explicitly written as

(11.56a) $\mathbf{X}_{40A}(m) = \begin{pmatrix} 0 \\ \mathbf{X}_{30}(m) \end{pmatrix}$,

(11.56b) $\mathbf{X}^{40A}(m) = \begin{pmatrix} 0 & \mathbf{X}^{30}(m) \end{pmatrix}$

and the full evolutionary forms for $\mathbf{X}_{40A}(m)$ and $\mathbf{X}^{40A}(m)$ are thus

(11.57a) $\mathbf{X}_{40A}(m) = -m \begin{pmatrix} 0 \\ \mathbf{X}_{3+} \end{pmatrix} + \begin{pmatrix} 0 \\ \mathbf{X}'_{30} \end{pmatrix}$,

(11.57b) $\mathbf{X}^{40A}(m) = -m \begin{pmatrix} 0 & \mathbf{X}^{3-} \end{pmatrix} + \begin{pmatrix} 0 & \mathbf{X}'^{30} \end{pmatrix}$.

Using these evolutionary forms, the inner product $\mathbf{X}^{40A}(m)\mathbf{X}_{40A}(m)$ comprises four separate URM3 products as follows:

(11.58a) $m^2 \begin{pmatrix} 0 & \mathbf{X}^{3-} \end{pmatrix} \begin{pmatrix} 0 \\ \mathbf{X}_{3+} \end{pmatrix}$

(11.58b) $-m \begin{pmatrix} 0 & \mathbf{X}^{3-} \end{pmatrix} \begin{pmatrix} 0 \\ \mathbf{X}'_{30} \end{pmatrix}$

(11.58c) $-m \begin{pmatrix} 0 & -\mathbf{X}'^{30} \end{pmatrix} \begin{pmatrix} 0 \\ \mathbf{X}_{3+} \end{pmatrix}$

(11.58d) $\begin{pmatrix} 0 & -\mathbf{X}'^{30} \end{pmatrix} \begin{pmatrix} 0 \\ \mathbf{X}'_{30} \end{pmatrix}$.

Using URM3 relations $\mathbf{X}^{3-}\mathbf{X}_{3+} = 0$ (F1), $\mathbf{X}^{3-}\mathbf{X}_{30} = 0$ (F5), and $\mathbf{X}^{30}\mathbf{X}_{3+} = 0$ (F5), the first three terms are all zero. The last term is just the initial $m = 0$ term, which, using the URM3 DCE $\mathbf{X}'^{30}\mathbf{X}'_{30} = +C^2$ (F3), is simply

(11.59) $\mathbf{X}^{40A}\mathbf{X}_{40A} = \begin{pmatrix} 0 & \mathbf{X}'^{30} \end{pmatrix} \begin{pmatrix} 0 \\ \mathbf{X}'_{30} \end{pmatrix} = C^2$, $m = 0$.

Therefore the inner product $\mathbf{X}^{40A}\mathbf{X}_{40A}$ is exactly as per URM3, i.e. the DCE (F3)

(11.60) $\mathbf{X}^{40A}\mathbf{X}_{40A} = +C^2$.

Also, using URM3 orthogonality relations $\mathbf{X}^{3-}\mathbf{X}_{3+} = 0$ (F1), $\mathbf{X}^{3-}\mathbf{X}_{30} = 0$ (F5), and $\mathbf{X}^{30}\mathbf{X}_{3+} = 0$ (F3), the mixed inner products of $\mathbf{X}_{40A}$ and $\mathbf{X}^{40B}$, and $\mathbf{X}_{40B}$ and $\mathbf{X}^{40A}$, are zero, showing that the '0A' vector is orthogonal to the '0B' vector in this case

(11.61a) $\mathbf{X}^{40B}(m)\mathbf{X}_{40A} = 0$

(11.61b) $\mathbf{X}^{40A}\mathbf{X}_{40B}(m) = 0$.

Furthermore, using $\mathbf{X}^{3-}\mathbf{X}_{3+} = 0$ (F1), and $\mathbf{X}^{30}\mathbf{X}_{3+} = 0$ (F5), the inner product $\mathbf{X}^{40A}(m)\mathbf{X}_{4+}$ is also zero, i.e.

(11.62) $\mathbf{X}^{40A}(m)\mathbf{X}_{4+} = 0$.

and, using $\mathbf{X}^{3-}\mathbf{X}_{3+} = 0$ (F1), the inner product $\mathbf{X}^{40B}\mathbf{X}_{4+}$ is seen to be zero

(11.63) $\mathbf{X}^{40B}\mathbf{X}_{4+} = 0$.

The inner product $\mathbf{X}^{4+}\mathbf{X}_{40B}$ is slightly more involved, but it is still zero as follows

(11.64) $\mathbf{X}^{4+}\mathbf{X}_{40B} = \left(2sC \quad -(s^2+m^2)\mathbf{X}^{3-} + 2m\mathbf{X}'^{30} + \mathbf{X}'^{3+}\right)\begin{pmatrix} +C \\ -s\mathbf{X}_{3+} \end{pmatrix}$

Since $\mathbf{X}^{3-}\mathbf{X}_{3+} = 0$ (F1) and $\mathbf{X}^{30}\mathbf{X}_{3+} = 0$ (F5), this evaluates to

(11.65) $\mathbf{X}^{4+}\mathbf{X}_{40B} = 2sC^2 - s\mathbf{X}'^{3+}\mathbf{X}_{3+}$

and because $\mathbf{X}^{3+}\mathbf{X}_{3+} = 2C^2$ (F4) then, once again,

(11.66) $\mathbf{X}^{4+}\mathbf{X}_{40B} = 0$.

The inner product $\mathbf{X}^{4+}\mathbf{X}_{40A}(m)$ is evaluated as follows:

$$\mathbf{X}^{4+}\mathbf{X}_{40A}(m) =$$

$$(11.67) \quad -m\left(2sC \quad -(s^2+m^2)\mathbf{X}^{3-} + 2m\mathbf{X}'^{30} + \mathbf{X}'^{3+}\begin{pmatrix} 0 \\ \mathbf{X}_{3+} \end{pmatrix}\right)$$

$$\left(2sC \quad -(s^2+m^2)\mathbf{X}^{3-} + 2m\mathbf{X}'^{30} + \mathbf{X}'^{3+}\begin{pmatrix} 0 \\ \mathbf{X}_{30} \end{pmatrix}\right)$$

Using URM3 relations $\mathbf{X}^{3-}\mathbf{X}_{3+}=0$ (F1) and $\mathbf{X}^{30}\mathbf{X}_{3+}=0$ (F5) in the first term, and $\mathbf{X}^{3-}\mathbf{X}_{30}=0$ (F5) and $\mathbf{X}^{3+}\mathbf{X}_{30}=0$ (F6) in the second term, this simplifies this to

$$(11.68) \quad \mathbf{X}^{4+}\mathbf{X}_{40A}(m) = -m\mathbf{X}'^{3+}\mathbf{X}_{3+} + 2m\mathbf{X}'^{30}\mathbf{X}_{30},$$

and with URM3 relations $\mathbf{X}^{3+}\mathbf{X}_{3+}=+2C^2$ (F4) and $\mathbf{X}^{30}\mathbf{X}_{30}=+C^2$ (F3), the right of (11.68) simply cancels to zero so that the inner product is zero

$$(11.69) \quad \mathbf{X}^{4+}\mathbf{X}_{40A}(m) = 0.$$

Hence $\mathbf{X}^{4+}$ is also orthogonal to $\mathbf{X}_{40A}(m)$ for arbitrary variations in m, and also s, since $\mathbf{X}_{40A}(m)$ has no s dependence.

Having gained a full URM4 eigenvector solution in terms of two parameters m and s, the next and final section of Part I moves on to discussing how the eigenvectors evolve as these parameters grow.

(12) Temporal Evolution

This section discusses temporal, evolutionary aspects of URMT, and serves primarily as a prelude to the work in Part II, which focuses on the extension of URMT to five and higher dimensions, with each dimension, three and higher, possessing its own evolutionary, temporal parameter.

Introduction

One of the most important aspects of URMT, as first detailed in URM3 [1],#3, is that the eigenvectors parametrically evolve, i.e. the complete eigenvector solution can be parametrically varied as a form of variational method, where the variational parameters are physically associated with time. The evolution of the URM3 eigenvector solution is given in Appendix (B) in terms of a single, temporal parameter m, ($\sim t_3$ in Part II). In earlier sections, from (2) onward, an additional, temporal parameter s is introduced, giving URM4 two temporal, evolutionary parameters m and s ($s \sim t_4$ in Part II). These two parameters are special cases of the more general form of variational parameter, δ_{ij} (8.16) for URM4 and $\eta, \delta, \varepsilon$ (8.4) for URM3. As explained in Section (8), such variational parameters enter into URMT indirectly via the Invariance Principle (8.7), which is essentially a statement that the dynamical equations remain invariant to arbitrary variations in these parameters.

(12-1) URM4 Evolution Conditions

The time evolution of URM4 is studied under quite specific conditions, including URM4 Pythagoras conditions (4.1), which are encountered when lifting URM3 solutions, Section (7). Furthermore, these conditions are sufficient to retain an invariant, zero Potential, Section (9), when subject to variations, Section (11), i.e. the Potential remains constant (and zero) with time. A summary of all these conditions is given below.

The embedding is done under URM4 Pythagoras conditions (4.1)

(4.1a) $\bar{S} = -S$, $\bar{T} = -T$, $\bar{U} = U$

(4.1b) $\bar{P} = P$, $\bar{Q} = Q$, $\bar{R} = -R$.

A zero, URM4 Potential is imposed

(4.16) $QT = (PS + RU) \Rightarrow V = 0$.

The conjugate parameter $\bar{s}$ in **A** (2.25) is related to s by

(7.4) $\bar{s} = -s$ (7.4)

The eigenvector **X** (1.5) has the first element w constrained to zero

(2.3) $w = 0$.

The conjugate vector $\overline{\mathbf{X}}$ (2.8) in **A** (2.25) satisfies

(2.24) $\overline{\mathbf{X}} = \begin{pmatrix} 0 & \mathbf{X}^{3-} \end{pmatrix}$

(A35c) $\mathbf{X}^{3-} = \begin{pmatrix} x & y & -z \end{pmatrix}$,

which means that the URM4-specific dynamical variables must be assigned according to **A** (2.25), as follows:

(7.11a) $\bar{S} = -S = sx$

(7.11b) $\bar{T} = -T = sy$

(7.11c) $\bar{U} = U = sz$.

(12-2) Temporal Parameters m and s

By associating the URMT evolutionary (variational) parameters m and s with time, this forces the three types of eigenvector (I11) to assume the physical units of second-order, first-order and constant derivatives with respect to time; see Appendix (G). These derivatives are commonly associated with acceleration, velocity and position in the standard physical interpretation of URMT, Appendix (J), making the DCE an energy conservation equation (per unit mass) in the process. These points form a strong reason to keep the physical association of the variational parameters m and s with units of time, as opposed to, say, length or other.

The parameter s is a new URM4 temporal parameter, and not in URM3 [1]. When it is zero, the entire URM4 almost reduces to URM3, barring a non-zero contribution to the fourth dimension comprising simply the constant eigenvalue C, as will be seen in Part II. Thus, URM3 can be crudely thought of as the initial state of URM4. On the other hand, parameter m controls URM3 only, and can evolve URM3 independently of UMR4's parameter s. Both parameters work independently and there are no mixed terms containing products of both m and s. A zero value for m represents the initial conditions of URM3, but unlike s in URM4, it does not reduce URM3 to a lower dimension, i.e. the realm of URM2, Appendix (H). Indeed, URM3 holds a special place in the evolutionary behaviour of URM4, and URMT in general, as will be seen in Part II where it is shown that URMT, when under invariant, zero Potential conditions, always evolves to converge on URM3 in the large evolutionary time limit.

For advance information on Part II, each jth dimension in URMn, $j = 3 \ldots n$, has a unique temporal parameter t_j, where $t_3 = m$ and $t_4 = s$ here in Part I, and it is the evolutionary behaviour of the complete, n-dimensional eigenvector space (I17), as these evolutionary times t_j grow, which is examined and shown to converge to the evolutionary behaviour of URM3.

(12-3) Real, Observer (lab) time

Whilst m and s are interpreted as temporal parameters in the standard URMT physical association, they are not necessarily the same as the familiar, observer (or laboratory) time t in The Special Theory of Relativity (STR). This STR time t is normally given in the scaled, length form as ct (little 'c' being the speed of light) in the four-vector position $(ct \quad x \quad y \quad z)$, or $(x \quad y \quad z \quad ct)$ according to convention. Such a time t might be considered as an interval (difference) of time rather than an absolute time, i.e. a difference of two absolute, evolutionary times with a common origin. Throughout this whole book, such a time has not required any consideration since it is evolutionary times (m and s so far), acting as variational parameters, that determine how the eigenvector solution evolves. Nevertheless, it is worth mentioning where such a laboratory time might fit into URMT. It is stressed, however, this is rather speculative and further work is required.

Note that the coordinates x, y, z in the aforementioned STR four-vector $(ct \quad x \quad y \quad z)$ are not the same as those in URMT, which are actually associated with accelerations in URMT. It is, in fact, the scale factors $\alpha_4, \beta_4, \gamma_4$ (3.56) which are associated with position, and represented by the four-vectors $\mathbf{X}^{4+}$ (3.60) and $\mathbf{X}_{4-}$, where

$\mathbf{X}_{4-} = \left(\mathbf{X}^{4+} \mathbf{T}_4 \right)^T$ (E7).

Equating the URM4 position vector $\mathbf{X}^{4+}$ and $\mathbf{X}_{4-}$ with the time coordinate in STR form, i.e.

(12.30) $\mathbf{X}^{4+} = \left(2sC \quad \alpha_4 \quad \beta_4 \quad \gamma_4 \right) \sim \left(- \quad - \quad - \quad +ct \right),$

(12.31) $\mathbf{X}_{4-} = \begin{pmatrix} 2sC \\ \alpha_4 \\ \beta_4 \\ -\gamma_4 \end{pmatrix} \sim \begin{pmatrix} - \\ - \\ - \\ -ct \end{pmatrix}$

implies that the scale factor γ_4 is actually associated with that of time

(12.32) $\gamma_4 = ct$.

In fact, there is some room for an additional, constant multiplying factor in $\mathbf{X}^{4+}$ (and $\mathbf{X}_{4-}$) in this expression, i.e. for some constant multiplier k, $\mathbf{X}^{4+} \rightarrow k\mathbf{X}^{4+}$, $\mathbf{X}_{4-} \rightarrow k\mathbf{X}_{4-}$ and, therefore, $\gamma_4 = kct$. This is so because $\mathbf{X}^{4+}$ (and $\mathbf{X}_{4-}$) are Pythagorean quadruples, and the inner product $k\mathbf{X}^{4+} k\mathbf{X}_{4-}$ is zero so that k^2 cancels, i.e.

(12.33) $k\mathbf{X}^{4+} k\mathbf{X}_{4-} = k^2 \left[(2sC)^2 + \alpha_4^2 + \beta_4^2 - \gamma_4^2 \right] = 0$.

This zero inner product physically represents a zero-norm, or zero interval in STR, i.e. it represents a null STR interval, where STR proper time τ is zero. In other words, if $\mathbf{X}^{4+}$ and $\mathbf{X}_{4-}$ are associated with a four-vector position, they represent a photon or graviton trajectory, and not that of an object with a finite mass, constrained to a speed less than that of the speed of light. This is of note since mass does not explicitly appear in URMT (yet), or at least cancels throughout an equation, just like k in (12.33) above.

Summarising the above, there are now three distinct, temporal coordinates, m, s and t in URM4. However, there does appear to be a notional difference between the two evolutionary parameters m and s, and the observer time parameter t. In STR, the velocity c (the speed of light) scales the observer time t to give a length coordinate ct. However, in URMT, m and s scale an acceleration coordinate, x, y and z, to give a velocity. For example, in URM3, parameter m scales the variational matrix Δ_3^P (8.133) to modify the dynamical variables in matrix $\mathbf{A}_3$, as in

(11.1) $\mathbf{A}_{30}(m) = \mathbf{A}_3' - m\Delta_3^P$, see also $\mathbf{A}_m = \mathbf{A}' - m\Delta^P$ (8.144).

As regards physical units, the variational matrix Δ_3^P (8.133) comprises coordinates x, y, z, i.e.

(J8) $units(\Delta_3^P) = units(x, y, z) = LT^{-2}$, acceleration

(J9) $units(m) = T$, time,

and so the variational product $m\Delta_3^P$ has units of velocity, i.e.

(12.35) $units(m\Delta_3^P) = LT^{-1}$, velocity,

which consistently matches the dynamical variables forming the elements of $\mathbf{A}_3$ or $\mathbf{A}_{30}(m)$ in (11.1)

(J6) $units(\mathbf{A}_3) = units(P, Q, R) = LT^{-1}$, velocity.

As regards parameter s, this enters URMT via the scaling of vectors $\mathbf{X}^{3-}$ and $\mathbf{X}_{3+}$ in the matrix $\mathbf{A}$ ($\sim \mathbf{A}_{40}$)

(7.5) $\mathbf{A}_{40} = \begin{pmatrix} 0 & -s\mathbf{X}^{3-} \\ s\mathbf{X}_{3+} & \mathbf{A}_3(m) \end{pmatrix}$.

(J3) $units(\mathbf{X}, \mathbf{X}_{3+}, \mathbf{X}^{3-}) = units(x, y, z) = LT^{-2}$, acceleration

(J10a) $units(s) = T$, time

and so

(J31) $units(s\mathbf{X}_3^-) = LT^{-1}$, velocity

which matches $units(\mathbf{A}_3) = LT^{-1}$ (J6) and $units(\mathbf{A})$.

There may well be a case for making all three parameters equal, or making m and s equal, leaving t distinct. With all three equal, this would reduce URM4 back to three spatial dimensions and one time dimension, as per STR. However, whilst we appear to live in a 3D spatial world with a single time parameter, a more symmetric viewpoint may well have three spatial and three time parameters. A full discussion on the physical nature of all temporal time parameters is, unfortunately, beyond the scope of this book, and will be dealt with in a later publication.

(12-4) URM4 Eigenvector Evolutionary Forms

To get an overview of the temporal evolution of the URM4 eigenvector solution, the full solution is reproduced below from Section (11), with the URM3 eigenvectors, also expanded in their full evolutionary forms, as given in Appendix (B). Note that Part II examines the temporal evolution of eigenvectors in far more detail, and this section primarily serves as an introduction to URM4's evolutionary behaviour, which is basically identical to URM3's own evolution over long evolutionary timescales.

The URM4 eigenvectors are written to show their full functional dependence on parameters m and s as follows, noting they all share the same physical units as their URM3 counterparts:

(2.4) $\mathbf{X}_{4+} = \begin{pmatrix} 0 \\ \mathbf{X}_{3+} \end{pmatrix}$, static, no m, s dependence, see also (B2a)

(11.37) $\mathbf{X}_{4-}(m,s) = \begin{pmatrix} 2sC \\ -(s^2 + m^2)\mathbf{X}_{3+} + 2m\mathbf{X}'_{30} + \mathbf{X}'_{3-} \end{pmatrix}$, see also (B2c)

(11.54a) $\mathbf{X}_{40A}(m) = \begin{pmatrix} 0 \\ -m\mathbf{X}_{3+} + \mathbf{X}'_{30} \end{pmatrix}$, see also (B2b)

(11.49) $\mathbf{X}_{40B}(s) = \begin{pmatrix} +C \\ -s\mathbf{X}_{3+} \end{pmatrix}$, no URM3 equivalent.

From a quick inspection of the standard forms of the above URM4 eigenvector solution, the following points are made:

$\mathbf{X}_{4+}$, like $\mathbf{X}_{3+}$, has no evolution and is static, remaining at its initial value.

$\mathbf{X}_{4-}$ is a function of both m and s.

$\mathbf{X}_{40A}$, like $\mathbf{X}_{30}$, is a function of m only, with no URM4 evolution.

$\mathbf{X}_{40B}$ is a function of s only, i.e. a URM4 vector with no URM3 equivalent.

(12-5) Large Evolution Limits

Of most interest, as in [1],#3, is the form of the eigenvector solution over long evolutionary times. Such times are simply defined here as

(12.50) $|m| \gg 0$ and or $|s| \gg 0$.

To keep things simple it will be assumed that m and s evolve in the forward, positive direction, i.e. they always increase and are much greater than zero

(12.51) $m \gg 0$ and or $s \gg 0$, with time increasing.

It is also assumed that all initial values for x, y, z, P, Q, R, α, β, γ and the eigenvalue C are all relatively small compared with m and s, and all terms dependent on m and s are much larger than any of these initial values. Of course, sufficiently large values for m and s can always be found to satisfy this assumption. This aspect is actually made more rigorous in Part II.

For such large evolutionary times, the URM4 eigenvectors approximate as follows in (12.52), below, where the highest order terms in either m or s, or both, have been retained, which could be linear or quadratic, or even static in the case of $\mathbf{X}_{4+}$ (12.52a). Naturally, only quadratic terms will dominate and it is clear, in particular from vector $\mathbf{X}_{4-}$ (12.52b), that this will become the largest magnitude vector, dwarfing all others in the large evolutionary limit. Nevertheless, proceeding in the aforementioned manner gives the dominant terms for each vector as

(12.52)

(12.52a) $\mathbf{X}_{4+} = \begin{pmatrix} 0 \\ \mathbf{X}_{3+} \end{pmatrix}$, static, no m, s dependence

$$(12.52\text{b}) \quad \mathbf{X}_{4-}(m,s) \approx -s^2 \begin{pmatrix} 0 \\ \mathbf{X}_{3+} \end{pmatrix} - m^2 \begin{pmatrix} 0 \\ \mathbf{X}_{3+} \end{pmatrix}, \quad m,s >> 0$$

$$(12.52\text{c}) \quad \mathbf{X}_{40A}(m) \approx -m \begin{pmatrix} 0 \\ \mathbf{X}_{3+} \end{pmatrix}, \quad m >> 0, \text{ no } s \text{ dependence}$$

$$(12.52\text{d}) \quad \mathbf{X}_{40B}(s) \approx -s \begin{pmatrix} 0 \\ \mathbf{X}_{3+} \end{pmatrix}, \quad s >> 0, \text{ no } m \text{ dependence.}$$

Every vector is seen to align parallel or anti-parallel to $\mathbf{X}_{4+}$ (12.52a) which itself is just the static URM3 vector $\mathbf{X}_{3+}$ with an additional, zero first element. Hence, for either large m or s, URM4 predominantly occupies the three-dimensional space of URM3, i.e. it looks just like URM3, as anticipated. This is no trivial result; effectively URM4 starts off as four-dimensional and evolves to become more and more three-dimensional, with the fourth dimension (first element of each vector) shrinking, relatively speaking, to zero. For $\mathbf{X}_{4+}$ and $\mathbf{X}_{40A}$ this is trivial in that the first element is always zero. However, vectors $\mathbf{X}_{4-}$ and $\mathbf{X}_{40B}$ have a non-zero, first element, albeit linear in their evolutionary parameters m or s, as opposed to quadratic in their other three elements, and are true 4D vectors. Nevertheless, the quadratic, evolutionary growth rapidly outstrips the linear growth, and the 4D contributions, once again, become insignificant.

The vector $\mathbf{X}_-$ is physically associated with a position, and $\mathbf{X}_{40B}$ with a velocity, and hence, by the above arguments, they have non-zero positions and velocities in their first element. These first elements decrease in magnitude relative to the other three elements, and become insignificant for large evolutionary times.

Since $\mathbf{X}_{4+}$ is physically associated with an acceleration, but has a zero first element regardless of evolutionary period, this means there is no acceleration component in the fourth dimension, and all acceleration is restricted to the 3D subspace occupied by the last three non-zero elements. Likewise, $\mathbf{X}_{0A}$ is physically associated with a velocity, which also has a zero first element, and thus has no velocity component in the fourth dimension either.

It is of note that the observer time t, Section (12-3), has not yet been linked in any way with the evolutionary times m or s, and this time t can change independently. Nevertheless, because it is assumed that all initial values, x, y, z, P, Q, R, α, β, γ and C are all relatively small compared with m or s, this means that the time t, as embedded within γ by the relation $\gamma = ct$, Section (12-3), is also relatively small compared to m or s. This then puts a condition on variable t (fixed c) linking it to m or s. It must be considered, however, that time is really only an interval, i.e. t is really 'Δt', as is often written in STR texts. This is also ultimately true of m or s, but intervals Δt take place in the context of an ever evolving (increasing) m or s. One might consider that the interval Δt is much smaller than m or s, i.e.

(12.53) $\Delta t \ll m, s$

Of course, in the very early stages of evolution, i.e. extremely small m or s, say $m, s < 10$, this is not so and all the above approximations become increasingly invalid as m and s approach zero. Physically, these small values are considered Planck times, i.e. a single unit interval of time is approximately $10^{-42} s$. Such Planck scales within URMT are discussed in [1],#3, to which the reader is referred.

(12-6) Eigenvector Calculus

The evolution of the eigenvector space, i.e. how it changes with time, is, of course, associated with the calculus of the eigenvectors, namely their derivatives with respect to time. This was studied in [1],#3, and lent considerable weight to the association of the eigenvectors with acceleration, velocity and position vectors – not only are they dimensionally correct, but they are also related to each other via straightforward calculus relations, e.g. the derivative of the position vector is the velocity vector (to within a scale factor of two), and the derivative of the velocity vector is the negative of the acceleration vector.

Appendix (G) reworks these calculus relations for the general, n-dimensional URMn case, showing that identical relations are preserved from URM3. The only key difference being that, in the case of URM4, there are now two temporal parameters m and s, as opposed to just m in URM3, and derivatives are now partial with respect to m or s. Therefore, from these results, it is concluded for URM4 that

- the calculus of URM4's $\mathbf{X}_{4+}$ is identical to that of URM3's $\mathbf{X}_{3+}$, i.e. a static acceleration with no m or s parametric dependence.

- the calculus of $\mathbf{X}_{4-}(m,s)$ is identical to that of URM3's $\mathbf{X}_{3-}$, i.e. its derivative is twice that of the velocity vectors $\mathbf{X}_{40A}(m)$ and $\mathbf{X}_{40B}(s)$.

- the calculus of URM4's $\mathbf{X}_{40A}(m)$ is identical to URM3's $\mathbf{X}_{30}(m)$, i.e. its derivative is the negative of the static acceleration vector $\mathbf{X}_{3+}$.

- the calculus of URM4's $\mathbf{X}_{40B}(s)$ is physically equivalent to that of URM4's $\mathbf{X}_{40A}(m)$, i.e. its derivative with respect to s is also the negative of the static acceleration vector $\mathbf{X}_{3+}$.

This similarity between URM4 and URM3 calculus is actually quite some result because it means that quantities such as curvature of the eigenvector trajectory with respect to the path parameter, i.e. the evolutionary time, will have the same form. In particular, it was shown in [1],#3 that the position vector $\mathbf{X}_{3-}$ has an inverse square law curvature with respect to evolutionary time m, and because $\mathbf{X}_{4-}(m,s)$ has the same functional behaviour as $\mathbf{X}_{3-}$, it too has the same inverse square law behaviour, but now with respect to both m and s.

(12-7) Geometric Evolution

Geometric evolution within URMT covers angular evolution of the axes set formed from the four standard eigenvectors, and the curvature of the path traced out by the vectors as they evolve. The four eigenvectors in URM4, as per URM3's three vectors, are linearly independent but highly oblique and of non-unit length, i.e. they are not orthogonal, unit vectors. Nevertheless, by virtue of their properties, they form a very interesting basis set, which parametrically evolves to show inverse square law trajectories and flattening behaviour as evolution proceeds, whilst still retaining, at every stage, the same set of conservation equations and scalar invariants.

Most of the these issues of curvature, flattening and invariants have already been discussed quite extensively in [1], particularly [1],#3, and there is very little change, if any, in URM4 and beyond. As seen from the URM4 eigenvector solution (12.52), the functional dependence of the eigenvectors with respect to the temporal parameters is as per URM3, i.e. quadratic in m and s for $\mathbf{X}_{4-}$ (cf. $\mathbf{X}_{3-}$), linear in m and s for $\mathbf{X}_{40A}(m)$ and $\mathbf{X}_{40B}(s)$ respectively (cf. $\mathbf{X}_{30}$), and static in $\mathbf{X}_{4+}$ (cf. $\mathbf{X}_{3+}$). Unsurprisingly then, the geometric properties of URM4, such as curvature, are the same as for URM3. With this in mind, the work in this section will largely be confined to showing URM4 properties of 'flatness' and curvature are virtually identical to URM3, but now with four, 4D vectors, i.e. URM4 inherits all the same, important geometric features in URM3, as discussed in [1],#3, to which the reader is referred for a more detailed analysis.

(12-8) Flattening

Flattening in URM3 is the term used to describe how the two eigenvectors $\mathbf{X}_{30}$ and $\mathbf{X}_{3-}$ align anti-parallel to $\mathbf{X}_{3+}$ as evolution progresses, i.e. the eigenvector basis is said to flatten (I5). The same behaviour is now shown to occur in URM4, using much the same definitions and language as in URM3, [1],#3.

Denoting the angle between $\mathbf{X}_{4+}$ and $\mathbf{X}_{4-}$ by θ_{+-}, the angle between $\mathbf{X}_{4+}$ and $\mathbf{X}_{40A}/\mathbf{X}_{40B}$ by $\theta_{+0A}/\theta_{+0B}$, the angle between $\mathbf{X}_{40A}/\mathbf{X}_{40B}$ and $\mathbf{X}_{4-}$ by $\theta_{0A-}/\theta_{0B-}$, then these angles are obtained by the following standard inner product relations, using the definition of vector magnitude $|\mathbf{X}| = \sqrt{\mathbf{X} \cdot \mathbf{X}}$ (I7),

(12.80)

(12.80a) $\cos\theta_{+-} = \mathbf{X}_{4+} \cdot \mathbf{X}_{4-} / |\mathbf{X}_{4+}| |\mathbf{X}_{4-}|$

(12.80b) $\cos\theta_{+0A} = \mathbf{X}_{4+} \cdot \mathbf{X}_{40A} / |\mathbf{X}_{4+}| |\mathbf{X}_{40A}|$

(12.80c) $\cos\theta_{+0B} = \mathbf{X}_{4+} \cdot \mathbf{X}_{40B} / \left| \mathbf{X}_{4+} \right\| \mathbf{X}_{40B} \right|$

(12.80d) $\cos\theta_{0A-} = \mathbf{X}_{40A} \cdot \mathbf{X}_{4-} / \left| \mathbf{X}_{40A} \right\| \mathbf{X}_{4-} \right|.$

(12.80e) $\cos\theta_{0B-} = \mathbf{X}_{40B} \cdot \mathbf{X}_{4-} / \left| \mathbf{X}_{40B} \right\| \mathbf{X}_{4-} \right|.$

Given the magnitudes are taken as positive, the sign of the angles is determined by the inner products on the right of (12.80). As a consequence, the behaviour of these inner products is examined to see how the angle between the axes evolves, generally for large evolutionary times.

Using the URM4 eigenvectors $\mathbf{X}_{4+}$ (12.52a) and $\mathbf{X}_{4-}$ (12.52b), the inner product $\mathbf{X}_{4+} \cdot \mathbf{X}_{4-}$ splits nicely into a URM4, s dependent component and a URM3, m dependent component,

(12.81) $\mathbf{X}_{4+} \cdot \mathbf{X}_{4-} \approx -(s^2 + m^2)\mathbf{X}_{3+} \cdot \mathbf{X}_{3+}$, $m, s \gg 0$.

As expected, if $s = 0$, the inner product reduces to its identical URM3, large m form $-m^2 \mathbf{X}_{3+} \cdot \mathbf{X}_{3+}$. The product is also symmetric in m and s, so its large m behaviour is identical to that for large s. Since $\mathbf{X}_{3+}$ is obviously parallel to itself, the $-(m^2 + s^2)$ factor, which is always negative, means the angle between $\mathbf{X}_{4+}$ and $\mathbf{X}_{4-}$ evolves anti-parallel as m or s, or both, grow large, flattening out to 180 deg in the large m, s limit.

(12.82) $\lim_{m \to \infty} \theta_{+-} = 180\,\text{deg}$, $\lim_{s \to \infty} \theta_{+-} = 180\,\text{deg}$.

For the angle θ_{+0A}, using the URM4 eigenvectors $\mathbf{X}_{4+}$ (12.52a) and $\mathbf{X}_{40A}$ (12.52c), then

(12.83) $\mathbf{X}_{4+} \cdot \mathbf{X}_{40A} \approx -m^2 \mathbf{X}_{3+} \cdot \mathbf{X}_{3+}.$

As expected, this inner product is identical to its URM3 form for large m, as per [1],#3, hence vectors $\mathbf{X}_{4+}$ and $\mathbf{X}_{40A}$ evolve anti-parallel and θ_{+0A} converges to 180 deg.

(12.84) $\lim_{m \to \infty} \theta_{+0A} = 180\,\text{deg}.$

For the angle θ_{+0B}, by the symmetry of the solution for $\mathbf{X}_{40A}$ (12.52c) and $\mathbf{X}_{40B}$ (12.52c) with regard to interchange of m and s, the behaviour of θ_{+0B} for large s is the same as the behaviour of θ_{+0A} for large m. Consequently, the angle θ_{+0B} between $\mathbf{X}_{4+}$ and $\mathbf{X}_{40B}$ converges to 180 deg in the large s limit.

(12.85) $\lim\limits_{s \to \infty} \theta_{+0B} = 180 \deg$.

Thus, both $\mathbf{X}_{40A}$ and $\mathbf{X}_{40B}$ align anti-parallel to $\mathbf{X}_{4+}$ in the large m, s limit, exactly as per the URM3 vector $\mathbf{X}_{30}$, which aligns anti-parallel to $\mathbf{X}_{3+}$.

For the angle θ_{0A-}, using large m and large s approximations for the URM4 eigenvectors $\mathbf{X}_{4-}$ (12.52b) and $\mathbf{X}_{40A}$ (12.52c), then the inner product $\mathbf{X}_{4-} \cdot \mathbf{X}_{40A}$ is approximated as

(12.86) $\mathbf{X}_{4-} \cdot \mathbf{X}_{40A} \approx m^3 \mathbf{X}_{3+} \cdot \mathbf{X}_{3+} + ms^2 \mathbf{X}_{3+} \cdot \mathbf{X}_{3+}$, $m \gg 0$, $s \gg 0$.

Regardless of the relative size of m and s, with $\mathbf{X}_{3+}$ obviously parallel to itself, then $\mathbf{X}_{4-}$ and $\mathbf{X}_{40A}$ align parallel to each other for large m or s, or both, i.e.

(12.87) $\lim\limits_{m \to \infty} \theta_{0A-} = 0 \deg$, $\lim\limits_{s \to \infty} \theta_{0A-} = 0 \deg$

For the angle θ_{0B-}, by the symmetry of the solution for $\mathbf{X}_{4-}$ (12.52b) and $\mathbf{X}_{40B}$ (12.52d) with regard to interchange of m and s, the behaviour of θ_{0B-} for large m or s, or both, is the same as the behaviour of θ_{0A-}. Consequently, $\mathbf{X}_{4-}$ and $\mathbf{X}_{40B}$ align parallel to each other for large m or s, or both, i.e.

(12.88) $\lim\limits_{m \to \infty} \theta_{0B-} = 0 \deg$, $\lim\limits_{s \to \infty} \theta_{0B-} = 0 \deg$

To summarise, in the large m or s limit:

$\mathbf{X}_{4+}$ and $\mathbf{X}_{4-}$ align anti-parallel

$\mathbf{X}_{4+}$ and $\mathbf{X}_{40A}$ align anti-parallel (no s dependence)

$\mathbf{X}_{4+}$ and $\mathbf{X}_{40B}$ align anti-parallel (no m dependence)

$\mathbf{X}_{4-}$ and $\mathbf{X}_{40A}$ align parallel

$\mathbf{X}_{4-}$ and $\mathbf{X}_{40B}$ align parallel

This is exactly the same alignment behaviour seen in URM3 between the equivalent vectors, with all angles limiting to 0 or 180 deg, and hence the flattening behaviour seen in URM3 is also seen in the 4D world of URM4.

(12-9) Curvature

The curvature properties of the evolving, URM3 eigenvector trajectories was assessed in [1],#3 by calculating the rate of change of the angle between a particular eigenvector and the static, reference eigenvector $\mathbf{X}_{3+}$, with respect to its path parameter m. Of course, in URM4, there are now two path parameters, m and s, but there remains the single, static reference vector $\mathbf{X}_{4+}$, which is basically the same as $\mathbf{X}_{3+}$ by its definition (2.4).

Given the various dependencies of the eigenvectors on m and s, there are potentially four possible, non-zero curvatures of note in URM4, denoted by generic symbol κ. These all measure the rate of change of the angle between $\mathbf{X}_{4-}$, $\mathbf{X}_{40A}$, $\mathbf{X}_{04B}$ and reference vector $\mathbf{X}_{4+}$.

(12.90a) $\kappa_{s-} = \partial_s \theta_{+-}$

(12.90b) $\kappa_{m-} = \partial_m \theta_{+-}$

(12.90c) $\kappa_{0A} = \partial_m \theta_{0A+}$

(12.90d) $\kappa_{0B} = \partial_s \theta_{0B+}$.

The full, accurate computations of curvature become fairly complex, even in URM3, albeit the usual approximations are made based upon large evolutionary times. Fortunately, however, this is all unnecessary because the calculus behaviour of the URM4 eigenvectors, Appendix (G), is seen to be identical to that of URM3, as is the geometric flattening behaviour as described in the previous section. Hence the curvature follows the same inverse square law (with respect to evolutionary time) as seen in URM3. Indeed, this is also confirmed by numerical results. To which end, the reader is referred to [1],#3 for full information.

The book now proceeds to Part II, which generalises the theory and results to URM5 and, indeed, any arbitrary number of dimensions n.

Part II

Part II

(13) URM5

(13-1) The General URM5 Formulation

Following on from the definition of the general, 4x4 unity root matrix **A** (1.3a), then the URM5, 5x5 unity root matrix **A** is defined similarly, as follows:

$$(13.1)\ \mathbf{A} = \begin{pmatrix} 0 & M & H & N & J \\ \overline{M} & 0 & S & T & U \\ \overline{H} & \overline{S} & 0 & R & \overline{Q} \\ \overline{N} & \overline{T} & \overline{R} & 0 & P \\ \overline{J} & \overline{U} & Q & \overline{P} & 0 \end{pmatrix},\ \text{URM5,}$$

comprising ten dynamical variables

$(13.2)\ M, H, N, J \in \mathbb{Z}$

$\quad S, T, U \in \mathbb{Z}$, URM4 (1.3b)

$\quad P, Q, R \in \mathbb{Z},\ (P, Q, R) \neq (0,0,0)$, URM3 (A1b)

and their conjugates

$(13.3)\ \overline{M}, \overline{H}, \overline{N}, \overline{J} \in \mathbb{Z}$

$\quad \overline{S}, \overline{T}, \overline{U} \in \mathbb{Z}$, URM4 (1.3b)

$\quad \overline{P}, \overline{Q}, \overline{R} \in \mathbb{Z},\ (\overline{P}, \overline{Q}, \overline{R}) \neq (0,0,0)$, URM3 (A1c)

Notes

The **A** matrix (13.1) embeds the existing URM4 matrix (1.3a) in its lower right, 4x4 sub-matrix, and adds four new dynamical variables M, H, N, J and their conjugates $\overline{M}, \overline{H}, \overline{N}, \overline{J}$.

Just as in the URM4 case for dynamical variables S, T, U, these new dynamical variables M, H, N, J are not defined as having the unity root property, which is strictly still reserved only for the original, URM3 dynamical variables P, Q, R, see (A44); the same remark applies to their conjugate forms. This loss of unity root property is discussed in Section (1-6) for the URM4 case, and is applicable to all higher dimensional variants of **A** above three-dimensions.

The usage of four non-consecutive capitals M, H, N, J in the top row and left column of **A** is unfortunate, but primarily due to the inability to find four such consecutive capitals that are not already reserved in URMT. The peculiar alphabetic ordering, i.e. M, H, N, J, is also legacy, and due to some other unpublished simplifications to the matrix.

The lead diagonal of **A** comprises all zeros, as for URM4 (1.3a), to retain the zero trace, $Tr(\mathbf{A}) = Tr(\mathbf{A}_3) = 0$ (1.2).

A single URM5 eigenvector **X** is defined comprising five coordinates v, w, x, y, z

(13.4)

(13.4a) $\mathbf{X} = \begin{pmatrix} v \\ w \\ x \\ y \\ z \end{pmatrix}$

(13.4b) $v, w, x, y, z \in \mathbb{Z}$, $(v, w, x, y, z) \neq (0,0,0,0,0)$

(13.4c) $\gcd(v, w, x, y, z) = 1$, see URM4 (1.6)

As always, **X** is an eigenvector of matrix **A**, eigenvalue C, i.e.

(1.7) $\mathbf{AX} = C\mathbf{X}$, $C \in \mathbb{Z}$, $C \geq 1$.

(13-2) URM5 Pythagoras Conditions

As for all forms of URMT, i.e. URMn, $n \geq 3$, the general case requires simplifying conditions to produce analytic solutions that are considered to be of physical relevance. There is a common set of conditions, termed 'Pythagoras conditions' (I13), which are present in every incarnation of URM n, and where the elements of every eigenvector $\mathbf{X}$, for a non-zero eigenvalue ($\pm C$), obey the Pythagoras equation, i.e. for URM5

(13.21) $\quad 0 = v^2 + w^2 + x^2 + y^2 - z^2$.

The URM5 Pythagoras conditions on the conjugate dynamical variables (13.3) are

(13.22)

(13.22a) $\overline{M} = -M$, $\overline{H} = -H$, $\overline{N} = -N$, $\overline{J} = J$

(13.22b) $\overline{S} = -S$, $\overline{T} = -T$, $\overline{U} = U$

(13.22c) $\overline{P} = P$, $\overline{Q} = Q$, $\overline{R} = -R$.

The Pythagoras conditions for URM4 are detailed earlier in Section (4), and those for URM3 are given in Appendix (A), equation (A18) onward. In general, Pythagoras conditions for URM n are a subset of URM $(n+1)$.

Because of the complexity and goals of URM5, it is currently only studied under Pythagoras conditions, noting that there are also a few additional conditions imposed to obtain some specific, highly desirable physical properties, e.g. an invariant, zero Potential, as justified in (9.0); see Section (12-1) for a summary.

All work hereafter will assume to be formulated under Pythagoras conditions.

When under these conditions, matrix $\mathbf{A}$ is relabelled $\mathbf{A}_{50}$, where the '5' in the subscript denotes URM5, and the '0' represents the standard form of the unity root matrix under Pythagoras conditions. From here onward, all matrices and eigenvectors are subscripted with an 'n', according to which n-dimensional incarnation of URM n they represent.

Using conditions (13.22), matrix **A** (13.1) becomes $\mathbf{A}_{50}$:

$$(13.23) \quad \mathbf{A}_{50} = \begin{pmatrix} 0 & M & H & N & J \\ -M & 0 & S & T & U \\ -H & -S & 0 & R & Q \\ -N & -T & -R & 0 & P \\ J & U & Q & P & 0 \end{pmatrix}.$$

A kinetic term K and Potential term V are defined as follows, whereby these forms are intentionally chosen to simplify the characteristic equation for $\mathbf{A}_{50}$

$$(13.24) \quad K = J^2 + P^2 + Q^2 + U^2 - (H^2 + M^2 + N^2 + R^2 + S^2 + T^2)$$

$$V = [QT - (PS + RU)]^2 + [NQ - (JR + HP)]^2 +$$
$$(13.25) \quad [HU - (JS + MQ)]^2 + [NU - (JT + MP)]^2 - \quad ,$$
$$[HT - (MR + NS)]^2$$

see point (13.27) below.

Using these two terms, K and V, the characteristic equation for matrix $\mathbf{A}_{50}$, eigenvalue λ, is

$$(13.26) \quad 0 = \lambda(-\lambda^4 + K\lambda^2 + V).$$

(13.27) The URM5 Potential V (13.25) is actually a quartic polynomial in the dynamical variables, and not the more usual quadratic form imposed in URM4 and URM3, enforced by division of C^2 and C respectively. It could be made quadratic by a similar division of the eigenvalue, but is left as is for now and, indeed, in this entire book. A chief reason being that the quartic simplifies to a quadratic in a quadratic eigenvalue expression, i.e. (13.2) above. Unsurprisingly, the form of the Potential is specific to each incarnation of URMn. Under URM3 Pythagoras conditions it (V_3) is always zero (A24). In URM4 it is given by just the first term in (13.25), i.e. $V = [QT - (PS + RU)]^2$ (4.3c), and is not generally zero under URM4 Pythagoras conditions without further simplifications, see (4.15).

(13-3) Invariant Zero Potential Conditions

An invariant, zero Potential solution is desirable in URMT on physical grounds, outlined at the beginning of Section (9), paragraph (9.0). Such a solution is obtained for URM5 and, indeed, URMn using the method of 'lifting', Section (7), when under Pythagoras conditions, and progressed using variational methods detailed in both Sections (9) and (11). However, it is the latter results in Section (11) that are used herein, themselves a culmination of several preceding sections, i.e. embedding work in Section (2), Pythagoras Conditions in Section (4), lifting methods in Section (7), and 'global' variational methods in Section (8-7).

The end result of applying these conditions is to obtain two symmetric, non-zero eigenvalues, with all others zero, i.e. $\lambda = \pm C, 0, 0, 0$ for URM5, (13.39) further below. This combination then makes the URMT eigenvector solution a quadratic function of its variables, and completely compatible with URM3.

To extend the URM4 work in Section (11) to a URM5 invariant, zero Potential solution requires the URM5 Pythagoras conditions, as given in the previous section, and some additional conditions as follows, with some explanation given shortly afterward. These conditions are equivalent to those detailed in Section (12-1).

The first two coordinates of $\mathbf{X}_{5+}$ are zeroed, i.e.

(13.31)

(13.31a) $v = 0$, $w = 0$,

and the URM4 and URM5 dynamical variables, S, T, U and M, H, N, J respectively, are assigned as scalar multiples of the eigenvector $\mathbf{X}_{5+}$ (13.4a), now with two zero coordinates v, w (above), where the scalars are the evolutionary parameters t_4 and t_5,

(13.31b) $M = 0$, $H = -t_5 x$, $N = -t_5 y$, $J = +t_5 z$

(13.31c) $S = -t_4 x$, $T = -t_4 y$, $U = +t_4 z$

(13.31d) $t_4, t_4 \in \mathbb{Z}$.

With conditions (13.31), the matrix $\mathbf{A}_{50}$ and eigenvector $\mathbf{X}$ (now relabelled $\mathbf{X}_{5+}$) become

$$(13.32)\ \mathbf{A}_{50} = \begin{pmatrix} 0 & 0 & -t_5 x & -t_5 y & +t_5 z \\ 0 & 0 & -t_4 x & -t_4 y & +t_4 sz \\ +t_5 x & +t_4 x & 0 & R & Q \\ +t_5 y & +t_4 y & -R & 0 & P \\ +t_5 z & +t_4 z & Q & P & 0 \end{pmatrix},\ \mathbf{X}_{5+} = \begin{pmatrix} 0 \\ 0 \\ x \\ y \\ z \end{pmatrix}$$

with eigenvector equation

$$(13.33)\ \mathbf{A}_{50}\mathbf{X}_{5+} = C\mathbf{X}_{5+}.$$

The eigenvector $\mathbf{X}_{5+}$ is simply the embedded (lifted) version of $\mathbf{X}_{3+}$ (equivalently $\mathbf{X}_{4+}$), and evidently only occupies the subspace of URM3 vector $\mathbf{X}_{3+}$. Noting that non-trivial (I10) 5D eigenvectors, e.g. $\mathbf{X}_{5-}$, will emerge as the work progresses; just as is the case for URM4 lifting in Sections (7) and (11).

The matrix $\mathbf{A}_{50}$ in (13.32) is basically an extension of the URM4 matrix $\mathbf{A}_{40}$ (7.10) where parameter s is replaced by t_4, and t_5 is the new, URM5 evolutionary parameter associated with the fifth dimension. The general, n-dimensional version $\mathbf{A}_{n0}$ is given in Appendix (C).

An explanation of conditions (13.31) follows shortly after a quick summary of their effect on the energy terms and eigenvalues.

Every bracketed term in the Potential V (13.25) is now zero and so too, therefore, the overall sum

$$(13.34)\ V = 0.$$

Substituting for the dynamical variables M, H, N, J, S, T, U from (13.31) into the kinetic term K (13.24), and using the Pythagoras equation (13.21), gives

$$(13.35)\ K = P^2 + Q^2 - R^2.$$

By associating K with the positive constant C^2, i.e.

$$(13.36)\ K = C^2,$$

then, for a zero Potential (13.34), the kinetic expression (13.35) becomes the familiar URM3 Dynamical Conservation Equation (DCE), as per URM3 Pythagoras conditions

(13.37) $C^2 = P^2 + Q^2 - R^2$, the DCE.

With a zero Potential, and a kinetic term (13.36), the characteristic equation (13.26) becomes

(13.38) $0 = \lambda^3 (C^2 - \lambda^2)$.

This characteristic equation factors with the following five eigenvalues as roots, three of which are zero

(13.39) $\lambda = \pm C, 0, 0, 0$.

As expected, when lifting a solution, each dimensional extension of URMT, i.e. $\mathrm{URM}\,n$ to $\mathrm{URM}(n+1)$, adds another zero eigenvalue, starting with one zero eigenvalue for URM3, i.e. $\lambda = \pm C, 0$, two for URM4, $\lambda = \pm C, 0, 0$, and three for URM5, $\lambda = \pm C, 0, 0, 0$ etc, with the only non-zero eigenvalues being $\lambda = \pm C$. The zero eigenvalue having a multiplicity [5] of $n-2$, albeit the matrix is not 'deficient' [4], or 'defective'.

An explanation on the choice of conditions (13.31), and the rather abstract form of $\mathbf{A}_{50}$ (13.32), is now given.

By writing $\mathbf{A}_{50}$ (13.32) in the following block matrix form, in terms of URM3 vectors $\mathbf{X}_{3+}$ (A33a), $\mathbf{X}^{3-}$ (A35c) and unity root matrix $\mathbf{A}_{30}(t_3)$ (A19), where $t_3 \sim m$,

(13.40) $\mathbf{A}_{50} = \begin{pmatrix} 0 & 0 & -t_5 \mathbf{X}^{3-} \\ 0 & 0 & -t_4 \mathbf{X}^{3-} \\ t_5 \mathbf{X}_{3+} & t_4 \mathbf{X}_{3+} & \mathbf{A}_{30}(t_3) \end{pmatrix}$, see (7.5) for example

(A3) $\mathbf{X}_{3+} = \begin{pmatrix} x \\ y \\ z \end{pmatrix}$, (A35c) $\mathbf{X}^{3-} = \begin{pmatrix} x & y & -z \end{pmatrix}$

$$\text{(A19)} \quad \mathbf{A}_{30}(t_3) = \begin{pmatrix} 0 & R & Q \\ -R & 0 & P \\ Q & P & 0 \end{pmatrix}, \text{(13.48)},$$

then $\mathbf{A}_{50}$ can be seen to be decomposed into three time-dependent matrix components $\mathbf{A}_{53}$, $\mathbf{\Delta}_{54}$ and $\mathbf{\Delta}_{55}$, i.e.

(13.41) $\mathbf{A}_{50} = \mathbf{A}_{53}(t_3) - t_4\mathbf{\Delta}_{54} - t_5\mathbf{\Delta}_{55}$. (13.49),

where $\mathbf{A}_{53}$, $\mathbf{\Delta}_{54}$ and $\mathbf{\Delta}_{55}$ are defined as follows, and $\mathbf{0}_{33}$ is defined as a 3×3 matrix of zeros,

$$\text{(13.42)} \quad \mathbf{A}_{53} = \begin{pmatrix} 0 & 0 & 0 \\ 0 & 0 & 0 \\ 0 & 0 & \mathbf{A}_{30}(t_3) \end{pmatrix},$$

$$\text{(13.43)} \quad \mathbf{\Delta}_{54} = \begin{pmatrix} 0 & 0 & 0 \\ 0 & 0 & \mathbf{X}^{3-} \\ 0 & -\mathbf{X}_{3+} & \mathbf{0}_{33} \end{pmatrix}, \quad \mathbf{\Delta}_{55} = \begin{pmatrix} 0 & 0 & \mathbf{X}^{3-} \\ 0 & 0 & 0 \\ -\mathbf{X}_{3+} & 0 & \mathbf{0}_{33} \end{pmatrix}, \text{(13.50)}$$

The matrices $\mathbf{\Delta}_{55}$ and $\mathbf{\Delta}_{54}$ are known as variational, 'delta' matrices in URMT (see Section (8)), and have the following annihilator property

(13.44) $\mathbf{\Delta}_{55}\mathbf{X}_{5+} = 0$ and $\mathbf{\Delta}_{54}\mathbf{X}_{5+} = 0$ (13.51).

The annihilation property works because the $\mathbf{X}^{3-}$ vector, embedded in the first and second rows of $\mathbf{\Delta}_{55}$ and $\mathbf{\Delta}_{54}$ respectively, is orthogonal to the $\mathbf{X}_{3+}$ vector embedded in the $\mathbf{X}_{5+}$ vector (13.32), i.e.

(13.45) $\quad \mathbf{X}^{3-}\mathbf{X}_{3+} = x^2 + y^2 - z^2 = 0$ Appendix (F1), Pythagoras, orthogonality.

The first and second row of matrix product $\mathbf{A}_{50}\mathbf{X}_{5+}$ (13.33) is simply equivalent to the inner product $\mathbf{X}^{3-}\mathbf{X}_{3+}$, which is just the Pythagoras equation and therefore zero.

Using this annihilator property, the eigenvector equation $\mathbf{A}_{50}\mathbf{X}_{5+}$ (13.33) becomes

(13.46) $\mathbf{A}_{50}\mathbf{X}_{5+} = \mathbf{A}_{53}\mathbf{X}_{5+}$

Writing $\mathbf{X}_{5+}$ (13.32) in block matrix form in terms of $\mathbf{X}_{3+}$

(13.47) $\mathbf{X}_{5+} = \begin{pmatrix} 0 \\ 0 \\ \mathbf{X}_{3+} \end{pmatrix},$

then, from the definition of $\mathbf{A}_{53}$ (13.42) in terms of $\mathbf{A}_{30}(t_3)$ (A19), the product $\mathbf{A}_{53}\mathbf{X}_{5+}$ in the eigenvector equation (13.46) is effectively the same as $\mathbf{A}_{30}\mathbf{X}_{3+}$ (disregarding dimensionality). Furthermore, since $\mathbf{A}_{30}\mathbf{X}_{3+} = C\mathbf{X}_{3+}$ by its URM3 eigenvector definition (A4), the original URM5 eigenvector equation (13.33) is restored, i.e. $\mathbf{A}_{50}\mathbf{X}_{5+} = C\mathbf{X}_{5+}$.

The important point here is that the eigenvector equation (13.33) is invariant to arbitrary variations $-t_5\Delta_{55}$ and $-t_4\Delta_{54}$ (13.41). The eigenvector equation holds in URM5, just as it does in URM3 (and also URM4 or URMn in general), invariant to any arbitrary variations $-t_5\Delta_{55}$ and $-t_4\Delta_{54}$ in matrix $\mathbf{A}_{50}$ (13.40). This might well seem pointless since nothing has been achieved. Which, in a sense, is the whole point of invariance transformations - to do nothing. But, and it's a big but, the transformations do not leave the other four eigenvectors $\mathbf{X}_{5-}$, $\mathbf{X}_{50A}$, $\mathbf{X}_{50B}$, $\mathbf{X}_{50C}$ (for eigenvalues, $\lambda = -C, 0, 0, 0$) invariant; on the contrary, they will change according to the values of t_4 and t_5. Consequently it is these latter four vectors that generate an evolving eigenvector space in URM5.

If anything, a fair criticism would be that $\mathbf{X}_{5+}$ (13.47) is nothing more than $\mathbf{X}_{3+}$ with a couple of zeros added to the front to extend it from three to five dimensions. However, not all eigenvectors are quite so simple - two of the eigenvectors, $\mathbf{X}_{50B}$ and $\mathbf{X}_{50C}$, see (13.52) further below, have four non-zero elements, and $\mathbf{X}_{5-}$ has a full five non-zero elements, making it a non-trivial (I10), five-dimensional vector. Although it may seem that such vectors are, therefore, only parameterised in terms of the two variational parameters t_4 and t_5, the solutions themselves are expressed in terms of the URM3 eigenvectors, which are fully parameterised by three parameters, t_3, k and l. Hence they are 5D vectors with a 5D parameterisation (t_3, t_4, t_5, k and l). This has the caveat that not all eigenvectors utilise the full parameterisation. For example, $\mathbf{X}_{50A}$ (13.52c) will be seen to be simply an embedding of the URM3 vector $\mathbf{X}_{30}$, so it is only actually parameterised by the three URM3 parameters, t_3, k and l. Nevertheless, in general, all five parameters are used in the complete solution for URM5 (14.56).

This completes the justification for the form of $\mathbf{A}_{50}$.

Notes

(13.48) Matrix $\mathbf{A}_{30}(t_3)$ is also a function of URM3 evolutionary parameter t_3, since the dynamical variables P, Q, R are functions of t_3, ($\sim m$ Appendix (B)). It is defined in [1],#1 as $\mathbf{A}_{30}(t_3) = \mathbf{A}'_{30} - t_3 \mathbf{\Delta}_3^P$, where the primed superscript denotes an initial value as in $\mathbf{A}'_{30} = \mathbf{A}_{30}(t_3 = 0)$, and $\mathbf{\Delta}_3^P$ is defined as $\mathbf{\Delta}^P$ (or $\mathbf{A}_+$) in [1],#1. See Sections (8-1) and (8-7) for a review of URM3 variational methods and $\mathbf{\Delta}^P$.

(13.49) The subscript '54' in $\mathbf{\Delta}_{54}$ denotes the 5×5 matrix for coefficient t_4. Likewise, the subscript '55' in $\mathbf{\Delta}_{55}$ denotes the 5×5 matrix for coefficient t_5.

(13.50) Since $\mathbf{X}_{3+}$ and $\mathbf{X}^{3-}$ are defined purely in terms of coordinates x, y, z, and are completely invariant to variations in t_3, (or t_4 and t_5 for that matter), then the matrices $\mathbf{\Delta}_{55}$ and $\mathbf{\Delta}_{54}$, which comprise $\mathbf{X}_{3+}$ and $\mathbf{X}^{3-}$, are also static, i.e. not a function of time.

(13.51) The $\mathbf{X}_{3+}$ vector embedded in the first column of $\mathbf{\Delta}_{55}$, and the second column of $\mathbf{\Delta}_{54}$, is seemingly useless since it only multiplies the first two, zero elements of $\mathbf{X}_{5+}$. Whilst it is intentional to have no effect, i.e. remain invariant, it raises the question as to why not use any three arbitrary elements and not, specifically, the $\mathbf{X}_{3+}$ vector? The answer is simple: the $-\mathbf{X}_{3+}$ in the column is the negative conjugate of $\mathbf{X}^{3-}$ in the row, and ultimately it means dynamical variables H, N, J and their conjugates $\overline{H}, \overline{N}, \overline{J}$ satisfy the Pythagoras conditions (4.2a). This is a must, and since $\mathbf{X}^{3-}$ cannot be chosen arbitrarily (it must satisfy orthogonality (2.9), (3.21)), it forces the two columns in $\mathbf{\Delta}_{55}$ and $\mathbf{\Delta}_{54}$ to embed $-\mathbf{X}_{3+}$, and not any just any arbitrary vector.

(13-4) URM5 Eigenvector Evolution Equations

Equipped with a URM5 unity root matrix $\mathbf{A}_{50}$ (13.40), and a definition for eigenvector $\mathbf{X}_{5+}$ (13.47), the other four eigenvectors $\mathbf{X}_{5-}$, $\mathbf{X}_{50A}$, $\mathbf{X}_{50B}$, $\mathbf{X}_{50C}$, (13.52b) to (13.52e) below, are determined using the Residual Matrix Method, detailed in Section (3) for URM4, and generalised in Appendix (C) for the n-dimensional case, URMn. For this reason, the URM5 eigenvector solution is quoted below, without explanation.

The eigenvector evolution equations for the URM5 eigenvectors, in terms of the URM3 eigenvectors, are as follows:

(13.52)

$$(13.52a) \quad \mathbf{X}_{5+} = \begin{pmatrix} 0 \\ 0 \\ \mathbf{X}_{3+} \end{pmatrix}$$

$$(13.52b) \quad \mathbf{X}_{5-} = -(t_5^2 + t_4^2)\begin{pmatrix} 0 \\ 0 \\ \mathbf{X}_{3+} \end{pmatrix} + 2t_5\begin{pmatrix} C \\ 0 \\ \mathbf{0}_3 \end{pmatrix} + 2t_4\begin{pmatrix} 0 \\ C \\ \mathbf{0}_3 \end{pmatrix} + \begin{pmatrix} 0 \\ 0 \\ \mathbf{X}_{3-} \end{pmatrix}$$

$$(13.52c) \quad \mathbf{X}_{50A} = \begin{pmatrix} 0 \\ 0 \\ \mathbf{X}_{30} \end{pmatrix}$$

$$(13.52\text{d}) \quad \mathbf{X}_{50B} = -t_4 \begin{pmatrix} 0 \\ 0 \\ \mathbf{X}_{3+} \end{pmatrix} + \begin{pmatrix} 0 \\ C \\ \mathbf{0}_3 \end{pmatrix}$$

$$(13.52\text{e}) \quad \mathbf{X}_{50C} = -t_5 \begin{pmatrix} 0 \\ 0 \\ \mathbf{X}_{3+} \end{pmatrix} + \begin{pmatrix} C \\ 0 \\ \mathbf{0}_3 \end{pmatrix}$$

Looking at this solution, there are no mixed t_4, t_5 terms, and it splits nicely into independent terms in t_4 and t_5. The vector $\mathbf{X}_{5-}$, that contains terms in t_4 and t_5, is split into its 4D and 5D components, denoted by $\mathbf{X}_{5-}(t_4)$ and $\mathbf{X}_{5-}(t_5)$, as follows:

(13.53)

$$(13.53\text{a}) \quad \mathbf{X}_{5-}(t_4) = -t_4^2 \begin{pmatrix} 0 \\ 0 \\ \mathbf{X}_{3+} \end{pmatrix} + 2t_4 \begin{pmatrix} 0 \\ C \\ \mathbf{0}_3 \end{pmatrix} + \frac{1}{2} \begin{pmatrix} 0 \\ 0 \\ \mathbf{X}_{3-} \end{pmatrix}$$

$$(13.53\text{b}) \quad \mathbf{X}_{5-}(t_5) = -t_5^2 \begin{pmatrix} 0 \\ 0 \\ \mathbf{X}_{3+} \end{pmatrix} + 2t_5 \begin{pmatrix} C \\ 0 \\ \mathbf{0}_3 \end{pmatrix} + \frac{1}{2} \begin{pmatrix} 0 \\ 0 \\ \mathbf{X}_{3-} \end{pmatrix}$$

$$(13.53\text{c}) \quad \mathbf{X}_{5-} = \mathbf{X}_{5-}(t_4) + \mathbf{X}_{5-}(t_5)$$

The zero eigenvectors $\mathbf{X}_{50B}$ and $\mathbf{X}_{50C}$ are also, rather conveniently, already separated into a 4D and 5D, time-dependent form (13.54), with each uniquely associated to its dimension, i.e. $\mathbf{X}_{40B}$ with the fourth, and $\mathbf{X}_{50C}$ with the fifth, as in

$$(13.53\text{d}) \quad \mathbf{X}_{50B}(t_4) \equiv \mathbf{X}_{50B}, \text{ see } (13.52\text{d})$$

$$(13.53\text{e}) \quad \mathbf{X}_{50C}(t_5) \equiv \mathbf{X}_{50C}, \text{ see } (13.52\text{e}).$$

(13.54) Note that $\mathbf{X}_{50A}$ (13.52c) is an embedding of the URM3 vector $\mathbf{X}_{30}$, with no non-zero elements in the excess dimensions, albeit $\mathbf{X}_{30}$ is a function of URM3's time t_3 in the first three dimensions of URM3, (B2b).

Not only does the URM5 solution separate into unique 4D and 5D terms, but these terms are identical, disregarding the particular excess dimension. They are identical in so far as the URM4 component contributes a linear term $2t_4 C$ in $\mathbf{X}_{5-}(t_4)$ and C in $\mathbf{X}_{50B}(t_4)$, and the URM5 component contributes a term $2t_5 C$ in $\mathbf{X}_{5-}(t_5)$ and C in $\mathbf{X}_{50C}(t_5)$, which are identical upon interchange of times t_4 and t_5. As will be seen in the next section, (14), since it is only the magnitude of the contribution in the excess dimension that matters in the analysis of the 'compactification ratio' χ (14.23), each dimension can be treated separately and, furthermore, identical findings for one dimension, apply to the other.

Concluding from the previous paragraph, the separability of the URM5 solution into 4D and 5D unique components, and the interchange symmetry of t_4 and t_5 between the two components (as regards calculating a compactification ratio χ), means that each dimension acts independently, with identical behaviour with respect to their individual evolutionary times.

(14) Compactification

(14-1) Overview

This last Section presents a physical application of the methods and results obtained so far in extending URMT to four and five dimensions, and also completes the final step of extending URMT to an arbitrary number of dimensions. The application is known as 'compactification', or 'dimensional reduction', in the scientific literature, and is the phenomena whereby an n-dimensional space appears of lower dimension by virtue that one or more of dimensions are unobservably small. Of course, the classic example comes from string theory, which is favoured to be formulated in eleven dimensions, seven of which are 'curled up' and seemingly invisible from our everyday experience. The compactification process presented here shows that in URMT it is a result of natural time evolution, and that in the earlier stages of evolution, i.e. small timescales, the dimensions may well have been more visible. With this in mind, the work in this section is fully titled as follows:

Compactification of an n-dimensional eigenvector

space over long evolutionary timescales.

The work shows how a discrete, n-dimensional eigenvector space can appear of lower dimension over long evolutionary timescales, ultimately compactifying to appear as a two-dimensional subspace within three dimensions.

Each excess dimension, i.e. any dimension higher than the third, has a unique, temporal coordinate, not necessarily associated with the familiar laboratory time, which controls the evolution of the dimension. Over a long evolutionary period, in a particular excess dimension, the dimension expands relative to all other excess dimensions, but appears to contract relative to the first three dimensions. Specifically, the entire n-dimensional space appears to align along a particular direction in the

245

three-dimensional space, the direction given by one of the eigenvectors, which is physically associated with an acceleration vector, and specified by two non-temporal, arbitrary parameters. The third parameter in the three-dimensional space is a temporal coordinate, which also controls the evolution of the three dimensions, and shows the same alignment behaviour as for the excess dimensions. The initial state of the entire space is specified by the initial values for the acceleration vector and, most importantly, a single energy-related constant controls the initial size of all excess dimensions.

The compactification process is detailed by way of a four and five-dimensional case, expressed in terms of the three-dimensional solution, with a full 5D numeric example provided in Appendix (D). A complete n-dimensional solution is given and the compactification arguments generalised for an arbitrary number of dimensions.

(14-2) The Compactification Ratio

Since the aim of this section is to show compactification occurs over long evolutionary timescales, in one or more excess dimensions, a quantitative measurement of the relative size of a dimension j, with respect to the first three dimensions of URM3, is required. Such a measurement is termed the compactification ratio χ_j, and defined further below in terms of the 'magnitude' of a dimension, defined next.

(14.21a) Definition: The **magnitude** (or size), symbol $|\mathbf{X}|_j$, of a particular, excess jth dimension, $j \geq 4$, is a measure (usually an approximation) of the dominant jth coordinate in that dimension. For advance information, this measure is invariably approximated as the time-scaled multiple of the eigenvalue, i.e. $|2t_j C| \approx 2t_j C$, where $t_j \geq 0$ by convention (14.32), and $C \geq 1$ by definition (1.7).

(14.21b) The magnitude of the first three dimensions, symbol $|\mathbf{X}|_3$, is a measure of the size of the URM3 3D subspace of the full n-dimensional, eigenvector space. This measure is invariably approximated from just the dominant $\mathbf{X}_{n-}$ vector, and then only using the quadratic term $|t_j^2 \mathbf{X}_{3+}|$, $j \geq 4$. That $j \neq 3$ here is intentional because the compactification behaviour is studied with respect to URM3.

(14.22) Definition: The **compactification ratio** of dimension j, denoted by χ_j, is the ratio of the magnitude $|\mathbf{X}|_j$ (14.21a) of the jth dimension to the magnitude $|\mathbf{X}|_3$ (14.21b) of the first three dimensions (URM3), i.e.

(14.23) $\chi_j = \dfrac{|\mathbf{X}|_j}{|\mathbf{X}|_3}$.

With a compactification ratio χ_j now defined, then showing that compactification occurs over evolutionary timescales translates to showing the ratio χ_j decreases to zero as the jth dimension's evolutionary time t_j increases without bound, i.e.

(14.24) $\lim\limits_{t_j \to \infty} \chi_j = 0$.

Since the above definitions for magnitudes $|\mathbf{X}|_j$ and $|\mathbf{X}|_3$ allude to the fact that they are approximated as follows:

(14.25) $|\mathbf{X}|_j \approx 2t_j C$

(14.26) $|\mathbf{X}|_3 \approx \left| t_j^2 \mathbf{X}_{3+} \right|$

then the compactification ratio χ_j (14.23) of dimension j is approximated by

(14.27) $\chi_j \approx \dfrac{2C}{t_j |\mathbf{X}_{3+}|}$, $t_j \neq 0$, use (14.23) when $t_j = 0$.

From this approximation it is seen that χ_j is inversely proportional to time t_j, hence the limit (14.24) is satisfied. Moving on to specifics, the calculation of the compactification ratio, and its behaviour for the URMT eigenvector solutions, is now the main focus of the work with regard to demonstrating compactification in URMT.

(14-3) Eigenvector Solutions

To study the compactification behaviour, three sets of eigenvector solution are examined, in order,

(14.31)

      1. URM4 in terms of URM3

      2. URM5 in terms of URM3

      3. URM n in terms of URM3, $n \geq 4$

The URM3 eigenvector solution is given in Appendices (A) and (B).

All solutions are given in block matrix form.

In all discussion, the first element of the vector is always the nth dimension in URM n, and the remaining $(n-1)$ elements represent the $(n-1)$ dimensions in URM $(n-1)$. The last three elements are always dimensions one to three, i.e. URM3, and referred to as 'the first three dimensions'.

To keep things simple, it will be assumed that all evolution proceeds in the forward, positive direction with time always increasing:

(14.32) $t_j \geq 0$, $j = 3 \ldots n$,

However, this is convention only, none of the work specifically requires such an assumption, and t_j can be positive or negative. Remember, t_j is a variational parameter and it certainly could be positive or negative, and proceeding forward or backward in time. Nevertheless, using the standard physical interpretation, Appendix (J), it always has physical units of time.

(14-4) URM4 Eigenvector Evolution Equations

The eigenvector evolution equations for the URM4 eigenvectors, in terms of the URM3 eigenvectors, are reproduced below from Section (11), with the following notational equivalences $t_4 \sim s$, $t_3 \sim m$, $\mathbf{X}_{3-} \sim \mathbf{X}_{3-}(m)$, $\mathbf{X}_{30} \sim \mathbf{X}_{30}(m)$

(14.41)

(14.41a) $\mathbf{X}_{4+} = \begin{pmatrix} 0 \\ \mathbf{X}_{3+} \end{pmatrix}$, (2.4)

(14.41b) $\mathbf{X}_{4-} = -t_4^2 \begin{pmatrix} 0 \\ \mathbf{X}_{3+} \end{pmatrix} + 2t_4 \begin{pmatrix} C \\ \mathbf{0}_3 \end{pmatrix} + \begin{pmatrix} 0 \\ \mathbf{X}_{3-} \end{pmatrix}$, (11.37)

(14.41c) $\mathbf{X}_{40A} = \begin{pmatrix} 0 \\ \mathbf{X}_{30} \end{pmatrix}$, (11.54a)

(14.41d) $\mathbf{X}_{40B} = -t_4 \begin{pmatrix} 0 \\ \mathbf{X}_{3+} \end{pmatrix} + \begin{pmatrix} C \\ \mathbf{0}_3 \end{pmatrix}$, (11.45)

Before proceeding with a more detailed analysis in Section (14-5), the following observations of this solution are made:

(14.42a) Regardless of the size, relative or absolute, of the fourth dimension's evolutionary time t_4, the only vectors that contribute anything to the size of the fourth dimension are $\mathbf{X}_{4-}$ and $\mathbf{X}_{40B}$ since they have a non-zero component, eigenvalue C, as their first element.

(14.42b) In fact, looking ahead to the general, n-dimensional solution, Section (14-9), C is the only quantity, other than evolutionary time, that is present in each excess dimension, and then it only appears in the linear product term $t_j C$, $j = 4 \ldots n$, or the constant term, as itself, C. Note too that C does not explicitly appear in any of the three URM3 eigenvectors; see (A33). The contribution of the jth dimension at the initial stage of evolution, i.e. $t_j = 0$, is thus governed by the magnitude of C compared to the magnitude of the URM3 eigenvectors. Given C is related to the total energy, $E = C^2$ (1.14b), it means that at $t_j = 0$ a comparatively large value for C would make for a sizeable excess dimension with a lot of energy in it, C being suitably chosen as an initial condition.

(14.43) For any sufficiently large evolutionary time t_4, (14.44) below, the URM3 vector $\mathbf{X}_{3+}$ dominates the entire solution. See, for example, $\mathbf{X}_{4-}$ (14.41b) with a quadratic term $-t_4^2$. Because $\mathbf{X}_{30}$ and $\mathbf{X}_{3-}$ are functions of the evolutionary time t_3 ($\sim m$), Appendix (B), they also grow with t_3. However, for a large URM3 evolutionary time t_3, regardless of time t_4, URM3 itself converges (flattens) to also align with $\mathbf{X}_{3+}$, and ultimately all evolution tends to align with $\mathbf{X}_{3+}$. Nevertheless, that any excess dimension is dominated by one or more of the URM3 vectors, $\mathbf{X}_{3+}$, $\mathbf{X}_{30}$ and $\mathbf{X}_{3-}$, only serves to bolster arguments that the higher, excess dimensions compactify to those of URM3. Suffice to note, it is not the relative size of URM3 vectors that matters with regard to compactification of excess dimensions, but the size of the excess dimensions relative to those spanned by the URM3 vectors.

(14.44) The caveat 'sufficiently large evolutionary time' appears repeatedly throughout. Loosely speaking, it means any time t large enough such that the approximation under discussion is valid. In actuality, it means the magnitude of the quadratic term t_j^2, in the $\mathbf{X}_{n-}$ eigenvector, for excess dimension j, $j = 4 \ldots n$, dominates all other terms in all eigenvectors. It is given a more rigorous definition in Section (14-6).

(14-5) Analysis of URM4 Compactification

Looking at the URM4 eigenvector solution (14.41) in more detail, the only two vectors in URM4 that contribute to the fourth dimension are $\mathbf{X}_{4-}$ and $\mathbf{X}_{40}$, i.e. they have a non-zero first element. In vector $\mathbf{X}_{4-}$, the size of the fourth dimension (first element) is controlled by the linear term $2t_4 C$, and in vector $\mathbf{X}_{40}$, the size of the fourth dimension is controlled by the constant term C. There are a few sensible ways to combine these two sizes, e.g. root sum squares or summation of the magnitude of the individual components, i.e.

$$(14.51) \quad |\mathbf{X}|_4 = \sqrt{(2t_4 C)^2 + C^2} \text{ or } |\mathbf{X}|_4 = |2t_4 C| + |C|,$$

but, given the analysis is primarily interested in large evolutionary times, i.e. $t_4 \gg 0$, it is clear that only the $2t_4 C$ term from $\mathbf{X}_{4-}$ will dominate, i.e.

$$(14.52) \quad 2t_4 C \gg C \text{ for } t_4 \gg 0,$$

and therefore the magnitude of the fourth dimension is simply approximated as the magnitude of the time-dependent component, i.e.

(14.53) $\left|\mathbf{X}\right|_4 \approx 2t_4 C$, $t_4 \gg 0$.

From here onward, to avoid repetition, the following two points are assumed throughout this section:

(14.54a) All calculations of $\left|\mathbf{X}\right|_3$ are restricted to their URM3 components only (last three elements).

(14.54b) All approximations for $\left|\mathbf{X}\right|_3$ are assumed valid for sufficiently large t_4 with justification given in Section (14-6).

Note that there is a subtle distinction in (14.54b) between a large time t_4, as in much greater than zero ($t_4 \gg 0$), and a 'sufficiently large' t_4 such that any approximation is actually valid. In the first case (14.53), of sizing the fourth dimensional component, the $t_4 \gg 0$ criterion is sufficient given (14.52). In the second case, of sizing the other three dimensions, having $t_4 \gg 0$ might not, by itself, be sufficient to justify the approximation, with a more exact definition required. This topic is considered again in Section (14-6).

The magnitude of the first three dimensions, denoted by $\left|\mathbf{X}\right|_3$, might seem a much messier affair because there are now three URM3 vectors $\mathbf{X}_{3+}$, $\mathbf{X}_{30}$ and $\mathbf{X}_{3-}$ embedded within the URM3 dimensions (last three elements) of URM4. Looking at the URM4 eigenvectors (14.41), the URM3 components for each vector are

(14.55)

(14.55a) $\mathbf{X}_{4+} = \mathbf{X}_{3+}$, URM3 components only

(14.55b) $\mathbf{X}_{4-} = -t_4^2 \mathbf{X}_{3+} + \mathbf{X}_{3-}$, ditto

(14.55c) $\mathbf{X}_{40A} = \mathbf{X}_{30}$, ditto

(14.55d) $\mathbf{X}_{40B} = -t_4 \mathbf{X}_{3+}$, ditto

A measure of $\left|\mathbf{X}\right|_3$ can be obtained by combining the magnitudes of these above components and, again, a root sum squares or sum of individual vector magnitudes are the two common methods of combination:

(14.56)

(14.56a) $\quad |\mathbf{X}|_3 = \sqrt{\left|\mathbf{X}_{4+}\right|^2 + \left|\mathbf{X}_{40A}\right|^2 + \left|\mathbf{X}_{40B}\right|^2 + \left|\mathbf{X}_{4-}\right|^2}$, root sum of squares.

(14.56b) $|\mathbf{X}|_3 = \left|\mathbf{X}_{4+}\right| + \left|\mathbf{X}_{40A}\right| + \left|\mathbf{X}_{40B}\right| + \left|\mathbf{X}_{4-}\right|$, sum of magnitudes

(14.57) It should be noted that all URM n vector spaces are, generally, highly oblique, i.e. the eigenvectors, as a basis, are far from orthogonal to each other, and neither are they are of unit magnitude. Therefore any such measures (root sum squares, etc.) are relatively basic estimates, but considered acceptable if consistently applied.

Of the four vectors (14.55) in these expressions (14.56), only $\mathbf{X}_{4-}$ is dominant because it is the only vector with a quadratic term in t_4. Therefore, $|\mathbf{X}|_3$ can be approximated by $\left|\mathbf{X}_{4-}\right|$ for some sufficiently large time t_4, i.e.

(14.58) $|\mathbf{X}|_3 \approx \left|\mathbf{X}_{4-}\right|$.

Furthermore, (14.55b) shows that $\mathbf{X}_{4-}$ is dominated by the term $-t_4^2 \mathbf{X}_{3+}$ when assuming the following, which is basically the criterion for a 'sufficiently large time t_4,

(14.59) $\left|t_4^2 \mathbf{X}_{3+}\right| \gg \left|\mathbf{X}_{3-}\right|$.

Under this assumption, $\mathbf{X}_{4-}$ is approximated by $-t_4^2 \mathbf{X}_{3+}$, and formalised as

(14.60) $\left|\mathbf{X}_{4-}\right| \approx \left|t_4^2 \mathbf{X}_{3+}\right|$.

(14.61) However, although not shown, $\mathbf{X}_{3-}$ also evolves with URM3's evolutionary time t_3, and it too can be approximated, for some sufficiently large t_3, by $\left|t_3^2 \mathbf{X}_{3+}\right|$, i.e. $\left|\mathbf{X}_{3-}\right| \approx \left|t_3^2 \mathbf{X}_{3+}\right|$. Thus, it is easily possible a time t_3 can be found such that the magnitudes $\left|\mathbf{X}_{3-}\right|$ and $\left|t_4^2 \mathbf{X}_{3+}\right|$ are comparable, i.e. $\left|t_4^2 \mathbf{X}_{3+}\right| \approx \left|\mathbf{X}_{3-}\right| \approx \left|t_3^2 \mathbf{X}_{3+}\right|$. It might then be better to use a combined estimate such as $\left|\sqrt{(t_4^2+t_3^2)}\mathbf{X}_{3+}\right|$ for $\left|\mathbf{X}_{4-}\right|$. In fact, the first, simpler approximation $\left|t_4^2\mathbf{X}_{3+}\right|$ will be used, ignoring t_3 completely by virtue of the explanation given next.

(14.62) Fortunately, these aforementioned concerns about t_3, and the comparative size of its evolving vectors, are all irrelevant for the following reason: given the definition of χ_j has the URM3 magnitude $\left|\mathbf{X}\right|_3$ in its denominator, a smaller estimate for $\left|\mathbf{X}\right|_3$ will give a larger ratio χ_j, i.e. a more pessimistic measure of compactification. If a pessimistic measure of χ_j can be shown to converge to zero for large evolutionary times, then it will also converge to zero quicker, i.e. for smaller evolutionary times, when the true magnitude of $\left|\mathbf{X}\right|_3$ is greater than that used in the calculation.

Since $\left|\mathbf{X}\right|_3$ is dominated by $\left|\mathbf{X}_{4-}\right|$ due to the domination of the term $-t_4^2\mathbf{X}_{3+}$, and disregarding any URM3 contribution due to t_3 for reasons given above (an increasing t_3 only grows the relative size of URM3, and betters the compactification), then the approximation $\left|-t_4^2\mathbf{X}_{3+}\right|$ will be used as measure $\left|\mathbf{X}\right|_3$ of the magnitude of the last three dimensions of URM4, i.e.

(14.63) $\left|\mathbf{X}\right|_3 \approx t_4^2\left|\mathbf{X}_{3+}\right|$.

Having established approximations for $\left|\mathbf{X}\right|_4$ (14.53) and $\left|\mathbf{X}\right|_3$ (14.63), the URM4 compactification ratio χ_4 can now be calculated.

The URM4 Compactification Ratio χ_4

Substituting the approximation for $\left|\mathbf{X}\right|_4$ (14.53), and $\left|\mathbf{X}\right|_3$ (14.63) into (14.23), for $j=4$, the compactification ratio χ_4 of the fourth dimension in URM4 is approximated as follows; see also the next Section, (14-6), which shows this approximation improves as t_4 increases.

(14.64) $\chi_4 \approx \dfrac{2C}{t_4\left|\mathbf{X}_{3+}\right|}$, for sufficiently large t_4,

Note that $\chi_4 \geq 0$ since $C > 0$ by definition (1.7), and $t_4 > 0$ by convention (14.32). This convention is strictly $t_4 \geq 0$, but the ratio is only calculated for $t_4 > 0$ for the obvious reason to avoid a zero divisor in (14.64). A different compactification ratio for t_4, at time zero, could be calculated but it is rather pointless since the focus is on large evolutionary times.

The vector magnitude $|\mathbf{X}_{3+}|$ (14.66), further below, is constant with respect to time because the vector $\mathbf{X}_{3+}$ is static (I16), i.e. it has no dependence on any evolutionary parameter, notably t_3; see also (B2a). So too is eigenvalue C also a constant and, additionally, an initial condition. The ratio χ_4 is therefore just inversely proportional to the time t_4, and so tends to zero as t_4 tends to infinity, i.e.

(14.65) $\lim_{t_4 \to \infty} \chi_4 = 0$.

To conclude then, in the four-dimensional vector space of URM4, the excess, fourth dimension is seen to contract as its evolutionary time t_4 grows ever larger, eventually appearing to have zero size as t_4 grows infinite. Hence, under assumption (14.54b), the four-dimensional vector space compactifies to that of the eigenvector space of URM3 as evolution progresses. Specifically, all 4D eigenvectors align with the single, static URM3 eigenvector $\mathbf{X}_{3+}$, which occupies a 2D subspace (14.98) of URM3; hence URM4 compactifies to appear two-dimensional.

As regards URM4, it remains to define what is meant by a 'sufficiently large evolutionary time', and justify the assumption (14.54b) made. This follows next, and is applicable to any excess dimension j, $j = 4 \ldots n$.

Following this, the same compactification analysis is performed on the 5D solution, which is seen to have evolutionary terms in both t_4 and t_5.

(14.66) By the definition (A3) of $\mathbf{X}_{3+}$ in terms of acceleration coordinates x, y, z (J3), the magnitude $|\mathbf{X}_{3+}|$ is actually $|\mathbf{X}_{3+}| = \sqrt{2}|z|$. By the Pythagorean relation (F1) between x, y, z, $|z|$ is always greater than zero, see (A26c), and increases with increasing values for parameters k and l. This simultaneously increases compactification by decreasing the compactification ratio as $\left|\mathbf{X}_{3+}\right|$, and hence also $\left|\mathbf{X}\right|_3$ (14.63), grows in magnitude.

(14-6) A Sufficiently Large Evolutionary Time

Until now, the term 'sufficiently large evolutionary time' has been considered as any evolutionary time t, large enough such that the approximation under discussion is valid. Now that a specific approximation for the size $\left|\mathbf{X}\right|_3$ of the URM3 dimensions embedded within URM4 has been given, i.e. $\left|t_4^2 \mathbf{X}_{3+}\right|$ (14.63), the term can be made more definitive.

Although only URM4 has been analysed, this section will generalise to the jth dimension, $j = 4 \ldots n$, for URMn, which basically means just replacing every subscript of '4' by 'j', as regards results obtained in the previous section.

Firstly, the same approximations used in URM4 are now generalised for an arbitrary dimension j, $j = 4 \ldots n$, as follows:

(14.67)

(14.67a) $\left|\mathbf{X}\right|_3 \approx \left|\mathbf{X}_{j-}\right|$, see (14.58) for URM4

(14.67b) $\left|\mathbf{X}_{j-}\right| \approx \left|t_j^2 \mathbf{X}_{3+}\right|$, see (14.60) for URM4.

The two above approximations are combined to give

(14.67c) $\left|\mathbf{X}\right|_3 \approx \left|t_j^2 \mathbf{X}_{3+}\right|$, see (14.63) for URM4.

Readers are also referred to the general, n-dimensional solution in Section (14-9) to see these approximations.

A measure of the relative error ε in the approximation of $\left|\mathbf{X}\right|_3$ by $\left|t_j^2 \mathbf{X}_{3+}\right|$ (14.67c), at any time t_j, is given by

$$(14.68)\ \varepsilon = \frac{\left\|\mathbf{X}\right|_3 - \left|t_j^2 \mathbf{X}_{3+}\right\|}{\left|\mathbf{X}\right|_3}, \text{ estimate of relative error in approximation at}$$

time t_j.

(14.69) It is at this stage that a potential problem appears. Whilst the approximation (14.67c) $\left|t_j^2\mathbf{X}_{3+}\right|$ is acceptable when calculating the compactification ratio χ_j (χ_4), using just time t_j (t_4) (for reasons outlined in (14.61) and (14.62)), it is not so good when calculating the relative error ε in (14.68). The idea behind the calculation of ε, as seen further below in this section, is that it removes the quadratic term t_j^2 from the numerator in (14.68), leaving only linear terms in t_j. However, by ignoring all other evolutionary times t_i (for the ith dimension, where $i \neq j$, $i, j = 3...n$), in the approximation $\left|\mathbf{X}\right|_3$ (14.67c), and using just t_j, leaves quadratic terms in t_i which can be as large as t_j, if not larger. In other words, the error ε is not small in these circumstances, i.e. when ignoring t_i, and it will not always converge to zero as t_j grows, even though the compactification ratio will converge, as per (14.65).

There are three methods to overcome this:

(14.69a) Replace the crude approximation $\left|t_j^2\mathbf{X}_{3+}\right|$ by a better approximation $\left|T_n^2\mathbf{X}_{3+}\right|$, where $T_n^2 = \sum_{k=3}^{n} t_k^2$. This is alluded to in point (14.61) where it is suggested that $\left|\sqrt{\left(t_4^2 + t_3^2\right)}\mathbf{X}_{3+}\right|$ be used instead of $\left|t_4^2\mathbf{X}_{3+}\right|$.

(14.69b) Assert $t_j \gg t_i$, $i \neq j$, i.e. make the jth evolutionary time much greater than all others t_i.

(14.69c) Set all $t_i = 0$, where $i \neq j$, $i, j = 3...n$, leaving $t_j \gg 0$.

The first (14.69a) seems a good, obvious choice since the original approximation is very crude and, in reality, this better approximation, using T_n^2, should always be used for both an accurate calculation of the compactification ratio and in any error analysis. Nevertheless, it will not be used here solely because it makes the analysis clumsy and, most importantly, it isn't really necessary because the third option, (14.69c) below, circumvents the problem.

The second choice (14.69b) will do the job, i.e. make the approximation $|\mathbf{X}|_3$ (14.67c) reasonable, but is disliked because it means the evolution times can never be comparable, i.e. it becomes a condition that the jth evolutionary time t_j is always much greater than every other, ith time t_i. Since the evolutionary times may well all be identical, this solution is not acceptable except when all other times t_i are zero.

The third choice (14.69c) is the preferred option because it makes the analysis simple and will make the approximation for $|\mathbf{X}|_3$ valid, even if it is an artificial condition. Although artificial in that all evolutionary times other than t_j are zero, the computation of χ_j remains unchanged and valid. As noted for URM4, points (14.61) and (14.62), ignoring non-zero t_i will give a worst-case estimation of χ_j, and any non-zero times t_i will only make χ_j better (smaller), i.e. faster compactification.

Lastly on this issue, if true accuracy is required, it is a simple matter to revert to method (14.69a), i.e. replace time t_j^2 with the combined, quadratic time T_n^2 in the approximation of $|\mathbf{X}|_3$, which can then be used to calculate χ_j.

To conclude the above, for the purposes of this section only, the calculations will assume all times t_i, other than t_j, are zero; t_j being both non-zero and likely 'large', i.e.

(14.70) $t_j \neq 0$, $t_i = 0$, where $i \neq j$, $i, j = 3 \ldots n$.

Returning then to the calculation of the relative error ε (14.68), under the assumption (14.70), this calculation requires a true (accurate) expression for $|\mathbf{X}|_3$. This was left undecided in the previous section, with one of two options, (14.56a) and (14.56b), available. Given $|\mathbf{X}_{j-}|$ is the dominant term in $|\mathbf{X}|_3$, then whichever one of the two options is chosen, they both approximate to (14.67b) and (14.67c). The simplest form for analysis is the sum of magnitudes, i.e. (14.56b), and is thus chosen as a measure of $|\mathbf{X}|_3$, now formally defined by

(14.71) $|\mathbf{X}|_3 = \sum_{i=1}^{n} |\mathbf{X}_i|$, for all eigenvectors $\mathbf{X}_i$ in the n-dimensional basis.

To see how the relative error ε behaves with respect to time t_j, it is also useful to define an absolute error ε_- for the approximation of $|\mathbf{X}_{j-}|$ (14.67b), calculated as follows,

(14.72) $\varepsilon_- = |\mathbf{X}_{j-}| - |t_j^2 \mathbf{X}_{3+}|$.

Using this, and the sum form (14.71) for $|\mathbf{X}|_3$, then the numerator of ε (14.68) is re-written as

(14.73) $\left\| \mathbf{X} \right\|_3 - \left| t_j^2 \mathbf{X}_{3+} \right| = \left(|\mathbf{X}_{n+}| + |\mathbf{X}_{n0}| + \sum_{i=1}^{n-3} |\mathbf{X}_{n0i}| \right) + \varepsilon_-$.

For example, using the URM4 eigenvectors (14.55), and assuming ε_- is small, this becomes

(14.74) $\left\| \mathbf{X} \right\|_3 - \left| t_4^2 \mathbf{X}_{3+} \right| \approx \left(|\mathbf{X}_{4+}| + |\mathbf{X}_{40A}| + |\mathbf{X}_{40B}| \right)$, URM4.

Looking at (14.73), and ahead to (14.92b) for $|\mathbf{X}_{j-}|$, then because the quadratic term in t_j^2 has been removed by the subtraction of $|t_j^2 \mathbf{X}_{3+}|$ on the left of (14.73), ε_- is only of linear order in time t_j (t_4), for large t_j, and ignoring t_i (t_3) by assumption (14.70), then

(14.75) $O(\varepsilon_-) = t_j$.

Likewise, from the general solution (14.92), or using the URM4 vectors (14.55) as an example, all vectors in the bracketed term on the right of (14.73) and (14.74) are also only of linear order, since $\mathbf{X}_{j-}$ $(\mathbf{X}_{4-})$ is the only vector with a quadratic factor in t_j (t_4), i.e.

(14.76) $O\left(|\mathbf{X}_{n+}| + |\mathbf{X}_{n0}| + \sum_{i=1}^{n-3} |\mathbf{X}_{n0i}| \right) = t_j$.

Therefore, the entire order of the numerator term (14.73) is linear in t_j.

(14.77) $\quad O\left\| \mathbf{X} \right\|_3 - \left| t_j^2 \mathbf{X}_{3+} \right\| = t_j$.

Conversely, the denominator $\left| \mathbf{X} \right|_3$ still contains $\left| \mathbf{X}_{j-} \right|$ and remains a quadratic function of t_j, i.e.

(14.78) $\quad O\left| \mathbf{X} \right|_3 = t_j^2$.

Therefore, inserting numerator (14.73) and denominator (14.71) into (14.68) shows that the error ε is now inversely proportional to time t_j, i.e.

(14.79) $\quad \varepsilon \propto \dfrac{1}{t_j}$, under assumption (14.69c), see also (14.80) below.

In other words, choosing $\left| \mathbf{X} \right|_3$ as the form (14.71), and using the quadratic approximation (14.67c), gives an estimate for the approximation error ε, which is inversely proportional to time, and thus decreases to zero as time increases.

This last result (14.79) is pivotal in defining the term 'sufficiently large' because, basically, it means that for any time greater than t_j the relative error ε will always be less than its value at time t_j, which is formalised next.

(14.80) If the assumption (14.69c) is unpalatable, then follow the suggestion in (14.69a), i.e. replace t_j^2 with a better, more accurate time, T_n^2, in both the calculation of χ and ε. Doing so will then give the same, inverse-time result for ε.

Finally then, by choosing a value of the maximum, permissible error ε_j as a pre-condition:

(14.81) $\quad \varepsilon_j$ = the maximum, permissible error for ε (14.68), e.g. 0.01 for 1% error, then a definition for 'sufficiently large evolutionary time' is given as follows:

(14.82) Definition: a **sufficiently large evolutionary time** is considered to be a time t_j, for a specific dimension j, $j = 4...n$, if, for all times t greater than t_j, the relative error ε (14.68) in the estimate for the size $|\mathbf{X}|_3$ of the first three dimensions, i.e. the size of URM3 embedded in URM n, is less than ε_j, i.e.

if $\varepsilon < \varepsilon_j$ for all $t > t_j$ then t_j is 'sufficiently large'.

Furthermore, the error decreases with increasing time such that it converges to zero, i.e.

(14.83) $\lim\limits_{t \to \infty} \varepsilon = 0$.

Admittedly, this does not give any actual sufficiency time, but merely shows that by approximating the magnitude $|\mathbf{X}|_3$ by selecting the dominant, quadratic, evolutionary terms in t_j, under certain assumptions (14.69c) (which can be rectified, see (14.80)), the error in this approximation converges to zero as evolutionary time progresses. The numerical example in Appendix (D) provides some values of ε versus t_j.

(14-7) URM5 Eigenvector Evolution Equations

The eigenvector evolution equations for the URM5 eigenvectors, in terms of the URM3 eigenvectors, are reproduced below, from Section (13-4), as follows:

(13.52)

$$\text{(13.52a)} \quad \mathbf{X}_{5+} = \begin{pmatrix} 0 \\ 0 \\ \mathbf{X}_{3+} \end{pmatrix}$$

$$\text{(13.52b)} \quad \mathbf{X}_{5-} = -(t_5^2 + t_4^2)\begin{pmatrix} 0 \\ 0 \\ \mathbf{X}_{3+} \end{pmatrix} + 2t_5\begin{pmatrix} C \\ 0 \\ \mathbf{0}_3 \end{pmatrix} + 2t_4\begin{pmatrix} 0 \\ C \\ \mathbf{0}_3 \end{pmatrix} + \begin{pmatrix} 0 \\ 0 \\ \mathbf{X}_{3-} \end{pmatrix}$$

$$\text{(13.52c)} \quad \mathbf{X}_{50A} = \begin{pmatrix} 0 \\ 0 \\ \mathbf{X}_{30} \end{pmatrix}$$

$$\text{(13.52d)} \quad \mathbf{X}_{50B} = -t_4\begin{pmatrix} 0 \\ 0 \\ \mathbf{X}_{3+} \end{pmatrix} + \begin{pmatrix} 0 \\ C \\ \mathbf{0}_3 \end{pmatrix}$$

$$\text{(13.52e)} \quad \mathbf{X}_{50C} = -t_5\begin{pmatrix} 0 \\ 0 \\ \mathbf{X}_{3+} \end{pmatrix} + \begin{pmatrix} C \\ 0 \\ \mathbf{0}_3 \end{pmatrix}$$

To reiterate some points in (13-4), there are no mixed t_4, t_5 terms in this solution, and it splits nicely into independent terms in t_4 and t_5. The vector $\mathbf{X}_{5-}$, that contains terms in t_4 and t_5, is split into its 5D and 4D components, denoted by $\mathbf{X}_{5-}(t_4)$ and $\mathbf{X}_{5-}(t_5)$, as follows:

(13.53)

(13.53a) $\mathbf{X}_{5-}(t_4) = -t_4^2 \begin{pmatrix} 0 \\ 0 \\ \mathbf{X}_{3+} \end{pmatrix} + 2t_4 \begin{pmatrix} 0 \\ C \\ \mathbf{0}_3 \end{pmatrix} + \frac{1}{2} \begin{pmatrix} 0 \\ 0 \\ \mathbf{X}_{3-} \end{pmatrix}$

(13.53b) $\mathbf{X}_{5-}(t_5) = -t_5^2 \begin{pmatrix} 0 \\ 0 \\ \mathbf{X}_{3+} \end{pmatrix} + 2t_5 \begin{pmatrix} C \\ 0 \\ \mathbf{0}_3 \end{pmatrix} + \frac{1}{2} \begin{pmatrix} 0 \\ 0 \\ \mathbf{X}_{3-} \end{pmatrix}$

(13.53c) $\mathbf{X}_{5-} = \mathbf{X}_{5-}(t_4) + \mathbf{X}_{5-}(t_5)$

The zero eigenvectors (I18) $\mathbf{X}_{50B}$ and $\mathbf{X}_{50C}$ are also already separated into a 4D and 5D, time-dependent form, with each uniquely associated to its dimension, i.e. $\mathbf{X}_{40B}$ with the fourth, and $\mathbf{X}_{50C}$ with the fifth, see (13.53d) and (13.53e).

These 4D and 5D unique terms are also are identical upon interchange of times t_4 and t_5, and since it is only the magnitude of the contribution in the excess dimension that matters in the analysis of the ratio χ, each dimension can be treated separately, and findings for one dimension apply identically to the other.

To conclude the above points, the separability of the URM5 solution into 4D and 5D unique components, and the interchange symmetry of t_4 and t_5 between the two components (as regards calculating the ratio χ), means that each dimension acts independently, with identical behaviour with respect to their individual evolutionary times.

The compactification behaviour of the fifth dimension, isolated from the behaviour of the fourth dimension, can easily be analysed by equating time t_4 to zero, thereby nullifying the 4D component and leaving only the 5D solution in terms of t_5. However, as stated, barring the fact that the 5D component affects the fifth dimension, and not the fourth, it is effectively the same solution as that for URM4 (14.41), by virtue of the interchange symmetry between t_4 and t_5. Thus the same arguments used for URM4 can be applied to URM5 and, most importantly, the expression for χ_5, i.e. the URM5 equivalent of

URM4's ratio χ_4 (14.64), is simply written down by interchange of t_4 with t_5, i.e.

(13.43) $\chi_5 \approx \dfrac{2C}{t_5 |\mathbf{X}_{3+}|}$, for sufficiently large t_5.

As regards URM5, it only remains to examine when both t_4 and t_5 are comparably large, i.e.

(13.44) $t_5 \approx t_4$

Unsurprisingly, given the above discussion, if either one of t_4 or t_5 is sufficiently large then compactification will still occur since the dimensions act independently. If both t_4 and t_5 are large, it will only serve to increase this compactification process further. However, this process is nicely illustrated with regard to what is really physically happening, and that is that URM3, from the perspective of the fourth or fifth dimension, appears to expand with a constant acceleration.

(14-8) An Expanding URM3 Eigenvector Space

Whilst the analysis and discussion is focussed on the concept of compactification, i.e. relative shrinkage of dimensions, it is not that any dimension actually shrinks, but rather that the URM3 dimensions appear to expand, and then with a constant acceleration (constant with respect to evolutionary time), that of the URM3 static vector $\mathbf{X}_{3+}$, see point (14.84) below. Simultaneously the excess dimensions expand linearly with a constant velocity (eigenvalue C). Because the 3D expansion is a static acceleration, URM3 spatially expands along $\mathbf{X}_{3+}$ (within $\mathbf{X}_{5-}$), quadratically with respect to the evolutionary times t_4 and t_5, as witnessed by the solution for vector $\mathbf{X}_{5-}$ (13.52b); there is also the linear expansion (due to velocity C) in the excess dimensions of both $\mathbf{X}_{5-}$ and the zero vectors $\mathbf{X}_{50B}$ (13.52d) and $\mathbf{X}_{50C}$ (13.52e). But, of course, these linear terms becomes less important relative to the quadratic term as the evolution progresses, hence the apparent contraction of the excess dimensions relative to URM3.

(14.84) This is the reason why all the other eigenvectors align with $\mathbf{X}_{3+}$ over times t_j, since they are related to $\mathbf{X}_{3+}$ by calculus relations, e.g.

the velocity $\mathbf{X}_{n0j}$ is the integral of acceleration $\mathbf{X}_{n+}$ ($\sim \mathbf{X}_{3+}$), and the position $\mathbf{X}_{n-}$ is the corresponding integral of the velocity (both to within a constant factor), see Section (12-6) and Appendix (G) for more details.

Looking at the URM5 eigenvector solution (13.52), it is clear, for sufficiently large t_4 and/or t_5, that the solution is dominated by the quadratic term in $\mathbf{X}_{5-}$ scaling the vector $\mathbf{X}_{5+}$. This vector $\mathbf{X}_{5+}$ is really just the URM3 vector $\mathbf{X}_{3+}$ embedded in URM5, with a zero fourth and fifth dimensional contribution. As noted earlier, (14.61) and (14.62), ignoring the vector $\mathbf{X}_{3-}$ gives a worst-case ('pessimistic') compactification ratio. Therefore, just concentrating on $\mathbf{X}_{5-}$, it is approximated as follows, for sufficiently large t_4 and/or t_5,

(14.85) $\mathbf{X}_{5-} \approx -(t_5^2 + t_4^2)\mathbf{X}_{5+}$.

Given $\mathbf{X}_{5+}$ is really just a 5D embedding of $\mathbf{X}_{3+}$, the URM3 vector space grows quadratically along $\mathbf{X}_{3+}$ with respect to either t_4 or t_5, and it matters not if t_4 is small relative to t_5 ($t_5 \gg t_4$), providing t_5 is sufficiently large. Likewise, in the converse case, $t_4 \gg t_5$, there is still quadratic growth in URM3 along $\mathbf{X}_{3+}$. Thus, either the fourth or fifth dimension can act in isolation to increase the size of the URM3 space, by growth in its evolutionary parameter, t_4 or t_5 respectively; both evolutionary times acting together can only increase this growth further. Given all excess dimensions only grow linearly with evolutionary time, the quadratic growth in $\mathbf{X}_{3+}$ will have the desired effect of making all the excess dimensions appear to shrink (compactify) and align along $\mathbf{X}_{3+}$, as also happens in URM3 for large evolutionary periods t_3.

To summarise, when both evolution times are comparable, and at least one of them is sufficiently large, then the compactification process still occurs. Thus, for any sufficiently large fourth or fifth-dimension evolutionary time, the five-dimensional vector space compactifies to that of the eigenvector space of URM3 as evolution progresses. Specifically, all 5D eigenvectors align with the single, static URM3 eigenvector $\mathbf{X}_{3+}$, which occupies the discrete, 2D, conical subspace of

URM3, hence URM5, like URM4, also compactifies to appear two-dimensional within the 3D space of URM3..

The final stage then is to show that this same compactification behaviour arises for any arbitrary, n-dimensional space, which is demonstrated using the general, n-dimensional solution.

(14-9) The General n-dimensional Solution

The general solution for URMn, $n \geq 4$, is actually obtained recursively by calculating the residual matrix for $\mathbf{A}_{n0}$ using an embedded matrix $\mathbf{A}_{(n-1)0}$; Appendix (C) gives an outline. However, the recursive solution is best given in a much simpler, unravelled form in terms of the URM3 vectors. Barring a single, linear Diophantine equation in URM3, see (A27), URMn is a completely solved problem, with an analytic solution for the eigenvectors parameterised by all n parameters, i.e. $n-2$ temporal parameters t_j, $j = 3 \ldots n$, and two non-temporal parameters k and l (A26).

With $\mathbf{0}_{33}$ defined as a 3×3 matrix of zeros, and $\mathbf{I}_{n-3}$ as the $(n-3) \times (n-3)$ identity matrix, then a constant, $n \times n$ matrix $\mathbf{M}_n$ is constructed as follows:

$$(14.91) \quad \mathbf{M}_n = C \begin{pmatrix} \mathbf{I}_{n-3} & 0 \\ 0 & \mathbf{0}_{33} \end{pmatrix}$$

The subscript n on $\mathbf{M}_n$ will be dropped from here onward and $\mathbf{M}$ assumed a square $n \times n$ matrix. Instead, where appropriate, the kth and ith column of $\mathbf{M}$ will be denoted by $\mathbf{M}_{ki}$.

The matrix $\mathbf{M}$ has a lead diagonal with all elements equal to eigenvalue C, except for the last three diagonal elements, which are zero. These last three zeros are, of course, so that the matrix $\mathbf{M}$ has no URM3, 3D contribution. Equally importantly, $\mathbf{M}$ is a constant matrix and has no time dependence; the ith column (index i) of $\mathbf{M}_{ki}$ (all k rows) is the equivalent of the initial value, zero eigenvector at $t_{j+3} = 0$, $j = 1 \ldots n - 3$, denoted by primed vector $\mathbf{X}'_{n0j}$, see further below.

Note that the index i, as in $i = n - (j+2)$, (14.92b) further below, goes from $n-3$ to 1 as j goes from 1 to $n-3$, and works across the first $n-3$ columns of $\mathbf{M}$, which are non-zero, unlike the last three columns.

The general solution for the vector $\mathbf{X}_{n+}$ is nothing more than the embedding of the static, URM3 vector $\mathbf{X}_{3+}$ in its last three elements, padded with $n-3$ leading zeros.

(14.92)

(14.92a) $\mathbf{X}_{n+} = \begin{pmatrix} \mathbf{0}_{n-3} \\ \mathbf{X}_{3+} \end{pmatrix}$

Denoting the kth element of $\mathbf{X}_{n-}$ by $\mathbf{X}_{n-,k}$, $k=1\ldots n$, then the general solution for vector $\mathbf{X}_{n-}$ is given by

(14.92b) $\mathbf{X}_{n-,k} = \begin{pmatrix} \mathbf{0}_{n-3} \\ \mathbf{X}_{3-} \end{pmatrix}_k + \sum_{j=1}^{n-3} \left(-t_{j+3}^2 \begin{pmatrix} \mathbf{0}_{n-3} \\ \mathbf{X}_{3+} \end{pmatrix}_k + 2t_{j+3}\mathbf{M}_{ki} \right),$

$k=1\ldots n$, $n \geq 4$, $i=n-(j+2)$.

The first zero vector $\mathbf{X}_{n0A}$ $(\sim \mathbf{X}_{n0j}, j=0)$ is just an embedding of the URM3 zero vector $\mathbf{X}_{30}$

(14.92c) $\mathbf{X}_{n0A} = \begin{pmatrix} \mathbf{0}_{n-3} \\ \mathbf{X}_{30} \end{pmatrix}$, $\mathbf{X}_{n0A} \sim \mathbf{X}_{n0j}, j=0$.

For all other zero vectors, denoting the kth element of $\mathbf{X}_{n0j}$ by $\mathbf{X}_{n0j,k}$, $k=1\ldots n$, then the kth element of the jth, zero eigenvector $\mathbf{X}_{n0j}$, $j=1\ldots n-3$, is given by

(14.92d) $\mathbf{X}_{n0j,k} = -t_{j+3}\begin{pmatrix} \mathbf{0}_{n-3} \\ \mathbf{X}_{3+} \end{pmatrix}_k + \mathbf{M}_{ki},$

$j=1\ldots n-3$, $k=1\ldots n$, $n \geq 4$, $i=n-(j+2)$.

$\mathbf{X}_{n0B} \sim \mathbf{X}_{n01}$ $(j=1)$, $\mathbf{X}_{n0C} \sim \mathbf{X}_{n02}$ $(j=2)$ etc.

By setting the evolutionary time t_{j+3} to zero, it is seen that the ith column of $\mathbf{M}$ is the initial, zero vector $\mathbf{X}'_{n0j}$, for the jth dimension, $j=1\ldots n-3$, $(i=n-3\ldots 1)$, i.e.

(14.93) $\mathbf{X}'_{n0j,k} = \mathbf{M}_{ki}$ at $t_{j+3} = 0$,

$j = 1 \ldots n-3$, $k = 1 \ldots n$, $n \geq 4$, $i = n - (j+2)$.

e.g. for URM4, $n = 4$, $j = 1$, $i = 1$, $k = 1 \ldots 4$, $\mathbf{X}'_{401,k} = \mathbf{M}_{k1}$

(14.94) $\mathbf{M} = \begin{pmatrix} C & 0 \\ \mathbf{0}_3 & \mathbf{0}_{33} \end{pmatrix}$, $\mathbf{M}_{k1} = \mathbf{X}'_{401} = \begin{pmatrix} C \\ \mathbf{0}_3 \end{pmatrix}$, $t_4 = 0$

e.g. for URM5, $n = 5$

(14.95a) $\mathbf{M} = \begin{pmatrix} C & 0 & 0 \\ 0 & C & 0 \\ \mathbf{0}_3 & \mathbf{0}_3 & \mathbf{0}_{33} \end{pmatrix}$

$j = 1$, $i = 2$, $k = 1 \ldots 5$, $\mathbf{X}'_{501,k} = \mathbf{M}_{k2}$

(14.95b) $\mathbf{X}'_{501,k} = \begin{pmatrix} 0 \\ C \\ \mathbf{0}_3 \end{pmatrix}_k$, $\mathbf{M}_{k2} = \mathbf{X}'_{501} \sim \mathbf{X}_{50B} (t_4 = 0)$

$j = 2$, $i = 1$, $k = 1 \ldots 5$, $\mathbf{X}'_{502,k} = \mathbf{M}_{k1}$

(14.95c) $\mathbf{X}'_{502,k} = \begin{pmatrix} C \\ 0 \\ \mathbf{0}_3 \end{pmatrix}_k$, $\mathbf{M}_{k1} = \mathbf{X}'_{502} \sim \mathbf{X}_{50C} (t_5 = 0)$.

Some points on this general solution are made following:

(14.96a) The two vectors $\mathbf{X}_{n+}$ (14.92a) and $\mathbf{X}_{n0A}$ (14.92c) remain, effectively, the URM3 vectors $\mathbf{X}_{3+}$ (B2a) and $\mathbf{X}_{30}$ (B2b), embedded as the last three elements within their n-dimensional counterparts, and padded with $n-3$ leading zeros.

(14.96b) All expressions for arbitrary dimension n (including $n = 3$) are quadratic only in the evolutionary time; there are no higher order terms. Furthermore, it is only the $\mathbf{X}_{n-}$ vector (14.92b), physically equated with a position (J2), which contains this quadratic term.

(14.97) Each additional dimension adds a new zero eigenvalue and associated eigenvector, e.g. $\mathbf{X}_{40B}$ (14.41d) for URM4, and $\mathbf{X}_{50C}$ (13.52e) for URM5.

(14.98) Eigenvector $\mathbf{X}_{n-}$ generally always comprises n non-zero elements. Because such vectors are Pythagorean n-tuples, the Pythagoras equation acts as a constraint, and so the n elements occupy an $n-1$, discrete hypersurface of the n-dimensional space. In the 3D, URM3 case, the eigenvectors $\mathbf{X}_{3+}$ and $\mathbf{X}_{3-}$ are both two-parameter families, parameters k and l (A26) for $\mathbf{X}_{3+}$, and s and t (A30) for $\mathbf{X}_{3-}$. Since the elements of the vectors satisfy the Pythagoras equation, the two-parameter surface of each vector is geometrically known as a discrete cone in [1],#3.

(14.99) The zero vectors $\mathbf{X}_{n0B}$, $\mathbf{X}_{n0C}$ etc., always comprise four or less (usually always exactly four), non-zero elements, but never more than four in any arbitrary dimension n, i.e. they contain at least $n-4$ zero elements, and generally occupy a 4D subspace of URMn. The zero vector $\mathbf{X}_{n0A}$, not mentioned in this point, is noted above in point (14.96a).

(14.100) The zero vectors $\mathbf{X}_{n0A}$, $\mathbf{X}_{n0B}$, $\mathbf{X}_{n0C}$ etc., with their reciprocals $\mathbf{X}^{n0A}$, $\mathbf{X}^{n0B}$ and $\mathbf{X}^{n0C}$ etc., Appendix (E), satisfy the same, hyperbolic DCE (13.37) as per URM3, which is given by the scalar product $\mathbf{X}^{n0}\mathbf{X}_{n0}$, see Appendix (F11).

(14.101) Each zero vector $\mathbf{X}_{n0A}$, $\mathbf{X}_{n0B}$, $\mathbf{X}_{n0C}$ etc., is implicitly parameterised in terms of the URM3, three-parameter solution k, l (A26) and t_3 ($\sim m$) (A28), by virtue that they all embed the URM3 vector $\mathbf{X}_{30}$ (A33b). However, eigenvalue C is also present in all these vectors, except $\mathbf{X}_{n0A}$ (14.92c), and acts as an initialisation parameter, dictating the total conserved energy (per unit mass) of the space, as in $E = C^2$ (J1).

Continuing with the analysis of the n-dimensional solution for large evolutionary times t_n, identical remarks made for the URM4 and URM5 solutions also apply to the general case URMn. Therefore, generalising these remarks to URMn:

The only time-dependent $(t_i,\ i = 4...n)$ vectors with a non-zero contribution to the excess dimensions are $\mathbf{X}_{n-}$ and the zero eigenvectors $\mathbf{X}_{n0j}$, $j = 1...n-3$, $\mathbf{X}_{n01} \sim \mathbf{X}_{n0B}$, $\mathbf{X}_{n02} \sim \mathbf{X}_{n0C}$ etc, i.e. all but $\mathbf{X}_{n+}$ and $\mathbf{X}_{n0A}$ $(\sim \mathbf{X}_{n0j}, j = 0)$. The zero eigenvectors $\mathbf{X}_{n0j}$ are also naturally separated into one vector for each dimension.

There are no mixed terms comprising products of two or more evolutionary times, e.g. $t_4 t_5$, and $\mathbf{X}_{n-}$ can be decomposed into independent terms in each evolutionary time t_i, hence the summation form of $\mathbf{X}_{n-}$ (14.92b).

Not only does the URM n solution separate into a vector term for each unique, excess dimension, but also the terms are identical, when disregarding the particular, excess dimension. They are identical in so far as the rth dimensional component $(r = 4...n)$ contributes a linear term $2t_r C$ in $\mathbf{X}_{n-}(t_r)$, and C in $\mathbf{X}_{n0(r-3)}(t_r)$, and the s dimensional component $(s = 4...n,\ r \neq s)$ contributes a term $2t_s C$ in $\mathbf{X}_{n-}(t_s)$ and C in $\mathbf{X}_{n0(s-3)}(t_s)$, which are identical upon interchange of times t_r and t_s. Since it is only the magnitude of the contribution in the excess dimension that matters in the analysis of the compactification ratio χ, Section (14-5), each dimension can be treated separately and, furthermore, identical findings for one specific dimension apply to all the others. The separability of the URM n solution into unique dimensional components, and the interchange symmetry between any two components (as regards ratio χ), means that each dimension acts independently of the other, with identical behaviour for equal evolutionary times; the same general expression for χ being used in each case.

The solution for a single, excess rth dimension can be isolated from all the other $n-4$ dimensions s, where $s = 4...n$, $s \neq r$, by setting all other evolutionary times to zero, i.e. $t_r > 0$, $t_s = 0$.

With these points in mind, then generalising to a specific, rth dimension, $r = 4 \ldots n$, evolutionary time t_r, for sufficiently large times t_r (14.82), the rth dimension's compactification ratio χ_r is given in both an exact and approximated form as follows, which are simply relabelled versions of (14.23) and (14.27) respectively,

$$(14.102) \quad \chi_r = \frac{|\mathbf{X}|_r}{|\mathbf{X}|_3}, \text{ exact form, (14.23)}$$

$$\chi_r \approx \frac{2C}{t_r |\mathbf{X}_{3+}|}, \text{ (14.27), } t_r \neq 0, \text{ use (14.23) when } t_r = 0.$$

Since χ_r is inversely proportional to time t_r, it will converge to zero for all times $t \geq t_r$, and by the preceding arguments in Section (14-5) onwards, the entire solution compactifies to URM3.

To summarise, when any one or more evolutionary times in any excess dimension is sufficiently large, the n-dimensional vector space compactifies to that of the eigenvector space of URM3. Specifically, all n-dimensional eigenvectors align with the single, static URM3 eigenvector $\mathbf{X}_{3+}$, which occupies a 2D conical subspace of URM3, hence URMn compactifies to appear two-dimensional. The conical subspace is actually a 2-parameter, discrete surface embedded within 3D, and termed the 'cone' in URMT, [1],#3.

This completes the general analysis of compactification for an n-dimensional space in URMT.

(14-10) Answers to Anticipated Questions

Before summarising and concluding the work, a few anticipated questions are answered.

(14.110) Why not set all evolutionary times equal?

As currently formulated, n-dimensional URMT has $n-2$ temporal parameters controlling the evolution, of which $n-3$ parameters control the evolution of each of the $n-3$ excess dimensions, one for each dimension; the last three dimensions, that of URM3, share the single evolution parameter t_3. As concluded, if any one or more of these $n-2$ parameters is sufficiently large (Section (14-6)), then the entire n-dimensional eigenvector space compactifies to a two dimensional, conical subspace of URM3, that of eigenvector $\mathbf{X}_{3+}$. Because of this, there is no compelling reason to make them all the same parameter, as doing so would only hasten the compactification, but not change the final result - with one small exception: if only the jth parameter of an excess dimension were made large, and all others, say, relatively small or zero, i.e. $t_j \gg t_i, i = 4 \ldots n - 3, i \neq j$, then the jth dimension would appear relatively larger than all other excess dimensions, by a factor of t_j / t_i, because the jth dimension is of size $t_j C$ and the ith dimension of size $t_i C$. Of course, relative to URM3 and $\mathbf{X}_{3+}$, and under the sufficiency condition (14.82), the jth dimension would still appear small. Thus, this exception might be justification to set all evolutionary parameters equal so that all excess dimensions appear of the same relative size; this issue remains open.

(14.111) Does it have to be in integers?

Integers are used throughout URMT but, as regards compactification, it is not currently known whether this is strictly necessary. Certainly URMT in [1],#1 can go quite a long way before integers are required, and then they only enter when gcd conditions are imposed, which is after transformation invariance is imposed. Provided some form of quantisation is mandated, it may well be feasible to broaden the compactification aspects to the real and complex domains. Nevertheless, URMT is currently formulated entirely in integers, and therefore the mathematics of URMT compactification is also formulated in integers. It is anticipated that complex integers may well enter in further development of the theory, but this matter is still in its infancy.

(14.112) Why stop at URM3, why not URM2 or lower?

Firstly, there is no meaningful URM1, but URM2 is perfectly plausible, see Appendix (H).

The compactification has been shown to terminate at the 2D conical subspace, represented by $\mathbf{X}_{3+}$, of URM3's three-dimensional eigenvector space. However, can the compactification continue within URM3 down to URM2?

The answer is yes, if a trivial (I10) initial solution for $\mathbf{X}_{3+}$ is acceptable, and no, if unacceptable - which is the eventual answer, i.e. it isn't acceptable on the grounds of being too simplistic. However, assuming yes for a while, such a trivial solution would be a vector $\mathbf{X}_{3+} = (0,1,1)^T$, where the x coordinate in $\mathbf{X}_{3+}$ is zero and $y = 1$, $z = 1$, giving the trivial Pythagorean twin $(1,1)$, where $1^2 = 1^2$! This twin pair $(1,1)$ is the realm of URM2, and the solution $\mathbf{X}_{3+} = (0,1,1)^T$ is basically a URM2 lifted solution $\mathbf{X}_{2+} = (1,1)^T$ (the only such primitive solution) embedded within URM3. Whilst this pair $(1,1)$ is considered rather simple, it can actually be used as the starting point to generate 3D and higher solutions, as illustrated by examples (7-1) to (7-3) in Section (7). In other words, URMT can start at URM2 and work upward, just like URM3 works upward to URM4 etc. However, because $\mathbf{X}_{2+} = (1,1)^T$ is the only primitive, non-trivial solution, it is not even 2D or 1D, but just a point in space. A non-primitive solution such as $\mathbf{X}_{2+} = (y,y)^T$, for parameter y, is mathematically acceptable, giving a 1D, straight line solution, but then the gcd criterion (13.4c) on $\mathbf{X}_{2+}$ is not satisfied since y is a common factor of both elements. More importantly, even this solution is limited, and lifting it to 3D only gives a subset of the full URM3 formulation.

As a consequence of the above, URM2 is considered too limited to be of physical use, at least at present, i.e. it has not been rejected and its place in URMT is left undecided - it does have a beauty in its simplicity, but perhaps just a bit too simple. A good reason not to dismiss it lightly is that does not (cannot) have its own evolutionary parameter t_2, and cannot, therefore, compactify from two to one dimension, at least not by the growth of an evolutionary parameter. Secondly, its only free parameter is the eigenvalue C, which relates to the total conserved energy $E = C^2$ (J1). In fact, eigenvalue C is an initial condition since it is conserved, i.e. invariant to arbitrary variations in all n free parameters of URMn (when under Pythagoras conditions - there are many more when not under these conditions).

Knowing there is the capability to reduce to URM2, the next question is, why doesn't this appear in the URM3 evolution equations, Appendix (B)?

Basically, the triad of URM3 eigenvectors flattens (or aligns) in the large evolutionary limit, $t_3 \rightarrow \infty$, to $\mathbf{X}_{3+}$, see (I5), but none of the elements of the vector $\mathbf{X}_{3+}$ shrink relative to each other because t_3 scales every element of the vector $\mathbf{X}_{3+}$ equally, and every element x, y, z of $\mathbf{X}_{3+}$ is generally non-zero. The vector $\mathbf{X}_{3+}$, as for all higher dimensional forms $\mathbf{X}_{n+}$, is static, i.e. not a function of any evolutionary time t_j, but it is still a function of two free parameters, k and l. Whilst a subset of solutions for $\mathbf{X}_{3+}$ could have one of k or l zero, but not both (A26d), neither are generally zero and $\mathbf{X}_{3+}$, therefore, comprises three non-zero elements and is classed as non-trivial. Alternatively stated, none of the two elements, x or y of $\mathbf{X}_{3+}$, is always zero, and z is never zero by Pythagoras (F1).

The non-triviality of $\mathbf{X}_{3+}$ is the absolute key as to why the example 5D formulation, vector $\mathbf{X}_{5+}$ (13.32) embeds $\mathbf{X}_{3+}$ as x, y, z, but is zero in its fourth and fifth dimensional elements (the first two elements of the vector). In effect, it is an embedding of a non-trivial 3D formulation, i.e. that of URM3. Conversely, a non-trivial 4D embedding, i.e. four non-zero elements w, x, y, z, would compactify to URM4 but not compactify further to URM3.

Concluding the above, the compactification can be made to stop with URM2 according to the initial conditions, i.e. by arranging the number of non-zero elements in the lowest dimension of $\mathbf{X}_{n+}$, but with the simplest and only primitive, 2D solution, $\mathbf{X}_{2+} = (1,1)^T$, it would compactify to the point (1,1) - perhaps this really is telling us something! In the work presented herein, it stops at three by virtue of three non-zero elements x, y, z in $\mathbf{X}_{3+}$. Therefore, the biggest reason not to stop at URM2 is one of too much simplicity and not enough complexity. Mathematically this is fine, but physically it is rather simplistic. However, this raises the next question.

(14.113) Why not stop at URM4 or higher?

From the arguments at the end of the previous question it would seem that (using URM5 as an example), setting all four elements w, x, y, z of $\mathbf{X}$ (13.4a) non-zero, and the first element v to zero, is sufficient to stop the compactification process at URM4 instead of URM3.

First and foremost, there is a strong case against this given the following physical constraints placed upon the solutions.

(14.113a) The solution for $\mathbf{X}_{n+}$ must be invariant to all arbitrary variations in all evolutionary parameters - this is a URMT imposed, general constraint arising from the Invariance Principle (8.7), which is essentially a postulate of URMT.

(14.113b) The Potential V must be zero and invariant for all arbitrary variations in all parameters. In URM4 and URM5, at least, this condition makes for only two non-zero eigenvalues $\pm C$, with all others zero; see (4.14a) and (13.39). To satisfy this invariant, zero Potential constraint, there are the two further constraints:

(14.113c) The matrix $\mathbf{A}_{n0}$ must satisfy Pythagoras conditions to give two eigenvectors, $\mathbf{X}_{n+}$ and $\mathbf{X}_{n-}$, which are Pythagorean n-tuples. Because they satisfy the n-dimensional Pythagoras equation, the eigenvectors have zero norm (I8) and, in the case of URM4, they thus form STR-like, zero norm, four-vectors. For physical reasons, URMT must accommodate STR somewhere.

(14.113d) The matrix $\mathbf{A}_{n0}$ must have a certain form whereby all the rows and columns, barring the last three (URM3) or four (URM4), are multiples of the eigenvectors $\mathbf{X}^{(n-1)-}$ (rows) and $\mathbf{X}_{(n-1)+}$ (columns) as per (C1). In the four-dimensional case of URM4, this point is discussed in more detail in Section (3-2), and is very important for the purposes of obtaining a quadratic polynomial expression for the eigenvectors. This then leads to the highest order terms in any expression being a quadratic function of the evolutionary parameters $t_j, j = 3...n$, and making all scalar invariants quadratic in the eigenvalue, i.e. C^2, which is important to physically associate the general DCE (F11) with an energy conservation equation, irrespective of dimension n.

To satisfy the above conditions for URM3 is relatively easy and leads to an infinite set of solutions for eigenvectors $\mathbf{X}_{3-}$ and $\mathbf{X}_{30}$, parameterised by t_3 ($\sim m$ or δ), for every static eigenvector $\mathbf{X}_{3+}$. As noted, $\mathbf{X}_{3+}$ is a full, two-parameter family, parameters k and l, see (A26) and (14.98).

All these above points are very important because it seems to be all change in URM4. Taken together, the above conditions (14.113a to 14.113d) are severe, and to meet them in URM4 it seems there is only one non-trivial solution for $\mathbf{X}_{4+}$, i.e. all four elements non-zero, which is the 'PS+RU' solution. This solution is first given in Section (5), and subject to variational methods in Section (9), where it is shown to be unique under an arbitrary, global variation. It is actually considered remarkable that there is exactly one solution and not two or more, or even no solutions - just one single solution? Even more curious, this solution is the simplest it could possibly be, namely the Pythagorean quadruple $(2,1,2,-3)$, with strict adherence to the sign of its elements. It also only arises by adding in a lot of simplifications. Whilst this particular solution appears to be the only solution under its PS+RU parameterisation scheme, no general proof that URM4 cannot satisfy the above conditions exists, without more simplifications. The most notable simplification is to change $\mathbf{X}_{4+}$ from four non-zero elements to three, and thereby reduce the theory to that of URM3 under the above conditions. Whilst no proof is offered, the author has not found any general solution, of any worth without having to add too many simplifications or reducing URM4 back to URM3.

With four-dimensional STR in mind, it might seem nicer to stop at URM4. This would mean all four eigenvectors of URM4 would, generally, be non-trivial (four non-zero elements). However, non-trivial 4D vectors, specifically $\mathbf{X}_{4-}$, can already be generated from the URM3 formulation, as can higher, n-dimensional vectors $\mathbf{X}_{n-}$, by embedding URM3 within URMn. Given $\mathbf{X}_{n-}$ can be physically associated with a position vector, it would seem URM3 will suffice, at least spatially speaking, to give a four-vector position. Of course, that's just position, and not the only four-vector, so it is nicer to have a bit more flexibility, which is why URMT was originally extended from its 3D origins in [1].

From another physical perspective though, nature is generally ternary, everything (almost) appears to come in threes, from spatial dimensions to families of particles. It is also of note that URM5 (a favourite of the author and a reason behind its usage herein), has three evolutionary time parameters, t_3, t_4 and t_5, which makes a nice symmetric triplet to go with the three spatial dimensions of URM3. Admittedly this is aesthetics; after all, this does not include the laboratory time t, which is considered to be an interval (in URMT anyhow), not an absolute evolutionary time; see Section (12-3).

However, the original URMT formulation in [1] seems at its best for URM3, at least in terms of results, physical interpretation and tractability. It has three sets of standard variables, plus their conjugate and dual forms, see [1],#5. Additionally, the dynamical variables P, Q, R and their conjugates $\overline{P}, \overline{Q}, \overline{R}$ are isomorphic to the complex roots of unity (A2), whereas in the most general, n-dimensional formulation of URMT, the dynamical variables in the excess dimension, e.g. S, T, U, M, H, N, J in the URM5 $\mathbf{A}_{50}$ matrix (13.1), lose this isomorphic nature; see Section (1-6). This may seem a whim, but it is a nod to the fact that mathematical physics uses complex numbers as a given, and it is felt URMT requires an equivalent algebra in its discrete formulation. So, whatever n-dimensional URMT formulation is used, it is desirable it to encompass URM3 as a subset, and this is one reason why the matrix $\mathbf{A}_{30}$ (A1a) is embedded in $\mathbf{A}_{50}$ (13.1).

To conclude these questions, starting the general, n-dimensional URMT formulation at URM3 seems to offer the best combination of being able to extend to any number of dimensions, whilst retaining all physical properties of URM3. Although stopping the formulation at URM3 means that compactification also stops at two spatial dimensions, it also ensures that it doesn't descend into triviality, i.e. too much physical simplicity.

(14.114) One last question

If a velocity eigenvector, e.g. $\mathbf{X}_{50B}$, grows linearly with time, by virtue of a constant acceleration $\mathbf{X}_{3+}$ as in $\mathbf{X}_{50B} = -t_4 \mathbf{X}_{3+}$ (13.52d), will it not at some stage exceed the speed of light, little c?

Strictly yes, but this is a 'space', and it is not yet physically clear what the expansion limit of the space is or, indeed, whether it has a physical presence (for want of a better term). It could simply be that the acceleration is so tiny that the evolutionary time has to be enormous to compensate, and no such evolutionary stage has yet been reached. Going back to URMT's roots in URM3 [1], the space is basically a discrete, infinite set of points, i.e. the eigenvector space [I17], also known as the lattice '**L**', [1],#3, and it is the underlying space upon which it is thought that physics plays-out as functions on this lattice.

(14-11) Summary

In brief, the entire compactification process shows that the first three dimensions expand relative to the excess dimensions, making the excess dimensions appear to shrink over time with respect to the first three. However, the expansion in the first three dimensions is in a particular vector direction, characterised by two free parameters. Hence the compactification is said to converge to a two-dimensional subspace of the first three dimensions.

Note, all quantities are in integers; all spaces are discrete sets of points.

It is important to note that the entire work focuses on the evolution of the eigenvectors relative to those of URM3. Keep in mind that the URM3 eigenvectors also evolve in an identical fashion, all converging on the single $\mathbf{X}_{3+}$ vector, see Appendix (B). It is chiefly this reason that it is not necessary to delve into URM3's eigenvector evolution herein; the compactification conclusions are the same. URM3's geometric evolution, in terms of 'flattening', cones and hyperboloids, is discussed in Section (12), and fully covered in [1],#3.

Eigenvectors and Eigenvalues

The general, n-dimensional unity root matrix theory, URMn, $n \geq 3$, for a square $n \times n$ matrix $\mathbf{A}_{n0}$, has n linearly independent eigenvectors, each with n elements. The first element is the nth dimensional coordinate, and the last three elements are the first three dimensions, and functions of the three URM3 eigenvectors $\mathbf{X}_{3+}$ (acceleration), $\mathbf{X}_{30}$ (velocity) and $\mathbf{X}_{3-}$ (position).

The eigenvectors split into three types, as per URM3, a single acceleration vector $\mathbf{X}_{n+}$ (eigenvalue C), a single position vector $\mathbf{X}_{n-}$ (eigenvalue $-C$), and $n-2$ velocity vectors $\mathbf{X}_{n0j}$, $j = 0...n-3$ (for $n-2$ zero eigenvalues). These types are also termed 'plus', 'minus' and 'zero' un URMT, see (I11), (I9) and (17) respectively.

The entire, discrete, eigenvector space (I17) is characterised by $n-2$ temporal parameters t_j, $j = 3...n$, and two non-temporal, k and l, i.e. n parameters in total.

The first three dimensions (last three elements of all vectors) have a single evolutionary time parameter t_3, the remaining $n-3$ excess dimensions each have their own temporal parameter, t_j, $j = 4...n$.

Each temporal parameter emerges from an invariance transformation on the elements (the dynamical variables) of the unity root matrix $\mathbf{A}_{n0}$, which leaves the eigenvector equation $\mathbf{A}_{n0}\mathbf{X}_{n+} = C\mathbf{X}_{n+}$ invariant to any arbitrary variation in these parameters.

The other two non-temporal parameters, k and l, control the URM3 vector solution, most notably the acceleration $\mathbf{X}_{3+}$ and hence also $\mathbf{X}_{n+}$.

Other than these n parameters ($k\,l$, t_j, $j = 3...n$,), there is a single, fundamental constant, the eigenvalue C (big C), which is the only contributor to all excess dimensions, excepting their respective evolutionary, temporal parameters, which also affect the first three dimensions.

The constant eigenvalue C is equated with a scalar velocity (speed), and controls the size of all excess dimensions. It is related to the total conserved energy $E = C^2$, and is also the single tuning constant for specifying the initial, time-zero, expansion velocity $\mathbf{X}_{n0j}$ of the excess dimensions, $j = 1...n-3$.

Eigenvector evolution

The first three dimensions (last three elements) of the acceleration vector $\mathbf{X}_{n+}$ (14.92a) are just the URM3 acceleration vector $\mathbf{X}_{3+}$, which is a non-trivial (three non-zero elements), static (no evolutionary time-dependence) acceleration vector.

The first three dimensions of the jth velocity vector $\mathbf{X}_{n0j}$, $j = 1...n-3$ (14.92d), are a linear function in time t_{j+3} of the URM3 acceleration vector $\mathbf{X}_{3+}$, i.e. the integral of acceleration to give velocity, $\mathbf{X}_{n0j} = -t_{j+3}\mathbf{X}_{n+}$. This is also true for the URM3 velocity vector $\mathbf{X}_{30}$ scaled by t_3, i.e. $j = 0$.

The first three dimensions of the position vector $\mathbf{X}_{n-}$ (14.92b) are a quadratic function in time of the URM3 acceleration vector $\mathbf{X}_{3+}$, with a contribution for each excess temporal coordinate t_j, $j = 1 \ldots n-3$, i.e. the sum $\mathbf{X}_{n-} = \sum -t_{j+3}^2 \mathbf{X}_{3+}$. The magnitude of this growth is approximated by $\left| t_{j+3}^2 \mathbf{X}_{3+} \right|$, e.g. $\left| t_5^2 \mathbf{X}_{3+} \right|$ (13.52b), for one specific evolutionary time t_j, a summation more appropriate for evolution in two or more dimensions - see (14.61) or (14.69a). This is also true for the URM3 velocity vector $\mathbf{X}_{3-}$ scaled by t_3^2, i.e. $j = 0$.

All excess dimensions (the first $n-3$ elements) of the acceleration vector $\mathbf{X}_{n+}$ (14.92a) are all zero, i.e. there is no accelerating expansion in the excess dimensions.

The jth excess dimension (element $n-(j+2)$) of each velocity vector $\mathbf{X}_{n0j}$, $j = 1 \ldots n-3$ (14.92d), comprises the constant eigenvalue C only, with no time-dependence, i.e. static in all excess dimensions. Therefore, the excess dimensions of the velocity vector remain a constant 'size', i.e. constant velocity (or speed) C.

The jth excess dimension of the position vector $\mathbf{X}_{n-}$ (14.92b) is a linear function in evolutionary time t_j of the velocity constant C, i.e. the integral $2t_j C$, e.g. $\left| 2t_5 C \right|$ in (13.52b).

All the above eigenvector integrals are really just Newton II, to within a sign and scale factor. But note that the excess dimensions are a first integral of the constant velocity C, whereas the first three dimensions are first and second integrals of a constant acceleration $\mathbf{X}_{3+}$.

Of all these eigenvector growth rates, only the first three dimensions of the position vector $\mathbf{X}_{n-}$ have a quadratic dependence on time. There is no quadratic time contribution in any excess dimensions, only linear. All quadratic growth is therefore in the $\mathbf{X}_{n-}$ vector, which aligns, over time, in the direction of the $\mathbf{X}_{3+}$ acceleration vector, embedded in the first three dimensions, i.e. the home of URM3.

Although URM3's own explicit evolution has not been detailed, see further above, the URM3 eigenvector evolution equations in Appendix (B) show identical evolutionary behaviour for all URM3 eigenvectors; in particular, quadratic growth with respect to time t_3 in the $\mathbf{X}_{3-}$ vector and in the direction of the acceleration $\mathbf{X}_{3+}$.

Summarising the above: for any single, sufficiently large (Section (14-6)), evolutionary time t_j, where t_j is any one of the $n-2$ evolutionary time parameters, the entire n-dimensional vector space converges on the $\mathbf{X}_{n-}$ position vector, which itself aligns and grows along URM3's $\mathbf{X}_{3+}$ acceleration vector. Simultaneously, the excess dimensions continue to grow, albeit as a linear function of time with a constant velocity C in each excess dimension.

Comparing the linear growth of the jth excess dimension $2|t_j C|$, for a specific evolutionary time t_j, with the simultaneous, quadratic growth $|t_{j+3}^2 \mathbf{X}_{3+}|$ of the first three dimensions, gives a measure of the relative size of the jth excess dimension with respect to the first three dimensions. This measure is termed the compactification ratio of the jth dimension, denoted by χ_j, and approximated as follows; see (14.23) for the exact form for χ_j.

(14.27) $\chi_j \approx \dfrac{2C}{t_j |\mathbf{X}_{3+}|}$, the compactification ratio of the jth dimension.

Note that χ_j is dimensionless, and the ratio is inversely proportional to the evolutionary time t_j such that it limits to zero, i.e.

(14.24) $\lim\limits_{t_j \to \infty} \chi_j = 0$.

The relative error ε (14.68) in the approximation also limits to zero, i.e. the approximation gets better with increasing time t_j.

As a consequence of χ_j limiting to zero, then over a sufficiently large evolutionary time t_j the size of the excess, jth dimension appears to shrink into insignificance with respect to the first three dimensions; concurrent growth in any other excess dimension only hastening the compactification.

Lastly though, the 'first three dimensions' are really just the single direction of the acceleration vector $\mathbf{X}_{3+}$. Whilst this might seem to be a compactification to one dimension, i.e. that linear direction in which $\mathbf{X}_{3+}$ points, the vector $\mathbf{X}_{3+}$ is arbitrarily specified by two other, non-temporal parameters k and l. In fact, $\mathbf{X}_{3+}$ is actually a Pythagorean triple, where the two parameters form the standard Pythagorean parameterisation. Thus, $\mathbf{X}_{3+}$ is really a 2D, discrete, conical surface, described as two cones, 'upper' and 'lower', in URMT [1],#3.

(14-12) Conclusion

The n-dimensional, discrete eigenvector space of Unity Root Matrix Theory appears to reduce its dimensionality, i.e. compactify, as its temporal evolution progresses, to a two-dimensional, discrete, conical surface embedded within a three-dimensional, discrete, eigenvector space. The conical surface is formed from the elements of a two-parameter, static acceleration eigenvector, to which all eigenvectors align in the limit as the evolutionary time, in one or more dimensions, tends to infinity.

Appendices

Appendix (A) URM3

This Appendix provides some basic background information on URM3, as fully detailed in [1].

The general, unity root matrix $\mathbf{A}_3$, comprising 'dynamical variables' P, Q, R and their conjugates $\overline{P}, \overline{Q}, \overline{R}$, is defined as

(A1)

(A1a) $\mathbf{A}_3 = \begin{pmatrix} 0 & R & \overline{Q} \\ \overline{R} & 0 & P \\ Q & \overline{P} & 0 \end{pmatrix}$

(A1b) $P, Q, R \in \mathbb{Z}, \ (P, Q, R) \neq (0,0,0)$

(A1c) $\overline{P}, \overline{Q}, \overline{R} \in \mathbb{Z}, \ (\overline{P}, \overline{Q}, \overline{R}) \neq (0,0,0)$

(A1d) Notation

$\mathbf{A}_3 \sim \mathbf{A}$ in [1] for general URM3

$\mathbf{A}_3 \sim \mathbf{A}_{30} \sim \mathbf{A}_0$ in [1],#2 under URM3 Pythagoras conditions, further below.

The dynamical variables have the following unity (or 'primitive' [6]) root properties for eigenvalue C and arbitrary, integer exponent n; see (A3) for the moduli x, y, z. Strictly speaking, they are only 'unity' roots if the eigenvalue is unity, i.e. $C = 1$, but nth order congruences, for arbitrary $C \neq 0$, can be dealt with under the same subject, and are more generally referred to as 'power residues' [6].

(A2) $n \geq 2$

(A2a) $P^n \equiv +C^n \ (\mathrm{mod}\, x)$

(A2b) $Q^n \equiv +C^n \pmod{y}$

(A2c) $R^n \equiv -C^n \pmod{z}$

(A2d) $\overline{P}^n \equiv +C^n \pmod{x}$

(A2e) $\overline{Q}^n \equiv +C^n \pmod{y}$

(A2f) $\overline{R}^n \equiv -C^n \pmod{z}$.

The conjugate variables $\overline{P}, \overline{Q}, \overline{R}$ relate to their standard forms P, Q, R via the following 'conjugate relations'

(A2g) $\overline{P} \equiv P^{n-1} \pmod{x}$

(A2h) $\overline{Q} \equiv Q^{n-1} \pmod{y}$

(A2i) $\overline{R} \equiv -R^{n-1} \pmod{z}$.

The conjugation of a dynamical variable is therefore one of raising it to the power $n-1$ and obtaining its residue, modulo a coordinate, excepting a specific choice of sign.

By their definitions (A2), for unity eigenvalue $C = 1$, the dynamical variables are isomorphic to the complex roots of unity, e.g. $P \sim Z$, $\overline{P} \sim Z^* \sim Z^{n-1}$, and $P\overline{P} \equiv P^n \equiv +1 \pmod{x}$, $ZZ^* \sim ZZ^{n-1} = Z^n = 1$.

An eigenvector $\mathbf{X}_3$ to matrix $\mathbf{A}_3$, for the single eigenvalue C, is defined as

(A3) $\mathbf{X}_3 = \begin{pmatrix} x \\ y \\ z \end{pmatrix}$

$x, y, z \in \mathbb{Z}$, $(x, y, z) \neq (0,0,0)$, $\gcd(x, y, z) = 1$,

$\gcd(x, y) = \gcd(x, z) = \gcd(y, z) = 1$, co-primality in pairs.

(A4) $\mathbf{A}_3 \mathbf{X}_3 = C\mathbf{X}_3$, $C \in \mathbb{Z}$, $C \geq 1$,

(A4b) Notation

$\mathbf{X}_3 \sim \mathbf{X}$ in [1] for general URM3

$\mathbf{X}_3 \sim \mathbf{X}_{3+} \sim \mathbf{X}_+$ in [1] under URM3 Pythagoras conditions (A18).

The dynamical equations are obtained by expanding out the eigenvector equation (A4)

(A5)

(A5a) $Cx = Ry + \overline{Q}z$

(A5b) $Cy = \overline{R}x + Pz$

(A5c) $Cz = Qx + \overline{P}y$.

The non-singular matrix condition for the eigenvector equation (A4) to have solutions is

(A6) $\det(\mathbf{A}_3 - C\mathbf{I}_3) = 0$,

and evaluating this using (A1a), gives the following Dynamical Conservation Equation (DCE) :

(A7) $+C^2 = P\overline{P} + Q\overline{Q} + R\overline{R} + (PQR + \overline{PQR})/C$, the URM3 DCE.

A kinetic energy term K_3 and Potential energy term V_3 are defined in terms of the dynamical variables as

(A8) $K_3 = P\overline{P} + Q\overline{Q} + R\overline{R}$

(A9) $V_3 = \dfrac{(PQR + \overline{PQR})}{C}$.

(A10) Notation: $K_3 \sim K$ in [1], $V_3 \sim \dfrac{V}{C}$ in [1].

Note that the URM3 Potential, symbol 'V_3' here, is actually the same as the URM3 'modified' Potential V' in [1],#6. However, this modified Potential V' is not used in this book, as it is not required: Most Potential terms used herein are quadratic, except that of URM5, which is quartic, see (13.27). This is contrary to [1] where V_3 is superficially a cubic, and V' is quadratic, and introduced specifically to make the

Potential a quadratic function of the velocity, i.e. with physical units of energy per unit mass (J14b).

The Potential V_3 is related to the determinant of $\mathbf{A}_3$ (A1a) by

(A11) $\det(\mathbf{A}_3) = CV_3$.

The DCE (A7) is now re-written in terms of the kinetic energy K_3 (A8) and Potential energy V_3 (A9) as

(A12) $C^2 = K_3 + V_3$, the DCE.

Associated with $\mathbf{X}_3$ ($\sim \mathbf{X}_{3+}$) is a reciprocal, row eigenvector $\mathbf{X}^3$ ($\sim \mathbf{X}^{3+}$) comprising scale factors $\alpha_3, \beta_3, \gamma_3$

(A13) $\mathbf{X}^3 = \begin{pmatrix} \alpha_3 & \beta_3 & \gamma_3 \end{pmatrix}$.

The vector $\mathbf{X}_3$ satisfies the reciprocal, row eigenvector equation

(A14) $\mathbf{X}^3 \mathbf{A}_3 = C\mathbf{X}^3$, notation $\mathbf{X}^3 \sim \mathbf{X}^{3+} \sim \mathbf{X}^+$.

The scale factors $\alpha_3, \beta_3, \gamma_3$ are related to the coordinates x, y, z and the dynamical variables, P, Q, R and $\overline{P}, \overline{Q}, \overline{R}$, by three key relations, also known as 'divisibility relations',

(A15)

(A15a) $(C^2 - P\overline{P}) = \alpha_3 x$

(A15b) $(C^2 - Q\overline{Q}) = \beta_3 y$

(A15c) $(C^2 - R\overline{R}) = \gamma_3 z$.

There are six additional, equivalent relations

(A16)

(A16a) $(C\overline{R} + PQ) = \alpha_3 y$

(A16b) $(CQ + \overline{RP}) = \alpha_3 z$

(A16c) $(CR + \overline{PQ}) = \beta_3 x$

(A16d) $(C\overline{P} + QR) = \beta_3 z$

(A16e) $(C\overline{Q} + RP) = \gamma_3 x$

(A16f) $(CP + \overline{Q}\overline{R}) = \gamma_3 y$.

Summing all three equations, (A15a) to (A15c), and using the Potential V_3 (A9), the following URM3 'Potential equation' is obtained:

(A17) $2C^2 + V_3 = \alpha_3 x + \beta_3 y + \gamma_3 z$, see also (F4).

This completes the overview of general URMT, the remainder of this Appendix is based upon the following Pythagoras conditions in the dynamical variables.

Pythagoras Conditions

The URM3 Pythagoras conditions relate the conjugate dynamical variables $\overline{P}, \overline{Q}, \overline{R}$ to their standard forms P, Q, R by

(A18) $\overline{P} = P$, $\overline{Q} = Q$, $\overline{R} = -R$.

With these conditions, the matrix $\mathbf{A}_3$ (now relabelled $\mathbf{A}_{30}$ to specifically denote it is the unity root matrix under URM3 Pythagoras conditions) becomes

$$\text{(A19)} \quad \mathbf{A}_{30} = \begin{pmatrix} 0 & R & Q \\ -R & 0 & P \\ Q & P & 0 \end{pmatrix}.$$

Note that the conjugate relations for exponent $n = 2$ are now identities, according to (A18), and not congruences. See [1],#3 for the derivation of the relations (A18).

There are now three distinct eigenvalues

(A20) $\lambda = +C$, $\lambda = 0$, $\lambda = -C$,

and, consequently, two additional eigenvectors $\mathbf{X}_{30}$ and $\mathbf{X}_{3-}$ defined by their eigenvector equations as

(A21) $\mathbf{A}_{30}\mathbf{X}_{30} = 0$, $\lambda = 0$

(A22) $\mathbf{A}_{30}\mathbf{X}_{3-} = -C\mathbf{X}_{3-}$, $\lambda = -C$.

The elements x, y, z of $\mathbf{X}_{3+}$ satisfy the Pythagoras equation, as evident in the parameterisation (A3) further below, and hence the reason behind the name 'Pythagoras conditions'.

(A23) $x^2 + y^2 - z^2 = 0$.

The elements (scale factors) $\alpha_3, \beta_3, \gamma_3$ of $\mathbf{X}_{3-}$ also satisfy the Pythagoras equation, see (A30), further below.

The Potential term V_3 is zero

(A24) $V_3 = 0$.

With the Potential zero, the DCE (A8) is just the constant, kinetic term K_3

(A25) $K_3 = C^2 = P^2 + Q^2 - R^2$.

Under Pythagoras conditions (A18), URM3 is a completely solved problem with an analytic solution for all variables. The x, y, z coordinates forming $\mathbf{X}_{3+}$ are parameterised by two arbitrary integers k and l:

(A26)

(A26a) $x = 2kl$

(A26b) $y = (l^2 - k^2)$

(A26c) $z = (l^2 + k^2)$

(A26d) $k, l \in \mathbb{Z}$, $(k, l) \neq (0, 0)$, $\gcd(k, l) = 1$.

The scale factors $\alpha_3, \beta_3, \gamma_3$ (A30) and dynamical variables P, Q, R (A29) are obtained by solving the following linear Diophantine equation, using Euclid's algorithm [6] for unknown integers s and t, given k and l

(A27) $+C = ks - lt$, $s, t \in \mathbb{Z}$.

Note that the usage of symbol t here is slightly unfortunate in that it is not a time parameter. It is retained for compatibility with all existing URMT literature. In fact, it is shown in [1] that t has units of $\sqrt{L}$ (the square root of length), but since it always multiplies k or l (units

$\sqrt{LT^{-1}}$), or appears in squared form, then potentially irrational quantities do not appear in the solution, in keeping with URMT's all-integer formulation.

Solving (A27) introduces some indeterminacy into URM3 in an otherwise, completely deterministic, analytic solution. It has to be solved algorithmically, with no analytic solution. Physically, this indeterminacy is very likely a good thing; without it, the entire n-dimensional URMT solution would be completely deterministic once initial conditions are imposed.

Once a particular solution s' and t' is obtained by algorithmic means, then an infinite family of solutions can be generated, denoted by integers s and t, and parameterised by another arbitrary integer t_3, as follows:

(A28)

(A28a) $s = s' + t_3 l$

(A28b) $t = t' + t_3 k$

(A28c) $t_3 \in \mathbb{Z}$, notation $t_3 \sim m, \delta$ in [1]

(A28d) $s', t' \in \mathbb{Z}$, $(s', t', t_3) \neq (0,0,0)$.

This parameter t_3 is none other than the URM3 evolutionary time, also denoted by m or δ in [1].

The dynamical variables P, Q, R are parameterised as follows, in terms of k, l and t_3, via the general solutions for s (A28a) and t (A28b), likewise for the URM3 scale factors (A15) :

(A29)

(A29a) $P = -(ks + lt)$

(A29b) $Q = (ls - kt)$

(A29c) $R = -(ls + kt)$

(A30)

(A30a) $\alpha_3 = -2st$

(A30b) $\beta_3 = (t^2 - s^2)$

(A30c) $\gamma_3 = (t^2 + s^2)$.

Note that the scale factors $\alpha_3, \beta_3, \gamma_3$ also form a Pythagorean triple.

(A31) $\alpha_3^2 + \beta_3^2 - \gamma_3^2 = 0$.

(A32) The following table gives the solutions in all variables for eigenvalue $C = 1$, $t_3 = 0$, and a few small values of the parameters k and l.

l	k	x	y	z	s	t	P	-Q	R	α_3	β_3	$-\gamma_3$
2	1	4	3	5	1	0	-1	-2	-2	0	-1	-1
3	2	12	5	13	2	1	-7	-4	-8	-4	-3	-5
4	1	8	15	17	1	0	-1	-4	-4	0	-1	-1
4	3	24	7	25	3	2	-17	-6	-18	-12	-5	-13
5	2	20	21	29	3	1	-11	-13	-17	-6	-8	-10
5	4	40	9	41	4	3	-31	-8	-32	-24	-7	-25
6	1	12	35	37	1	0	-1	-6	-6	0	-1	-1
6	5	60	11	61	5	4	-49	-10	-50	-40	-9	-41
7	2	28	45	53	4	1	-15	-26	-30	-8	-15	-17
7	4	56	33	65	2	1	-15	-10	-18	-4	-3	-5
7	6	84	13	85	6	5	-71	-12	-72	-60	-11	-61
8	1	16	63	65	1	0	-1	-8	-8	0	-1	-1
8	3	48	55	73	3	1	-17	-21	-27	-6	-8	-10
8	5	80	39	89	5	3	-49	-25	-55	-30	-16	-34
8	7	112	15	113	7	6	-97	-14	-98	-84	-13	-85
9	2	36	77	85	5	1	-19	-43	-47	-10	-24	-26
9	4	72	65	97	7	3	-55	-51	-75	-42	-40	-58
9	8	144	17	145	8	7-	127	-16	-128	-112	-15	-113

The standard eigenvectors $\mathbf{X}_{3+}$, $\mathbf{X}_{30}$ and $\mathbf{X}_{3-}$ are defined in terms of the coordinates x, y, z, dynamical variables P, Q, R and scale factors $\alpha_3, \beta_3, \gamma_3$ respectively, as follows:

(A33)

$$(A33a) \ \mathbf{X}_{3+} = \begin{pmatrix} x \\ y \\ z \end{pmatrix}, \ (A33b) \ \mathbf{X}_{30} = \begin{pmatrix} P \\ -Q \\ R \end{pmatrix}, \ (A33c) \ \mathbf{X}_{3-} = \begin{pmatrix} \alpha_3 \\ \beta_3 \\ -\gamma_3 \end{pmatrix}.$$

Reciprocal Forms

The reciprocal eigenvectors, $\mathbf{X}^{3+}$, $\mathbf{X}^{30}$ and $\mathbf{X}^{3-}$, are 1x3 row-vectors, see (A35) further below, and obtained from the standard forms (A33) using the URM3 $\mathbf{T}_3$ operator relations:

(A34)

$$(A34a) \ \mathbf{T}_3 = \mathbf{T}^3 = \begin{pmatrix} +1 & 0 & 0 \\ 0 & +1 & 0 \\ 0 & 0 & -1 \end{pmatrix}$$

$$(A34b) \ \mathbf{X}^{3+} = \left(\mathbf{T}^3 \mathbf{X}_{3-} \right)^T$$

$$(A34c) \ \mathbf{X}^{30} = \left(\mathbf{T}^3 \mathbf{X}_{30} \right)^T$$

$$(A34e) \ \mathbf{X}^{3-} = \left(\mathbf{T}^3 \mathbf{X}_{3+} \right)^T$$

$$(A34f) \ \mathbf{X}_{3+} = \left(\mathbf{X}^{3-} \mathbf{T}_3 \right)^T$$

$$(A34g) \ \mathbf{X}_{30} = \left(\mathbf{X}^{30} \mathbf{T}_3 \right)^T$$

$$(A34h) \ \mathbf{X}_{3-} = \left(\mathbf{X}^{3+} \mathbf{T}_3 \right)^T$$

Using the forms (A33), the reciprocal vectors are expanded in full as

(A35)

$$(A35a) \ \mathbf{X}^{3+} = \begin{pmatrix} \alpha_3 & \beta_3 & \gamma_3 \end{pmatrix}$$

$$(A35b) \ \mathbf{X}^{30} = \begin{pmatrix} P & -Q & -R \end{pmatrix}$$

$$(A35c) \ \mathbf{X}^{3-} = \begin{pmatrix} x & y & -z \end{pmatrix}.$$

The reciprocal vectors satisfy the following eigenvector equations by definition

(A36)

(A36a) $\mathbf{X}^{3+}\mathbf{A}_{30} = C\mathbf{X}^{3+}$

(A36b) $\mathbf{X}^{30}\mathbf{A}_{30} = 0$

(A36c) $\mathbf{X}^{3-}\mathbf{A}_{30} = -C\mathbf{X}^{3-}$.

Conjugate Forms

Conjugate forms are much the same as reciprocal forms (see above), and the difference is mainly just notational, i.e. conjugate forms have an over-struck bar instead of a raised index. However, the notion of conjugacy in URMT also extends to elements of the vectors, see Appendix (E) for more details.

The conjugate vector forms are:

(A37)

(A37a) $\mathbf{X}^{3-} = \overline{\mathbf{X}}_{3+}$, $\mathbf{X}^{30} = \overline{\mathbf{X}}_{30}$, $\mathbf{X}^{3+} = \overline{\mathbf{X}}_{3-}$,

(A37b) $\overline{\mathbf{X}}^{3-} = \mathbf{X}_{3+}$, $\overline{\mathbf{X}}^{30} = \mathbf{X}_{30}$, $\overline{\mathbf{X}}^{3+} = \mathbf{X}_{3-}$

Note that the conjugate of a conjugate eigenvector is the original eigenvector. For example, by (A37a) and (A37b),

(A37c) $\overline{\mathbf{X}}_{3+} = \mathbf{X}^{3-} \Rightarrow \overline{\overline{\mathbf{X}}}_{3+} = \overline{\mathbf{X}}_{3-} \Rightarrow \overline{\overline{\mathbf{X}}}_{3+} = \mathbf{X}_{3+}$.

Combining (A34) and (A37) relates the conjugate forms to their standard forms as follows:

(A38)

(A38a) $\overline{\mathbf{X}}_{3+} = (\mathbf{T}\mathbf{X}_{3+})^{T}$

(A38b) $\overline{\mathbf{X}}_{30} = (\mathbf{T}\mathbf{X}_{30})^{T}$

(A38c) $\overline{\mathbf{X}}_{3-} = (\mathbf{T}\mathbf{X}_{3-})^{T}$

Comparing the elements of the standard (A33) and conjugate (A35) eigenvectors, the standard and conjugate variables are related as follows, see [1],#5.

(A39)

(A39a) $\bar{x} = x$, $\bar{y} = y$, $\bar{z} = -z$

(A39b) $\bar{P} = P$, $\bar{Q} = Q$, $\bar{R} = -R$, 'Pythagoras conditions' (A18)

(A39c) $\bar{\alpha} = \alpha$, $\bar{\beta} = \beta$, $\bar{\gamma} = -\gamma$

(A39d) $\bar{C} = -C$.

(A40) As for the eigenvectors, the conjugate of all conjugate variables returns the original variable. For example, by (A39a),

$$\bar{x} = x \Rightarrow \bar{\bar{x}} = \bar{x} \Rightarrow \bar{\bar{x}} = x,$$

or, by (A39b),

$$\bar{P} = P \Rightarrow \bar{\bar{P}} = \bar{P} \Rightarrow \bar{\bar{P}} = P.$$

Hence, in full,

(A40a) $\bar{\bar{x}} = x$, $\bar{\bar{y}} = y$, $\bar{\bar{z}} = z$,

(A40b) $\bar{\bar{P}} = P$, $\bar{\bar{Q}} = Q$, $\bar{\bar{R}} = R$.

The conjugate of a vector/matrix in URMT is the transpose of the vector/matrix with all their elements conjugated, hence the conjugate vector relations (A37) can be verified using the elemental definitions of the eigenvectors (A33) and (A35), and the relations (A39). Given these facts, it is also easy to see that the unity root matrix has the Hermitian property, i.e. it is equal to its transpose conjugate, since

$$(\text{A41}) \quad \left(\bar{\mathbf{A}}_3\right)^T = \begin{pmatrix} 0 & \bar{\bar{R}} & \bar{\bar{Q}} \\ \bar{\bar{R}} & 0 & \bar{\bar{P}} \\ \bar{\bar{Q}} & \bar{\bar{P}} & 0 \end{pmatrix} = \begin{pmatrix} 0 & R & \bar{Q} \\ \bar{R} & 0 & P \\ Q & \bar{P} & 0 \end{pmatrix} = \mathbf{A}_3, \text{ by (A40b).}$$

Divisibility Relations

All three sets of variables x, y, z, P, Q, R, $\alpha_3, \beta_3, \gamma_3$ and eigenvalue C are related by relations (A15), using (A18),

(A42)

(A42a) $C^2 - P^2 = \alpha_3 x$

(A42b) $C^2 - Q^2 = \beta_3 y$

(A42c) $C^2 + R^2 = \gamma_3 z$.

These relations (A42) satisfy the following congruences, as expected from definitions (A2), using (A18),

(A43)

(A43a) $P^2 \equiv C^2 \pmod{x}$

(A43b) $Q^2 \equiv C^2 \pmod{y}$

(A43c) $R^2 \equiv -C^2 \pmod{z}$.

If the eigenvalue is unity, i.e. $C = 1$, then P, Q, R are the square (or primitive) roots of unity in modulo arithmetic.

(A44)

(A44a) $P^2 \equiv +1 \pmod{x}$

(A44b) $Q^2 \equiv +1 \pmod{y}$

(A44c) $R^2 \equiv -1 \pmod{z}$.

The congruences (A43) and (A44) are the realm of 'quadratic residues' in number theory, see [6].

Appendix (B) URM3 Eigenvector Evolution

The URM3 eigenvector evolution equations are reproduced below from [1],#3, where the initial values at URM3 time zero, $m = 0$, are superscripted with a prime. Note that symbol m in Part I and [1] is replaced by t_3 in Part II of this book, i.e. $t_3 \sim m$.

(B1)

(B1a) $\mathbf{X}'_{3+} = \mathbf{X}_{3+} (m = 0)$

(B1b) $\mathbf{X}'_{30} = \mathbf{X}_{30} (m = 0)$

(B1c) $\mathbf{X}'_{3-} = \mathbf{X}_{3-} (m = 0)$.

The reader is referred to [1],#3 for a full account of the URM3 eigenvector evolution.

The evolution equations in standard vector form are:

(B2)

(B2a) $\mathbf{X}_{3+} = \mathbf{X}'_{3+}$, static, no m dependence

(B2b) $\mathbf{X}_{30} (m) = -m\mathbf{X}_{3+} + \mathbf{X}'_{30}$

(B2c) $\mathbf{X}_{3-} (m) = -m^2 \mathbf{X}_{3+} + 2m\mathbf{X}'_{30} + \mathbf{X}'_{3-}$,

and their reciprocal forms, using (A34),

(B2d) $\mathbf{X}^{3-} = \mathbf{X}'^{3-}$, static, no m dependence

(B2e) $\mathbf{X}^{30} (m) = -m\mathbf{X}^{3-} + \mathbf{X}'^{30}$

(B2f) $\mathbf{X}^{3+} (m) = -m^2 \mathbf{X}^{3-} + 2m\mathbf{X}'^{30} + \mathbf{X}'^{+}$.

Appendix (C) The n-dimensional Residual Method

This Appendix extends the residual matrix method, as detailed in Section (3), to the n- dimensional case, when used in the context of lifting solutions, and as first detailed in Section (7) and later expanded upon in Section (11).

(C1) $\mathbf{A}_{n0} = \begin{pmatrix} 0 & -t_n \mathbf{X}^{(n-1)-} \\ t_n \mathbf{X}_{(n-1)+} & \mathbf{A}_{(n-1)0} \end{pmatrix}$, $t_n \sim s$, $\bar{s} = -s = -t_n$.

The parameter s in (7.2b) is replaced by the evolutionary, temporal parameter t_n, $n \geq 3$, one for each excess dimension. Since $\bar{s} = -s$ (7.4), then $\bar{s}$ is simply equivalent to $-t_n$.

All equations are given under URMn Pythagoras conditions (I13), with the additional constraint of an invariant, zero Potential. For examples of Pythagoras conditions, see URM3 (A18), URM4 (4.1), URM5 (13.22). For examples of an invariant, zero Potential solution, see URM4 Section (11) and URM5 Section (13-3).

The n-dimensional residual matrices, $\mathbf{E}_{n+}$ and $\mathbf{E}_{n0}$, are defined in terms of the n-dimensional unity root matrix $\mathbf{A}_{n0}$ as the following quadratic polynomials in the eigenvectors, for eigenvalues $\lambda = C$ and $\lambda = 0$.

(C2) $\mathbf{E}_{n+} = (\mathbf{A}_{n0}^2 + C\mathbf{A}_{n0}) = \mathbf{X}_{n+}\mathbf{X}^{n+}$, $\lambda = C$

(C3) $\mathbf{E}_{n0} = (\mathbf{A}_{n0}^2 - C^2\mathbf{I}_n) = -\mathbf{X}_{n0}\mathbf{X}^{n0}$, $\lambda = 0$.

The outer products, $\mathbf{X}_{n+}\mathbf{X}^{n+}$ and $\mathbf{X}_{n0}\mathbf{X}^{n0}$, on the right of these definitions, are $n \times n$ matrices, identical to the residual matrices $\mathbf{E}_{n+}$ and $\mathbf{E}_{n0}$ respectively. See comment (C18), further below, on why the $\mathbf{E}_-$ residual matrix is not required.

The quadratic polynomial forms, (C2) and (C3), for the residual matrices, $\mathbf{E}_{n+}$ and $\mathbf{E}_{n0}$ respectively, are the same for all URMn, $n \geq 3$, and determined by carefully selecting the conditions such that the eigenvalues are always the same two non-zero values $\pm C$, with all the other eigenvalues zero; see also Section (3-2).

Using $\mathbf{A}_{n0}$ (C1) the residual matrix $\mathbf{E}_{n+}$ is calculated from (C2) as

$$(\text{C4}) \quad \mathbf{E}_{n+} = \begin{pmatrix} 0 & 0 \\ 2t_n C\mathbf{X}_{(n-1)+} & -t_n^2 \mathbf{X}_{(n-1)+} \mathbf{X}^{(n-1)-} + \mathbf{E}_{(n-1)+} \end{pmatrix}.$$

Given that $\mathbf{E}_{(n-1)+}$ can be expressed as the following outer product

$$(\text{C5}) \quad \mathbf{E}_{(n-1)+} = \mathbf{X}_{(n-1)+} \mathbf{X}^{(n-1)+}.$$

and $\mathbf{X}_{n+}$ is defined in terms of $\mathbf{X}_{(n-1)+}$ as

$$(\text{C6}) \quad \mathbf{X}_{n+} = \begin{pmatrix} 0 \\ \mathbf{X}_{(n-1)+} \end{pmatrix}, \quad (14.92\text{a}),$$

then comparing $\mathbf{E}_{n+} = \mathbf{X}_{n+} \mathbf{X}^{n+}$ (C2) with $\mathbf{E}_{n+}$ (C4), the vector $\mathbf{X}^{n+}$ is deduced to be

$$(\text{C7}) \quad \mathbf{X}^{n+} = -t_n^2 \begin{pmatrix} 0 & \mathbf{X}^{(n-1)-} \end{pmatrix} + 2t_n \begin{pmatrix} C & 0 \end{pmatrix} + \begin{pmatrix} 0 & \mathbf{X}^{(n-1)+} \end{pmatrix}.$$

Using $\mathbf{A}_{n0}$ (C1), the residual matrix $\mathbf{E}_{n0}$ (C3) is calculated as

$$(\text{C8}) \quad \mathbf{E}_{n0} = \begin{pmatrix} -C^2 & t_n C\mathbf{X}^{(n-1)-} \\ t_n C\mathbf{X}_{(n-1)+} & -t_n^2 \mathbf{X}_{(n-1)+} \mathbf{X}^{(n-1)-} + \mathbf{E}_{(n-1)0} \end{pmatrix}.$$

The matrix $\mathbf{E}_{n0}$ is now split into two components, $\mathbf{E}_{n0}(n)$ and $\mathbf{E}_{n0}(n-1)$,

$$(\text{C9}) \quad \mathbf{E}_{n0}(n) = \begin{pmatrix} -C^2 & t_n C\mathbf{X}^{(n-1)-} \\ t_n C\mathbf{X}_{(n-1)+} & -t_n^2 \mathbf{X}_{(n-1)+} \mathbf{X}^{(n-1)-} \end{pmatrix},$$

$$(\text{C10}) \quad \mathbf{E}_{n0}(n-1) = \begin{pmatrix} 0 & 0 \\ 0 & \mathbf{E}_{(n-1)0} \end{pmatrix}.$$

The second component $\mathbf{E}_{n0}(n-1)$ is recursively defined in terms of $\mathbf{E}_{(n-1)0}$, which is expressed as an outer product in the usual way, i.e.

(C11) $\mathbf{E}_{(n-1)0} = -\mathbf{X}_{(n-1)0}\mathbf{X}^{(n-1)0}$.

Armed with the knowledge that $\mathbf{X}^{n0} = (\mathbf{T}^n\mathbf{X}_{n0})^T$ (E6), and $\mathbf{E}_{n0} = -\mathbf{X}_{n0}\mathbf{X}^{n0}$ (C3), then $\mathbf{X}_{n0}$ and $\mathbf{X}^{n0}$ are derived from (C9) and (C11), and expressed recursively as

(C12) $\mathbf{X}_{n0} = -t_n\begin{pmatrix} 0 \\ \mathbf{X}_{(n-1)+} \end{pmatrix} + \begin{pmatrix} C \\ \mathbf{0}_{n-1} \end{pmatrix} + \begin{pmatrix} 0 \\ \mathbf{X}_{(n-1)0} \end{pmatrix}$

(C13) $\mathbf{X}^{n0} = -t_n\begin{pmatrix} 0 & \mathbf{X}^{(n-1)-} \end{pmatrix} + \begin{pmatrix} C & \mathbf{0}^{n-1} \end{pmatrix} + \begin{pmatrix} 0 & \mathbf{X}^{(n-1)0} \end{pmatrix}$.

Recursing down from n to $n = 4$, gives $n-3$, zero eigenvectors, one for each extra dimension. This is in addition to the single, zero eigenvector $\mathbf{X}_{30}$ (A33b) in the URM3 solution, so there are actually $n-2$, zero eigenvectors in total. The reader is referred to the URM5 solution (13.52) for an example of the two extra zero eigenvectors, $\mathbf{X}_{50B}$ and $\mathbf{X}_{50C}$ in URM5, where the URM3 zero eigenvector $\mathbf{X}_{30}$ is equivalent to URM5 eigenvector $\mathbf{X}_{50A}$.

Using $\mathbf{X}^{n+}$ (C7), $\mathbf{X}_{n-}$ is then obtained from the $\mathbf{T}$ operator relation (E7) to give

(C14) $\mathbf{X}_{n-} = -t_n^2\begin{pmatrix} 0 \\ \mathbf{X}_{(n-1)+} \end{pmatrix} + 2t_n\begin{pmatrix} C \\ \mathbf{0}_{n-1} \end{pmatrix} + \begin{pmatrix} 0 \\ \mathbf{X}_{(n-1)-} \end{pmatrix}$.

Using $\mathbf{X}_{n+}$ (C6), $\mathbf{X}^{n-}$ is then obtained from the $\mathbf{T}$ operator relation (E5) to give

(C15) $\mathbf{X}^{n-} = \begin{pmatrix} 0 & \mathbf{X}^{(n-1)-} \end{pmatrix}, = ... \begin{pmatrix} 0 & \mathbf{X}^{3-} \end{pmatrix}$

For reference, the two vectors $\mathbf{X}_{n-}$ (C14) and $\mathbf{X}^{n+}$ (C7) are usefully written in terms of the scale factors α_n, β_n and γ_n as

$$(\text{C16}) \quad \mathbf{X}_{n-} = \begin{pmatrix} - \\ \alpha_n \\ \beta_n \\ -\gamma_n \end{pmatrix}, \quad \mathbf{X}^{n+} = \begin{pmatrix} - & \alpha_n & \beta_n & \gamma_n \end{pmatrix}, \quad n \geq 4$$

where the first, blanked element '-' in $\mathbf{X}_{n-}$ is actually an $n-3$ element vector, given by the summation term in (14.92b), and involving the first $n-3$ rows and columns of the matrix $\mathbf{M}$. The scale factors α_n, β_n and γ_n are defined recursively as

(C17)

$$\alpha_n = -t_n^2 x + \alpha_{n-1}, \quad n \geq 4$$

$$\beta_n = -t_n^2 y + \beta_{n-1}$$

$$\gamma_n = t_n^2 z + \gamma_{n-1}.$$

(C18). The $\mathbf{E}_-$ residual matrix is not required here to obtain the scale factors (C16) since $\mathbf{X}_{n-}$ and $\mathbf{X}^{n-}$ are obtained from the $\mathbf{T}$ operator relations (E5) and (E7). Given $\mathbf{X}_{n+}$ is pre-defined (14.92a), $\mathbf{X}^{n-}$ is obtained without any recourse to the residual method using the $\mathbf{T}$ operator relation (E5). On the other hand, $\mathbf{X}_{n-}$ is obtained from the $\mathbf{T}$ operator relation (E7) using $\mathbf{X}^{n+}$ (C7), itself obtained from residual matrix $\mathbf{E}_{n+}$ (C4).

The three equations (C17) can be written in a vector form as follows:

$$(\text{C19}) \quad \begin{pmatrix} \alpha_n \\ \beta_n \\ -\gamma_n \end{pmatrix} = -t_n^2 \begin{pmatrix} x \\ y \\ z \end{pmatrix} + \begin{pmatrix} \alpha_{n-1} \\ \beta_{n-1} \\ -\gamma_{n-1} \end{pmatrix}.$$

Looking at the definition of URM3 vectors $\mathbf{X}_{3+}$ (A33a) and $\mathbf{X}_{3-}$ (A33c), it is clear that the last three elements (first three dimensions) of $\mathbf{X}_{n-}$, for URM n, can be written in an evolutionary, recursive form as

(C20) $\mathbf{X}_{n-}(t_n) = -t_n^2 \mathbf{X}_{3+} + \mathbf{X}_{(n-1)-}$, **last three elements only**.

This completes the n-dimensional solution for URM n, under Pythagoras conditions.

Appendix (D) URM5 Example

This numeric example illustrates the compactification behaviour of the fifth dimension of URM5 as its evolution progresses.

In this example the evolutionary time t_5, of the fifth dimension in URM5, is varied from zero upward, whilst leaving all other times t_3 and t_4 at their initial, zero value, in accordance with (14.70), and done primarily for ease of understanding. In other words, there is no evolution in the first four dimensions, only the fifth.

(D1) $t_3 = 0$, $t_4 = 0$.

The example uses (embeds) the standard URM3 Pythagorean (4,3,5) solution, for unity eigenvalue, as given in the first row of table (A32) in Appendix (A), and is reproduced below.

(D2)

$C = 1$, unity eigenvalue.

$x = 4$, $y = 3$, $z = 5$

$P = -1$, $Q = +2$, $R = -2$

$\alpha = 0$, $\beta = -1$, $\gamma = 1$.

Using these values, the URM3 eigenvector solution is, by (A33),

$$(D3) \quad \mathbf{X}_{3+} = \begin{pmatrix} 4 \\ 3 \\ 5 \end{pmatrix}, \quad \mathbf{X}'_{30} = \begin{pmatrix} -1 \\ -2 \\ -2 \end{pmatrix}, \quad \mathbf{X}'_{3-} = \begin{pmatrix} 0 \\ -1 \\ -1 \end{pmatrix}.$$

These three URM3 eigenvectors are all static by virtue of evolutionary time t_3 constrained to zero (D1), hence they all remain at their initial value and are superscripted with a prime to denote this, excepting $\mathbf{X}_{3+}$, which is always static and the superscript prime is therefore omitted as superfluous.

The URM5 eigenvector $\mathbf{X}_{5+}$ (13.52a) is also static and remains at its initial, $\mathbf{X}_{3+}$ value, i.e.

$$(13.52a) \quad \mathbf{X}_{5+} = \begin{pmatrix} 0 \\ 0 \\ \mathbf{X}_{3+} \end{pmatrix}.$$

Because $t_3 = 0$ (D1), then $\mathbf{X}_{30} = \mathbf{X}'_{30}$ as a consequence, and the URM5 eigenvector $\mathbf{X}_{50A}$ (13.52c) is also therefore static and remains at its initial, $\mathbf{X}'_{30}$ value, i.e.

$$(D4) \quad \mathbf{X}_{50A} = \begin{pmatrix} 0 \\ 0 \\ \mathbf{X}'_{30} \end{pmatrix}.$$

Lastly, on the subject of time-independent, static vectors, because $\mathbf{X}_{50B}$ is a function only of evolutionary time t_4, and since $t_4 = 0$ (D1), then $\mathbf{X}_{50B}$ reduces to the following static vector

$$(D5) \quad \mathbf{X}_{50B} = \begin{pmatrix} 0 \\ C \\ \mathbf{0}_3 \end{pmatrix}, \ t_4 = 0.$$

Note that this vector $\mathbf{X}_{50B}$ has a constant component, eigenvalue C, in its fourth dimension; the eigenvalue being unity (D2) in this example. Ordinarily, C can be made as large as desired, see the comment (14.42b).

Using initial values $\mathbf{X}_{3+}$ and $\mathbf{X}'_{30}$ (D3), and $C=1$ (D2), the static URM5 eigenvectors $\mathbf{X}_{5+}$, $\mathbf{X}_{50A}$ and $\mathbf{X}_{50B}$ are thus

$$\text{(D6)} \ \ \mathbf{X}_{5+} = \begin{pmatrix} 0 \\ 0 \\ 4 \\ 3 \\ 5 \end{pmatrix}, \ \mathbf{X}_{50A} = \begin{pmatrix} 0 \\ 0 \\ -1 \\ -2 \\ -2 \end{pmatrix}, \ \mathbf{X}_{50B} = \begin{pmatrix} 0 \\ 1 \\ 0 \\ 0 \\ 0 \end{pmatrix}, \ t_3 = 0, \ t_4 = 0.$$

This leaves just two time-dependent URM5 eigenvectors $\mathbf{X}_{5-}$ (13.52b) and $\mathbf{X}_{50C}$ (13.52e), which, for $t_3 = 0$ and $t_4 = 0$ (D1), become

$$\text{(D7)} \ \ \mathbf{X}_{5-} = -t_5^2 \begin{pmatrix} 0 \\ 0 \\ \mathbf{X}_{3+} \end{pmatrix} + 2t_5 \begin{pmatrix} C \\ 0 \\ \mathbf{0}_3 \end{pmatrix} + \begin{pmatrix} 0 \\ 0 \\ \mathbf{X}'_{3-} \end{pmatrix}.$$

$$\text{(D8)} \ \ \mathbf{X}_{50C} = -t_5 \begin{pmatrix} 0 \\ 0 \\ \mathbf{X}_{3+} \end{pmatrix} + \begin{pmatrix} C \\ 0 \\ \mathbf{0}_3 \end{pmatrix}.$$

Using initial values $\mathbf{X}_{3+}$ and $\mathbf{X}'_{3-}$ (D3), and $C=1$ (D2), then the initial values for the time-dependent vectors are:

$$\text{(D9)} \ \ \mathbf{X}'_{5-} = \begin{pmatrix} 0 \\ 0 \\ 0 \\ -1 \\ -1 \end{pmatrix}, \ \mathbf{X}'_{50C} = \begin{pmatrix} 1 \\ 0 \\ 0 \\ 0 \\ 0 \end{pmatrix}, \ t_3, t_4, t_5 = 0.$$

Calculations

For the purposes of URMT compactification, Section (14-1), the magnitude of the sum of two or more vectors is defined here as the sum of the individual magnitudes, which gives a maximum estimate of their size, i.e.

$$\text{(D10) if} \ \mathbf{X} = \sum_i \mathbf{X}_i \ \text{then} \ |\mathbf{X}| = \sum_i |\mathbf{X}_i|, \ i \geq 2.$$

This does not apply to any individual eigenvector, whose magnitude is calculated as the positive, root sum of squares of its elements, as per (I7).

All magnitudes, except $|\mathbf{X}|_5$ (the magnitude of the fifth dimension), are calculated from the three-element vector of the first three dimensions, which correspond to the last three elements of the five-element, URM5 vectors, i.e. the URM3 subspace.

The magnitudes of the static URM3 vectors are

(D11)

(D11a) $|\mathbf{X}_{3+}| = 5\sqrt{2}$, (D3)

(D11b) $|\mathbf{X}'_{30}| = 3$, (D3)

(D11c) $|\mathbf{X}'_{3-}| = \sqrt{2}$, (D3).

The magnitudes of the static, URM5 vectors are as follows, **first three dimensions only**

(D12)

(D12a) $|\mathbf{X}_{5+}| = 5\sqrt{2}$, (D6)

(D12b) $|\mathbf{X}_{50A}| = 3$, (D6)

(D12c) $|\mathbf{X}_{50B}| = 0$, (D6).

The two magnitudes, $|\mathbf{X}_{5-}|$ and $|\mathbf{X}_{50C}|$, **first three dimensions** only, are time-dependent, as follows

(D13)

(D13a) $|\mathbf{X}_{5-}| = \left| -t_5^2 \mathbf{X}_{3+} + \mathbf{X}'_{3-} \right|$, exact, (13.52b)

(D13b) $|\mathbf{X}_{5-}| = \sqrt{(50t_5^4 + 16t_5^2 + 2)}$, exact, from (D7) using (D3)

(D13c) $|\mathbf{X}_{5-}| \approx t_5^2 |\mathbf{X}_{3+}|$, $t_5 \gg 0$, large t_5 approximation of (D13a)

(D13d) $|\mathbf{X}_{5-}| \approx 5\sqrt{2}t_5^2$, $t_5 \gg 0$, using (D11a)

(D13e) $|\mathbf{X}_{50C}| = t_5 |\mathbf{X}_{3+}|$, exact, from (D8)

(D13f) $\left|\mathbf{X}_{50C}\right| = t_5 5\sqrt{2}$, exact, using (D11a).

The magnitude of the fifth dimension $\left|\mathbf{X}\right|_5$ (14.21a)

(D14)

(D14a) $\left|\mathbf{X}\right|_5 = \sqrt{(2t_5 C)^2 + C^2}$, exact, URM4 equivalent (14.51)

(D14b) $\left|\mathbf{X}\right|_5 = \sqrt{4t_5^2 + 1}$, $C = 1$ (D2)

(D14c) $\left|\mathbf{X}\right|_5 \approx 2t_5$, $t_5 \gg 0$, large t_5 approximation, URM4 equivalent (14.53)

The magnitude of the first three dimensions of URM5 $\left|\mathbf{X}\right|_3$ (14.21b),

(D15)

(D15a) $\left|\mathbf{X}\right|_3 = \left|\mathbf{X}_{5+}\right| + \left|\mathbf{X}_{50A}\right| + \left|\mathbf{X}_{50B}\right| + \left|\mathbf{X}_{50C}\right| + \left|\mathbf{X}_{5-}\right|$, definition, (14.21b)

Defining constants a and b

(D15b) $a = 5\sqrt{2} = 7.071068$ to 6dps.

(D15c) $b = a + 3 = 10.071068$ to 6dps.

using (D12) and (D13) for the individual, URM5 vector magnitudes gives

(D15d) $\left|\mathbf{X}\right|_3 = b + at_5 + \left|\mathbf{X}_{5-}\right|$, exact, use (D13b) for $\left|\mathbf{X}_{5-}\right|$

Compactification Ratio

The true compactification ratio χ_5 (14.23)

(D16a) $\chi_5 = \dfrac{\left|\mathbf{X}\right|_5}{\left|\mathbf{X}\right|_3}$, definition (14.23)

(D16d) $\chi_5 = \dfrac{\sqrt{4t_5^2 + 1}}{b + at_5 + \left|\mathbf{X}_{5-}\right|}$, exact, using (D14b) and (D15d).

The approximated compactification ratio χ_5 (14.27)

(D17a) $\chi_5 \approx \dfrac{2C}{t_5 |\mathbf{X}_{3+}|}$, (6.7), $t_5 \neq 0$: use (D16a) for $t_5 = 0$.

(D17b) $\chi_5 \approx \dfrac{\sqrt{2}}{5t_5}$, using (D11a), $t_5 \neq 0$ - see (D17a).

It is confirmed in both (D16d) and (D17b) that the compactification ratio limits to zero as t_5 grows to infinity since, for large t_5, both expressions are inversely proportional to t_5. Thus the fifth dimension shrinks to zero relative to the first three dimensions, i.e.

(D18) $\lim\limits_{t_5 \to \infty} \chi_5 = 0$ (14.24).

Error analysis

The absolute error ε_- (14.72)

(D19a) $\varepsilon_- = \left| \mathbf{X}_{5-} \right| - \left| t_5^2 \mathbf{X}_{3+} \right|$, definition, (14.72)

(D19b) $\varepsilon_- = \left| \mathbf{X}_{5-} \right| - 5\sqrt{2}t_5^2$, exact, using (D11a)

Expanding (D13b) $\left| \mathbf{X}_{5-} \right| = \sqrt{\left(50t_5^4 + 16t_5^2 + 2 \right)}$ binomially, to first order in $1/t_5$, gives

(D19c) $\left| \mathbf{X}_{5-} \right| = 5\sqrt{2}t_5^2 + 4\sqrt{2}/5 + O(1/t_5^2)$

Note that this binomial expansion (D19c) was not made in the main text, Section (14-6).

Substituting (D19c) into (D19b) approximates the absolute error ε_- as follows, to first order in $1/t_5$,

(D19d) $\varepsilon_- \approx \dfrac{4\sqrt{2}}{5} \approx 1.131371$ to 6dps.

Since the absolute error ε_- (D19d) is constant, it will shrink rapidly into insignificance compared with the quadratic term $5\sqrt{2}t_5^2$ in (D19c).

The relative error estimate ε (14.68) is

(D20a) $\varepsilon = \dfrac{\left\| \mathbf{X} \right|_3 - \left| t_5^2 \mathbf{X}_{3+} \right\|}{\left| \mathbf{X} \right|_3}$, (14.68)

Using (D15d) for $\left| \mathbf{X} \right|_3$ and definition (D19b) for the absolute error ε_-, the relative error estimate ε (D20a) becomes, after some additional algebraic manipulation and using a for the constant $5\sqrt{2}$ in (D19b),

(D20b) $\varepsilon = \dfrac{b + at_5 + \varepsilon_-}{b + at_5 + at_5^2}$.

From the approximation (D19d) for ε_-, the numerator is, to the highest order, linear in evolutionary time t_5. Hence, with a quadratic term at_5^2 in the denominator, it is confirmed from (D20b) that the error estimate ε limits to zero as t_5 grows to infinity, i.e.

(D20c) $\lim\limits_{t_5 \to \infty} \varepsilon = 0$, (14.83)

(D21) Tabulated Data

The following four sets of data are provided on the next few pages:

(D22) Time-dependent Vector $\mathbf{X}_{5-}$

(D23) Time-dependent Vector $\mathbf{X}_{50C}$

(D24) Time-dependent Magnitudes $\left| \mathbf{X}_{5-} \right|, \left| \mathbf{X}_{50C} \right|, \left| \mathbf{X} \right|_5, \left| \mathbf{X} \right|_3$

(D25) Compactification Ratios and Errors $\chi_5, \varepsilon, \varepsilon_-$.

The column headings are defined as follows:

t5 : evolutionary time t_5

X-(1) to X-(5) : five elements of time-dependent vector $\mathbf{X}_{5-}$ (D7)

|X-| : time-dependent magnitude $|\mathbf{X}_{5-}|$ (D13b), **first three dimensions** only

X0C(1) to X0C(5) : five elements of time-dependent vector $\mathbf{X}_{50C}$ (D8)

|X0C| : time-dependent magnitude $|\mathbf{X}_{50C}|$ (D13f), **first three dimensions** only

|X|5 : The magnitude of the fifth dimension $|\mathbf{X}|_5$ (D14b)

|X|3 : The magnitude of the first three dimensions $|\mathbf{X}|_3$ (D15d),

chi5 : true compactification ratio χ_5 (D16d)

chi5 app : approximated compactification ratio χ_5 (D17b)

chi5%err : percentage error in χ_5 approximation, i.e. 100×(true-approx)/true

eps : relative error estimate ε (D20b)

eps- : absolute error ε_- (D19b)

(D22) Time-dependent Vector X_{5-}

```
t5,  X-(1),  X-(2),   X-(3),   X-(4),   X-(5),    |X-|,
--,-----,-------,------,------,------,-------,
 0,     0,     0,      0,     -1,     -1,     1.4,
 1,     2,     0,     -4,     -4,     -6,     8.2,
 2,     4,     0,    -16,    -13,    -21,    29.4,
 3,     6,     0,    -36,    -28,    -46,    64.8,
 4,     8,     0,    -64,    -49,    -81,   114.3,
 5,    10,     0,   -100,    -76,   -126,   177.9,
 6,    12,     0,   -144,   -109,   -181,   255.7,
 7,    14,     0,   -196,   -148,   -246,   347.6,
 8,    16,     0,   -256,   -193,   -321,   453.7,
 9,    18,     0,   -324,   -244,   -406,   573.9,
10,    20,     0,   -400,   -301,   -501,   708.2,
11,    22,     0,   -484,   -364,   -606,   856.7,
12,    24,     0,   -576,   -433,   -721,  1019.4,
13,    26,     0,   -676,   -508,   -846,  1196.1,
14,    28,     0,   -784,   -589,   -981,  1387.1,
15,    30,     0,   -900,   -676,  -1126,  1592.1,
16,    32,     0,  -1024,   -769,  -1281,  1811.3,
24,    48,     0,  -2304,  -1729,  -2881,  4074.1,
32,    64,     0,  -4096,  -3073,  -5121,  7241.9,
48,    96,     0,  -9216,  -6913, -11521, 16292.9,
64,   128,     0, -16384, -12289, -20481, 28964.2,
```

(D23) Time-dependent Vector X_{50C}

```
t5,X0C(1),X0C(2),X0C(3),X0C(4),X0C(5),  |X0C|,
--,-----,-------,------,------,------,------,
 0,    1,      0,     0,     0,     0,   0.0,
 1,    1,      0,    -4,    -3,    -5,   7.1,
 2,    1,      0,    -8,    -6,   -10,  14.1,
 3,    1,      0,   -12,    -9,   -15,  21.2,
 4,    1,      0,   -16,   -12,   -20,  28.3,
 5,    1,      0,   -20,   -15,   -25,  35.4,
 6,    1,      0,   -24,   -18,   -30,  42.4,
 7,    1,      0,   -28,   -21,   -35,  49.5,
 8,    1,      0,   -32,   -24,   -40,  56.6,
 9,    1,      0,   -36,   -27,   -45,  63.6,
10,    1,      0,   -40,   -30,   -50,  70.7,
11,    1,      0,   -44,   -33,   -55,  77.8,
12,    1,      0,   -48,   -36,   -60,  84.9,
13,    1,      0,   -52,   -39,   -65,  91.9,
14,    1,      0,   -56,   -42,   -70,  99.0,
15,    1,      0,   -60,   -45,   -75, 106.1,
16,    1,      0,   -64,   -48,   -80, 113.1,
24,    1,      0,   -96,   -72,  -120, 169.7,
32,    1,      0,  -128,   -96,  -160, 226.3,
48,    1,      0,  -192,  -144,  -240, 339.4,
64,    1,      0,  -256,  -192,  -320, 452.5,
```

(D24) Time-dependent Magnitudes $|\mathbf{X}_{5-}|$, $|\mathbf{X}_{50C}|$, $|\mathbf{X}|_5$, $|\mathbf{X}|_3$

t5,	|X-|,	|X0C|,	|X|5,	|X|3,
0,	1.4,	0.0,	1.0,	11.5,
1,	8.2,	7.1,	2.2,	25.4,
2,	29.4,	14.1,	4.1,	53.6,
3,	64.8,	21.2,	6.1,	96.1,
4,	114.3,	28.3,	8.1,	152.6,
5,	177.9,	35.4,	10.0,	223.3,
6,	255.7,	42.4,	12.0,	308.2,
7,	347.6,	49.5,	14.0,	407.2,
8,	453.7,	56.6,	16.0,	520.3,
9,	573.9,	63.6,	18.0,	647.6,
10,	708.2,	70.7,	20.0,	789.0,
11,	856.7,	77.8,	22.0,	944.6,
12,	1019.4,	84.9,	24.0,	1114.3,
13,	1196.1,	91.9,	26.0,	1298.1,
14,	1387.1,	99.0,	28.0,	1496.1,
15,	1592.1,	106.1,	30.0,	1708.3,
16,	1811.3,	113.1,	32.0,	1934.5,
24,	4074.1,	169.7,	48.0,	4253.8,
32,	7241.9,	226.3,	64.0,	7478.3,
48,	16292.9,	339.4,	96.0,	16642.4,
64,	28964.2,	452.5,	128.0,	29426.8,

(D25) Compactification Ratios and Errors χ_5, ε, ε_-

t5,	chi5,	chi5 app,	chi5%err,	eps,	eps-,
--,	-------,	--------,	--------,	------,	-----,
0,	0.087,	0.000,	100.000,	0.000,	----,
1,	0.088,	0.283,	-221.140,	0.721,	1.175,
2,	0.077,	0.141,	-83.987,	0.473,	1.144,
3,	0.063,	0.094,	-48.891,	0.338,	1.137,
4,	0.053,	0.071,	-33.863,	0.259,	1.135,
5,	0.045,	0.057,	-25.711,	0.208,	1.133,
6,	0.039,	0.047,	-20.650,	0.174,	1.133,
7,	0.034,	0.040,	-17.221,	0.149,	1.132,
8,	0.031,	0.035,	-14.752,	0.130,	1.132,
9,	0.028,	0.031,	-12.893,	0.116,	1.132,
10,	0.025,	0.028,	-11.445,	0.104,	1.132,
11,	0.023,	0.026,	-10.286,	0.094,	1.132,
12,	0.022,	0.024,	-9.339,	0.086,	1.132,
13,	0.020,	0.022,	-8.550,	0.079,	1.132,
14,	0.019,	0.020,	-7.882,	0.074,	1.132,
15,	0.018,	0.019,	-7.311,	0.069,	1.132,
16,	0.017,	0.018,	-6.817,	0.064,	1.132,
24,	0.011,	0.012,	-4.419,	0.043,	1.131,
32,	0.009,	0.009,	-3.267,	0.032,	1.131,
48,	0.006,	0.006,	-2.147,	0.021,	1.131,
64,	0.004,	0.004,	-1.598,	0.016,	1.131,

Appendix (E) Reciprocal and Conjugate Eigenvectors

Reciprocal Eigenvectors

In the most general form of URMT, a 'standard' eigenvector $\mathbf{X}$ (1.5), for eigenvalue C, unity root matrix $\mathbf{A}$ (1.3a), is an n-element column vector which satisfies the eigenvector equation

(1.7) $\mathbf{AX} = C\mathbf{X}$, $C \in \mathbb{Z}$, $C \geq 1$.

Given this definition of a standard eigenvector $\mathbf{X}$, then the eigenvector reciprocal to $\mathbf{X}$ is an n-element row vector $\mathbf{X}^{+}$, which satisfies the reciprocal, row eigenvector equation

(A14) $\mathbf{X}^{+}\mathbf{A} = C\mathbf{X}^{+}$, notation $\mathbf{X}^{+} \sim \mathbf{X}^{3}, \mathbf{X}^{3+}$.

The name derives from the fact that the reciprocal vector $\mathbf{X}^{+}$ is a vector in the basis known as 'reciprocal' (or 'dual') to $\mathbf{X}$. However, the word 'dual' in URMT has a slightly different meaning, and is only currently used in URM3 in [1],#5, but is not used in this book. Basically, in URMT, dual eigenvectors are defined in the same basis set as the standard or reciprocal forms, and only under Pythagoras conditions. For example, in [1], the dual of $\mathbf{X}_{+}$ is $\mathbf{X}_{-}$ and vice versa, and $\mathbf{X}_{0}$ is its own dual, i.e. self-dual. Likewise, regarding URM3 reciprocal vectors, the dual of $\mathbf{X}^{+}$ is $\mathbf{X}^{-}$, and $\mathbf{X}^{0}$ is self-dual.

Frequently in the literature the reciprocal vector $\mathbf{X}^{+}$ is denoted by the 'bra' $\langle \mathbf{X} |$, and the standard vector $\mathbf{X}$ represented by the 'ket' $| \mathbf{X} \rangle$, see [5] for example. The vector product $\langle \mathbf{X} | \mathbf{X} \rangle$ is then simply the inner product $\mathbf{X}^{+}\mathbf{X}$ resulting in a scalar. There are four equivalent ways this can be written in URMT:

$$\langle \mathbf{X} | \mathbf{X} \rangle \sim \mathbf{X}^{+}\mathbf{X} \sim \mathbf{X} \cdot \mathbf{X}^{+} \sim \mathbf{X} \cdot \mathbf{X}^{+}.$$

Reciprocal eigenvectors are always denoted by a raised 'index' in URMT, e.g. $\mathbf{X}^+$, whereas the standard eigenvectors are denoted by a lowered index, e.g. $\mathbf{X}_+$, which is often dropped by assumption that URMT refers to standard, column eigenvectors, e.g. $\mathbf{X}$, unless specified otherwise.

Since the reciprocal eigenvector $\mathbf{X}^+$ is an n-element row eigenvector, and the standard vector $\mathbf{X}$ is an n-element column vector, then the product $\mathbf{X}^+\mathbf{X}$ is the equivalent of the scalar (or inner) vector product, i.e.

(E1) $\mathbf{X}^+\mathbf{X} \sim \mathbf{X}^+ \cdot \mathbf{X}$ = scalar (or inner) vector product.

Conversely, the product $\mathbf{XX}^+$ is the equivalent of the outer vector product, i.e.

(E2) $\mathbf{XX}^+ \sim \mathbf{X} \times \mathbf{X}^+$ = outer (or dyadic) vector product.

In general, multiplication of a reciprocal vector by a standard vector is the equivalent of an inner product operation, and multiplication of a standard vector by a reciprocal vector is the equivalent of an outer product operation.

Under general URMT, only the single eigenvector $\mathbf{X}$ (1.5), for eigenvalue C, is strictly defined, and so too therefore $\mathbf{X}^+$. However, throughout this book, and URMT in general, the majority of interest focuses on a complete set of n, linearly independent eigenvectors and, consequently, n related, reciprocal eigenvectors. Such a complete set of eigenvectors has only been evaluated under the specific conditions, summarised in Section (12-1), with the general, n-dimensional, eigenvector solution supplied in Section (14-9). This complete set comprises two standard eigenvectors, $\mathbf{X}_{n+}$ and $\mathbf{X}_{n-}$, for eigenvalues C and $-C$ respectively, and all the other $n-2$ eigenvectors are 'zero' eigenvectors, denoted by $\mathbf{X}_{n0j}$, $j = 0 \dots n-3$ (or $\mathbf{X}_{n0A}$, $\mathbf{X}_{n0B}$ etc.) for the eigenvalue zero, multiplicity $n-2$. Each of these n standard eigenvectors has its equivalent reciprocal eigenvector, $\mathbf{X}^{n+}$, $\mathbf{X}^{n-}$ and $\mathbf{X}^{n0j}$.

The remainder of this Appendix is assumed to be under the conditions summarised in Section (12-1), and therefore a complete set of eigenvectors exists, as described above and given in Section (14-9).

The set of reciprocal eigenvectors are obtained from the standard eigenvectors by use of the **T** operator $\mathbf{T}_n$ ($\sim \mathbf{T}^n$), $n \geq 2$ (this includes URM2), defined by

(E3) $\mathbf{T}_n = \mathbf{T}^n = \begin{pmatrix} \mathbf{I}_{n-1} & 0 \\ 0 & -1 \end{pmatrix}$, $\mathbf{I}_{n-1} = (n-1) \times (n-1)$ identity matrix,

which is its own inverse and transpose, i.e.

(E4) $\mathbf{T} = \mathbf{T}^{-1} = \mathbf{T}^T$.

The **T** operator is of the same form as the familiar Minkowski metric of STR η_{ij}, disregarding sign convention. Within URMT, the reciprocal eigenvectors, i.e. those with raised indices, e.g. $\mathbf{X}^{5-}$, are, indeed, related to the lowered index eigenvectors, $\mathbf{X}_{5+}$ in this example, via the relation

$\mathbf{X}^{5-} = \left(\mathbf{T}^5 \mathbf{X}_{5+}\right)^T$ and, conversely, $\mathbf{X}_{5+} = \left(\mathbf{X}^{5-} \mathbf{T}_5\right)^T$.

The reciprocal forms of the eigenvectors are not explicitly required in this paper for its central purpose, since all working can be done with the standard forms of eigenvectors. Nevertheless, most URMT conservation equations and scalar invariants arise from the inner products between the reciprocal and standard forms, as follows below. See the general solution, Section (14-9), for the standard forms of the URMn vectors from which the following reciprocal forms are obtained.

(E5) $\mathbf{X}^{n-} = \left(\mathbf{T}^n \mathbf{X}_{n+}\right)^T$, $\mathbf{X}_{n+} = \left(\mathbf{X}^{n-} \mathbf{T}_n\right)^T$

(E6) $\mathbf{X}^{n0j} = \left(\mathbf{T}^n \mathbf{X}_{n0j}\right)^T$, $j = 0 \ldots n-1$,

  where $\mathbf{X}^{n00} \sim \mathbf{X}^{n0A}$,

  and $\mathbf{X}^{n01} \sim \mathbf{X}^{n0B}$, etc

(E7) $\mathbf{X}^{n+} = \left(\mathbf{T}^n \mathbf{X}_{n-}\right)^T$, $\mathbf{X}_{n-} = \left(\mathbf{X}^{n+} \mathbf{T}_n\right)^T$.

See further below for some examples of **T** operator usage.

The reciprocal of a reciprocal vector returns the original vector, i.e. reciprocation is its own inverse, which is identical in behaviour to conjugation, see further below.

It is important to note, to avoid confusion, that the reciprocal of $\mathbf{X}_{n+}$ is considered to be $\mathbf{X}^{n+}$. Likewise, the reciprocal of $\mathbf{X}_{n-}$ is $\mathbf{X}^{n-}$ and the reciprocal of $\mathbf{X}_{n0j}$ is $\mathbf{X}^{n0j}$. However the above $\mathbf{T}$ operator relations link $\mathbf{X}_{n+}$ with $\mathbf{X}^{n-}$ (E5) and $\mathbf{X}_{n-}$ with $\mathbf{X}^{n+}$ (E7), so it is not strictly true to say (E5) is the reciprocal relation for $\mathbf{X}_{n+}$, and (E7) is the reciprocal relation for $\mathbf{X}_{n-}$. Suffice to note that $\mathbf{X}_{n0j}$ does relate directly to $\mathbf{X}^{n0j}$. This terminological inconsistency is avoided by using 'conjugate' forms instead, and defined next.

Conjugate Eigenvectors

The conjugate forms of the eigenvectors are denoted by an over-struck bar, and defined as follows:

(E8)

(E8a) $\overline{\mathbf{X}}_{n+} = \mathbf{X}^{n-}$,

(E8b) $\overline{\mathbf{X}}_{n0j} = \mathbf{X}^{n0j}$,

(E8c) $\overline{\mathbf{X}}_{n-} = \mathbf{X}^{n+}$,

(E8d) $\overline{\mathbf{X}}^{n-} = \mathbf{X}_{n+}$

(E8e) $\overline{\mathbf{X}}^{n0j} = \mathbf{X}_{n0j}$

(E8f) $\overline{\mathbf{X}}^{n+} = \mathbf{X}_{n-}$

Note that the conjugate of a conjugate eigenvector is the original eigenvector. For example, by (E8a) and by (E8d),

(E9) $\overline{\mathbf{X}}_{n+} = \mathbf{X}^{n-} \Rightarrow \overline{\overline{\mathbf{X}}}_{n+} = \overline{\mathbf{X}}^{n-} = \mathbf{X}_{n+}$.

From (E8) it is seen that the conjugate of a standard vector is a reciprocal vector. Notationally, the only difference between conjugate and reciprocal vectors appears to be the usage of an over-struck bar instead of a raised index. Superficially this is the case, but conjugation (or conjugacy) applies to all scalars, vector and matrices in URMT, and not just eigenvectors. First in URMT was the appearance of conjugate dynamical variables $\overline{P}, \overline{Q}, \overline{R}$, i.e. the conjugates of P, Q, R, and strictly defined by the conjugate relations (A2g) to (A2i). However, under a more unified scheme presented in [1],#5, coordinates x, y, z and scale factors $\alpha_3, \beta_3, \gamma_3$ also had conjugate forms - all defined strictly only under URM3 Pythagoras conditions. With a complete set of conjugate variables, all eigenvectors could then be given consistent conjugate forms when the conjugate of a vector or matrix is defined as the transpose of the vector with all its elements simultaneously conjugated, i.e. what is commonly known as the transpose conjugate. One can now see where Hermitian properties enter into URMT and, in fact, how a consistent isomorphism between complex numbers and dynamical variables and their conjugates, can be applied in URMT.

To recap, reciprocal and conjugate vectors are much the same, i.e. they are vectors defined in the basis that is reciprocal (or dual) to the standard basis. But reciprocal terminology is more loosely defined in that, when talking in terms of reciprocal vectors, any vector in the reciprocal, basis set, i.e. a vector denoted with a raised index is a reciprocal vector. On the other hand, conjugate vectors, whilst also forms of reciprocal vectors, are more precisely defined by consistent mathematical definitions between the standard and conjugate forms of their elements. The rules apply to all URMT variables of any rank, i.e. scalars, vectors and matrices. See Appendix (A) for URM3 specifics on reciprocal and conjugate forms.

This Appendix completes with some examples of **T** operator usage.

Examples

For URM5 the 5×5 matrix operator $\mathbf{T}_5$ $(= \mathbf{T}^5)$ is defined in block matrix form, using the 4×4 identity matrix $\mathbf{I}_4$, as

$$(E10) \quad \mathbf{T}_5 = \mathbf{T}^5 = \begin{pmatrix} \mathbf{I}_4 & 0 \\ 0 & -1 \end{pmatrix}.$$

Under Pythagoras conditions there is a reciprocal vector $\mathbf{X}^{5-}$, or conjugate vector $\overline{\mathbf{X}}_{5+}$, defined in terms of $\mathbf{X}_{5+}$ by

$$(E11) \quad \mathbf{X}^{5-} = \overline{\mathbf{X}}_{5+} = \left(\mathbf{T}^5 \mathbf{X}_{5+}\right)^T.$$

Using the $\mathbf{T}^5$ operator (E10) and $\mathbf{X}_{5+}$ (13.32), then $\mathbf{X}^{5-}$ (or $\overline{\mathbf{X}}_{5+}$) is thus,

(E12) $\mathbf{X}^{5-} = \overline{\mathbf{X}}_{5+} = \begin{pmatrix} 0 & 0 & x & y & -z \end{pmatrix}$.

The Pythagoras equation (13.21) is now expressed as the following inner product of $\mathbf{X}^{5-}$ (or $\overline{\mathbf{X}}_{5+}$) and $\mathbf{X}_{5+}$, which is a conservation equation in URMT with a scalar invariant quantity of zero, see (F7),

(E13) $\mathbf{X}^{5-}\mathbf{X}_{5+} = \overline{\mathbf{X}}_{5+}\mathbf{X}_{5+} = x^2 + y^2 - z^2 = 0$.

The reciprocal form of the URM4 eigenvector solution (14.41) is given below in terms of the URM3 reciprocal eigenvectors

(E14)

$$\mathbf{X}^{4-} = \begin{pmatrix} 0 & \mathbf{X}^{3-} \end{pmatrix}$$

$$\mathbf{X}^{4+} = -t_4^2 \begin{pmatrix} 0 & \mathbf{X}^{3-} \end{pmatrix} + 2t_4 \begin{pmatrix} C & \mathbf{0}^3 \end{pmatrix} + \begin{pmatrix} 0 & \mathbf{X}^{3+} \end{pmatrix}$$

$$\mathbf{X}^{40A} = \begin{pmatrix} 0 & \mathbf{X}^{30} \end{pmatrix}$$

$$\mathbf{X}^{40B} = -t_4 \begin{pmatrix} 0 & \mathbf{X}^{3-} \end{pmatrix} + \begin{pmatrix} C & \mathbf{0}^3 \end{pmatrix}$$

The conjugate form of the URM4 eigenvector solution (14.41) is given below in terms of the URM3 conjugate eigenvectors

(E15)

$$\overline{\mathbf{X}}_{4+} = \begin{pmatrix} 0 & \overline{\mathbf{X}}_{3+} \end{pmatrix}$$

$$\overline{\mathbf{X}}_{4-} = -t_4^2 \begin{pmatrix} 0 & \overline{\mathbf{X}}_{3+} \end{pmatrix} + 2t_4 \begin{pmatrix} C & \mathbf{0}^3 \end{pmatrix} + \begin{pmatrix} 0 & \overline{\mathbf{X}}_{3-} \end{pmatrix}$$

$$\overline{\mathbf{X}}_{40A} = \begin{pmatrix} 0 & \overline{\mathbf{X}}_{30} \end{pmatrix}$$

$$\overline{\mathbf{X}}_{40B} = -t_4 \begin{pmatrix} 0 & \overline{\mathbf{X}}_{3+} \end{pmatrix} + \begin{pmatrix} C & \mathbf{0}^3 \end{pmatrix}$$

The reciprocal form of the URM5 eigenvector solution (13.52) is given below in terms of the URM3 reciprocal eigenvectors

(E16)

$$\mathbf{X}^{5+} = -(t_5^2 + t_4^2)\begin{pmatrix} 0 & 0 & \mathbf{X}^{3-} \end{pmatrix} +$$
$$2t_5\begin{pmatrix} C & 0 & \mathbf{0}^3 \end{pmatrix} +$$
$$2t_4\begin{pmatrix} 0 & C & \mathbf{0}^3 \end{pmatrix} +$$
$$\begin{pmatrix} 0 & 0 & \mathbf{X}^{3+} \end{pmatrix}$$

$$\mathbf{X}^{5-} = \begin{pmatrix} 0 & 0 & \mathbf{X}^{3-} \end{pmatrix}$$

$$\mathbf{X}^{50A} = \begin{pmatrix} 0 & 0 & \mathbf{X}^{30} \end{pmatrix}$$

$$\mathbf{X}^{50B} = -t_4\begin{pmatrix} 0 & 0 & \mathbf{X}^{3-} \end{pmatrix} + \begin{pmatrix} 0 & C & \mathbf{0}^3 \end{pmatrix}$$

$$\mathbf{X}^{50C} = -t_5\begin{pmatrix} 0 & 0 & \mathbf{X}^{3-} \end{pmatrix} + \begin{pmatrix} C & 0 & \mathbf{0}^3 \end{pmatrix}$$

The conjugate form of the URM5 eigenvector solution (13.52) is given below in terms of the URM3 conjugate eigenvectors

(E17)

$$\overline{\mathbf{X}}_{5-} = -(t_5^2 + t_4^2)\begin{pmatrix} 0 & 0 & \overline{\mathbf{X}}_{3+} \end{pmatrix} +$$
$$2t_5\begin{pmatrix} C & 0 & \mathbf{0}^3 \end{pmatrix} +$$
$$2t_4\begin{pmatrix} 0 & C & \mathbf{0}^3 \end{pmatrix} +$$
$$\begin{pmatrix} 0 & 0 & \overline{\mathbf{X}}_{3-} \end{pmatrix}$$

$$\overline{\mathbf{X}}_{5+} = \begin{pmatrix} 0 & 0 & \overline{\mathbf{X}}_{3+} \end{pmatrix}$$

$$\overline{\mathbf{X}}_{50A} = \begin{pmatrix} 0 & 0 & \overline{\mathbf{X}}_{30} \end{pmatrix}$$

$$\overline{\mathbf{X}}_{50B} = -t_4\begin{pmatrix} 0 & 0 & \overline{\mathbf{X}}_{3+} \end{pmatrix} + \begin{pmatrix} 0 & C & \mathbf{0}^3 \end{pmatrix}$$

$$\overline{\mathbf{X}}_{50C} = -t_5\begin{pmatrix} 0 & 0 & \overline{\mathbf{X}}_{3+} \end{pmatrix} + \begin{pmatrix} C & 0 & \mathbf{0}^3 \end{pmatrix}$$

Appendix (F) Conservation Equations and Invariants

The six key conservation equations of URMT, as obtained from the inner product relations between the eigenvectors, are given in this Appendix for both URM3 and the general, n-dimensional case.

It is stressed that all URM3 equations listed in this Appendix are those under URM3 Pythagoras conditions. Likewise, their n-dimensional counterparts are all obtained under n-dimensional Pythagoras conditions. As a consequence, the most general conservation equation that started it all, i.e. the DCE Section (1-5), is not listed since no general solution in the dynamical variables has been given, without recourse to Pythagoras conditions.

The URM3 conservation equations are, **under URM3 Pythagoras conditions,**

(F1) $\mathbf{X}^{3-}\mathbf{X}_{3+} = \overline{\mathbf{X}}_{3+}\mathbf{X}_{3+} = x^2 + y^2 - z^2 = 0$ Pythagoras

(F2) $\mathbf{X}^{3+}\mathbf{X}_{3-} = \overline{\mathbf{X}}_{3-}\mathbf{X}_{3-} = \alpha_3^2 + \beta_3^2 - \gamma_3^2 = 0$ Pythagoras

(F3) $\mathbf{X}^{30}\mathbf{X}_{30} = \overline{\mathbf{X}}_{30}\mathbf{X}_{30} = P^2 + Q^2 - R^2 = +C^2$ the DCE

(F4) the Potential Equation, see (A17) for $V_3 = 0$

$$\mathbf{X}^{3+}\mathbf{X}_{3+} = \mathbf{X}^{3-}\mathbf{X}_{3-} = \alpha_3 x + \beta_3 y + \gamma_3 z = +2C^2$$

$$\overline{\mathbf{X}}_{3-}\mathbf{X}_{3+} = \overline{\mathbf{X}}_{3+}\mathbf{X}_{3-} = \alpha_3 x + \beta_3 y + \gamma_3 z = +2C^2$$

(F5) the Delta equation

$$\mathbf{X}^{30}\mathbf{X}_{3+} = \mathbf{X}^{3-}\mathbf{X}_{30} = xP - yQ - zR = 0$$

$$\overline{\mathbf{X}}_{30}\mathbf{X}_{3+} = \overline{\mathbf{X}}_{3+}\mathbf{X}_{30} = xP - yQ - zR = 0$$

325

(F6) the Dual Delta equation

$$\mathbf{X}^{30}\mathbf{X}_{3-} = \mathbf{X}^{3+}\mathbf{X}_{30} = \alpha_3 P - \beta_3 Q + \gamma_3 R = 0$$

$$\overline{\mathbf{X}}_{30}\mathbf{X}_{3-} = \overline{\mathbf{X}}_{3-}\mathbf{X}_{30} = \alpha_3 P - \beta_3 Q + \gamma_3 R = 0$$

The n-dimensional forms of these equations for URMn, $n \geq 4$, give exactly the same invariants. They are stated below, with an example given following of how they are proved by induction. Note that the scale factors α_n, β_n and γ_n are defined recursively in (C17).

(F7) $\mathbf{X}^{n-}\mathbf{X}_{n+} = \overline{\mathbf{X}}_{n+}\mathbf{X}_{n+} = 0$, Pythagoras

(F8) $\mathbf{X}^{n+}\mathbf{X}_{n-} = \overline{\mathbf{X}}_{n-}\mathbf{X}_{n-} = \displaystyle\sum_{j=4}^{n}(2t_j C)^2 + \alpha_n^2 + \beta_n^2 - \gamma_n^2 = 0$, Pythagoras

(F9) the Potential Equation ($V = 0$)

$$\mathbf{X}^{n+}\mathbf{X}_{n+} = \mathbf{X}^{n-}\mathbf{X}_{n-} = \alpha_n x + \beta_n y + \gamma_n z = +2C^2$$

$$\mathbf{X}^{n+}\mathbf{X}_{n+} = \overline{\mathbf{X}}_{n-}\mathbf{X}_{n-} = \alpha_n x + \beta_n y + \gamma_n z = +2C^2$$

For each of the zero eigenvectors $\mathbf{X}_{n0j}$, $j = 0 \dots n-3$, where $\mathbf{X}_{n00} \sim \mathbf{X}_{n0A}$, $\mathbf{X}_{n01} \sim \mathbf{X}_{n0B}$, $\mathbf{X}_{n02} \sim \mathbf{X}_{n0C}$, and $\mathbf{X}_{300} = \mathbf{X}_{30A} \sim \mathbf{X}_{30}$ for $n = 3$, $j = 0$ (URM3), then

(F10) $\mathbf{X}^{n0j}\mathbf{X}_{n0j} = \overline{\mathbf{X}}_{n0j}\mathbf{X}_{n0j} = P^2 + Q^2 - R^2 = C^2$, the DCE.

Note the following orthogonality relation holds between different, zero eigenvectors:

(F10b) $\mathbf{X}^{n0j}\mathbf{X}_{n0i} = C^2$ if $i = j$, and $\mathbf{X}^{n0j}\mathbf{X}_{n0i} = 0$ if $i \neq j$.

Defining a total sum, zero vector $\mathbf{X}_{n0}$, as given by $\mathbf{X}_{n0} = \displaystyle\sum_{j=0}^{n-3}\mathbf{X}_{n0j}$, etc.,

and related reciprocal vector $\mathbf{X}^{n0} = \left(\mathbf{T}^n\mathbf{X}_{n0}\right)^T$, then (F10) is simply the sum of $(n-2)$ DCE's:

(F11) $\mathbf{X}^{n0}\mathbf{X}_{n0} = \overline{\mathbf{X}}_{n0}\mathbf{X}_{n0} = P^2 + Q^2 - R^2 = (n-2)C^2$.

(F12) the Delta equation

$$\mathbf{X}^{n0}\mathbf{X}_{n+} = \mathbf{X}^{n-}\mathbf{X}_{n0} = 0$$

$$\overline{\mathbf{X}}_{n0}\mathbf{X}_{n+} = \overline{\mathbf{X}}_{n+}\mathbf{X}_{n0} = 0.$$

(F13) the Dual Delta equation

$$\mathbf{X}^{n0}\mathbf{X}_{n-} = \mathbf{X}^{n+}\mathbf{X}_{n0} = \alpha_n P - \beta_n Q + \gamma_n R = 0$$

$$\overline{\mathbf{X}}_{n0}\mathbf{X}_{n-} = \overline{\mathbf{X}}_{n-}\mathbf{X}_{n0} = \alpha_n P - \beta_n Q + \gamma_n R = 0.$$

Those inner products between eigenvectors and their reciprocals, for different eigenvalues, are always zero by the rules of matrix algebra, see (F25) further below. The inner products can also be proved inductively using a recursive form of the general solutions provided in Appendix (C). An example of this is given following, to prove the inner product relation (F9) inductively, starting with the general solutions for $\mathbf{X}_{n+}$ (14.92a) and $\mathbf{X}^{n+}$ (C7), reproduced below,

(F14)

$$\mathbf{X}_{n+} = \begin{pmatrix} 0 \\ \mathbf{X}_{(n-1)+} \end{pmatrix} \text{ (14.92a)},$$

$$\mathbf{X}^{n+} = \left(2t_n C \quad -t_n^2 \mathbf{X}^{(n-1)-} + \mathbf{X}^{(n-1)+} \right) \text{(C7)},$$

The inner product $\mathbf{X}^{n+}\mathbf{X}_{n+}$ is given by

(F15) $\mathbf{X}^{n+}\mathbf{X}_{n+} = -t_n^2 \mathbf{X}^{(n-1)-}\mathbf{X}_{(n-1)+} + \mathbf{X}^{(n-1)+}\mathbf{X}_{(n-1)+}.$

The right of this product is given purely in terms of the $n-1$ dimensional eigenvectors, albeit there are now two $n-1$ dimensional terms instead of one, n-dimensional term on the left.

If it can be shown that the first term on the right, $\mathbf{X}^{(n-1)-}\mathbf{X}_{(n-1)+}$, is zero (F24), then the product will reduce to $\mathbf{X}^{n+}\mathbf{X}_{n+} = \mathbf{X}^{(n-1)+}\mathbf{X}_{(n-1)+}$, i.e. the desired, inductive reduction of $\mathbf{X}^{n+}\mathbf{X}_{n+}$ to $\mathbf{X}^{(n-1)+}\mathbf{X}_{(n-1)+}$ will have been achieved. Really, of course, the inductive argument shows that, if it is true for the $n-1$ case then it is true for the n case - this argument is given at the end.

It is relatively trivial to prove $\mathbf{X}^{(n-1)-}\mathbf{X}_{(n-1)+}$ is zero because the vector $\mathbf{X}_{(n-1)+}$ is static, as seen by the recursive formula for $\mathbf{X}_{n+}$

(F16) $\mathbf{X}_{n+} = \mathbf{X}_{(n-1)+} = \mathbf{X}_{(n-2)+} = ... \begin{pmatrix} \mathbf{0}_{n-3} \\ \mathbf{X}_{3+} \end{pmatrix},$

and likewise for $\mathbf{X}^{(n-1)-}$, since it is simply obtained from the relation $\mathbf{X}^{(n-1)-} = \left(\mathbf{T}^{n-1}\mathbf{X}_{(n-1)+}\right)^{T}$, i.e.

(F17) $\mathbf{X}^{(n-1)-} = \left(\mathbf{0}^{n-3} \quad \left(\mathbf{T}^{3}\mathbf{X}_{3+}\right)^{T}\right),$

Using the URM3 relation $\left(\mathbf{T}^{3}\mathbf{X}_{3+}\right)^{T} = \mathbf{X}^{3-}$ (A34e), then $\mathbf{X}^{(n-1)-}$ is given in terms of the URM3 vector $\mathbf{X}^{3-}$ as

(F18) $\mathbf{X}^{(n-1)-} = \left(\mathbf{0}^{n-3} \quad \mathbf{X}^{3-}\right).$

Thus, using this and $\mathbf{X}_{(n-1)+}$ (F16), the inner product $\mathbf{X}^{(n-1)-}\mathbf{X}_{(n-1)+}$ is given in terms of the URM3 vectors as

(F19) $\mathbf{X}^{(n-1)-}\mathbf{X}_{(n-1)+} = \mathbf{X}^{3-}\mathbf{X}_{3+}.$

Since $\mathbf{X}^{3-}\mathbf{X}_{3+}$ is zero by Pythagoras (F1), then it is proved that the first term on the right of (F15) is zero, i.e.

(F20) $\mathbf{X}^{(n-1)-}\mathbf{X}_{(n-1)+} = 0,$

and the inner product (F15) becomes

(F21) $\mathbf{X}^{n+}\mathbf{X}_{n+} = \mathbf{X}^{(n-1)+}\mathbf{X}_{(n-1)+}.$

Retracing the argument inductively, if $\mathbf{X}^{(n-1)+}\mathbf{X}_{(n-1)+} = +2C^{2}$, then so too $\mathbf{X}^{n+}\mathbf{X}_{n+} = +2C^{2}$ by (F21). Since it is true for $n = 3$, i.e. $\mathbf{X}^{3+}\mathbf{X}_{3+} = +2C^{2}$ (F4), then it is also therefore true for $n = 4,5...$ etc, hence (F9) is proven for all $n \geq 3$.

(F22) Commentary

The n-dimensional vector space is characterised by n, independent parameters (k, l and t_{j}, $j = 3...n$), all but two (k and l, see (A26)) are physically interpreted as temporal coordinates. Each jth dimension, three and higher, has evolutionary behaviour governed by its jth temporal coordinate t_j. Thus, with n independent parameters, the eigenvector space is a discrete set of points in n dimensions, termed the 'lattice' (symbol **L**) in [1],#3. A point in the lattice is therefore uniquely specified by the n-element, coordinate vector $(k \quad 1 \quad t_j)$, $j = 3...n$. With every lattice point there is also an associated set of invariants $0, \pm C^2, \pm 2C^2$, (see (F23) below), given by the scalar products between the eigenvectors, (F7) to (F13). For the unity eigenvalue, $C = 1$, this gives the set of integers $0, \pm 1, \pm 2$. Ratios of these (except zero) may also be considered. Regardless of the size of the eigenvectors, and their respective elements, which could easily be $O(10^1)$ to $O(10^{80})$ or even larger, the same three numbers 0, 1 and 2 appear at every lattice position. What do these integer invariants physically represent? Are their ratios meaningful?

(F23) From (F22) above: the minus sign can be selected using a different sign convention for the eigenvectors.

(F24) Since $\mathbf{X}^{(n-1)-}$ is a reciprocal eigenvector for eigenvalue $-C$, and $\mathbf{X}_{(n-1)+}$ is a standard eigenvector for eigenvalue $+C$, then by the rules of matrix algebra these eigenvectors will be orthogonal, i.e. their inner product zero. This is usually described in the literature [5] under the subject of 'orthogonality' of eigenvectors to different eigenvalues. Note that this orthogonality is between reciprocal (row-vector) and standard (column-vector) forms, but not between standard-standard or reciprocal-reciprocal vector forms. In these two latter cases, as noted in URMT (14.57), the standard vectors form a highly oblique, non-orthogonal basis and so too, therefore, the reciprocal vectors. The orthogonality is more precisely stated as now follows.

(F25) Orthogonality of eigenvectors for unique (distinct) eigenvalues

If $\mathbf{X}_a$ is an eigenvector to matrix $\mathbf{A}$ for eigenvalue λ_a, i.e. $\mathbf{AX}_a = \lambda_a \mathbf{X}_a$, and $\mathbf{X}^b$ is a reciprocal eigenvector to the same matrix $\mathbf{A}$ for a different eigenvalue λ_b, i.e. $\mathbf{X}^b \mathbf{A} = \lambda_b \mathbf{X}^b$, where $\lambda_a \neq \lambda_b$, then $\mathbf{X}_a$ and $\mathbf{X}^b$ are orthogonal, i.e. $\mathbf{X}^b \mathbf{X}_a = \mathbf{X}_a \cdot \mathbf{X}^b = 0$.

The proof is very straightforward, as follows:

By definition

(F25a) $\mathbf{A}\mathbf{X}_a = \lambda_a \mathbf{X}_a$

(F25b) $\mathbf{X}^b \mathbf{A} = \lambda_b \mathbf{X}^b$.

Multiplying (F25a) on the left by $\mathbf{X}^b$, and (F25b) on the right by $\mathbf{X}_a$, gives

(F25c) $\mathbf{X}^b \mathbf{A}\mathbf{X}_a = \lambda_a \mathbf{X}^b \mathbf{X}_a$

(F25d) $\mathbf{X}^b \mathbf{A}\mathbf{X}_a = \lambda_b \mathbf{X}^b \mathbf{X}_a$.

Subtracting (F25d) from (X25c) gives

(F25c) $0 = (\lambda_a - \lambda_b)\mathbf{X}^b \mathbf{X}_a$,

and since the eigenvalues are distinct, by definition $\lambda_a \neq \lambda_b$, then the vector product $\mathbf{X}^b \mathbf{X}_a$ must be zero, i.e.

(F25d) $\mathbf{X}^b \mathbf{X}_a = 0$.

But since $\mathbf{X}^b \mathbf{X}_a$ is just the scalar, inner product of $\mathbf{X}_a$ and $\mathbf{X}^b$, also written as $\mathbf{X}_a \cdot \mathbf{X}^b$ or $\mathbf{X}^b \cdot \mathbf{X}_a$ (where the vector multiplication order does not matter), then the inner product is zero and, hence, the two eigenvectors are orthogonal.

Note that this does rely heavily on distinct (unique) eigenvalues, and that $\mathbf{X}^b$ is a vector reciprocal to $\mathbf{X}_a$. In this case $\mathbf{X}^b$ is considered a row vector, and $\mathbf{X}_a$ a column vector, hence the product $\mathbf{X}^b \mathbf{X}_a$ is equivalent to the scalar, inner product. Conversely, the product $\mathbf{X}_a \mathbf{X}^b$ is an outer product resulting in a matrix, not a scalar. The 'dot' notation, e.g. $\mathbf{X}^b \cdot \mathbf{X}_a$, circumvents this multiplication ordering issue by specifying the product is 'element-wise', e.g. element one of $\mathbf{X}^b$ multiplies element one of $\mathbf{X}_a$ etc., which is, of course, the same as multiplying element one of $\mathbf{X}_a$ by element one of $\mathbf{X}^b$, irrespective of multiplicative order.

Appendix (G) Calculus Properties of URMT

Whilst URM3 vectors $\mathbf{X}_{3+}$, $\mathbf{X}_{30}$, $\mathbf{X}_{3-}$ can be consistently interpreted in terms of their physical units, with an acceleration, velocity and position vector respectively, they are also related via the following calculus relations, further justifying the standard physical interpretation as given in Appendix (J).

The standard calculus derivative $\dfrac{d}{dt_3}$ ($\sim \dfrac{d}{dm}$ in [1],#3) is used as a good, large t_3 approximation for discrete differences, i.e.

(G1) $\quad \dfrac{d}{dt_3} \approx \dfrac{\delta}{\delta t_3} \sim \Delta t_3$, $t_3 >> 0$, $\Delta t_3 = \delta t_3 = 1$,

The following calculus relations have the caveat of valid only for large evolutionary times, as above. This caveat is only required in so far as the continuous derivative is used as an approximation for the discrete difference for any large, jth dimensional, evolutionary time t_j, $t_j >> 0$, $j \geq 3$, but is not particularly related to the 'sufficiently large' criterion for evolutionary times mentioned in (14-6).

(G2) $\quad \dfrac{d\mathbf{X}_{3-}}{dt_3} = 2\mathbf{X}_{30}$, derivative of position = twice velocity

(G3) $\quad \dfrac{d\mathbf{X}_{30}}{dt_3} = -\mathbf{X}_{+}$, derivative of velocity = negative of acceleration

(G4) $\quad \dfrac{d\mathbf{X}_{3+}}{dt_3} = 0$, derivative of acceleration = zero (constant acceleration)

(G5) $\dfrac{d^2\mathbf{X}_{3-}}{d^2 t_3} = -2\mathbf{X}_{3+}$, second derivative of position = - twice acceleration

These URM3 derivative relationships amongst the eigenvectors are maintained, as follows, in the general, n-dimensional case of URM n.

Calculus Properties of URMn

With more than one evolutionary parameter for four and higher dimensions, the standard calculus partial derivative $\dfrac{\partial}{\partial t_i}$ is now used in

place of $\dfrac{d}{dt_3}$ for derivatives with respect to evolutionary time t_i, i.e.

(G6) $\dfrac{\partial}{\partial t_i} \approx \dfrac{\delta}{\delta t_i} \sim \Delta t_i$, $t_i \gg 0$, $\sim \Delta t_i = \delta t_i = 1$.

For URM5, the partial derivatives in the excess dimensions, i.e. the fourth and fifth, are

(G7) $\dfrac{\partial \mathbf{X}_{5-}}{\partial t_4} = 2\mathbf{X}_{50B}$, derivative of position = twice velocity

(G8) $\dfrac{\partial \mathbf{X}_{5-}}{\partial t_5} = 2\mathbf{X}_{50C}$, ditto

(G9) $\dfrac{\partial \mathbf{X}_{50A}}{\partial t_3} = -\mathbf{X}_{5+}$, derivative of velocity = negative of acceleration

(G10) $\dfrac{\partial \mathbf{X}_{50B}}{\partial t_4} = -\mathbf{X}_{5+}$, ditto

(G11) $\dfrac{\partial \mathbf{X}_{50C}}{\partial t_5} = -\mathbf{X}_{5+}$, ditto

(G12) $\dfrac{\partial \mathbf{X}_{5+}}{\partial t_4} = 0$, $\dfrac{\partial \mathbf{X}_{5+}}{\partial t_5} = 0$, constant acceleration

(G13) $\dfrac{\partial^2 \mathbf{X}_{5-}}{\partial^2 t_4} = -2\mathbf{X}_{5+}$, $\dfrac{\partial^2 \mathbf{X}_{5-}}{\partial^2 t_5} = -2\mathbf{X}_{5+}$ second derivative of position

= -twice acceleration

For URM n, dimension $j+3$, $j = 0 \ldots n-3$, ($j = 0$ includes URM3 here) the same general relations hold true for evolutionary parameter t_{j+3}, e.g. t_3 for $j = 0$, t_4 for $j = 1$ and, in general, t_n for $j = n-3$.

(G14) $\dfrac{\partial \mathbf{X}_{n-}}{\partial t_{j+3}} = 2\mathbf{X}_{n0j}$, derivative of position = twice velocity

(G15) $\dfrac{\partial \mathbf{X}_{n0j}}{\partial t_{j+3}} = -\mathbf{X}_{n+}$, derivative of velocity = negative of acceleration

(G16) $\dfrac{\partial^2 \mathbf{X}_{n-}}{\partial^2 t_{j+3}} = -2\mathbf{X}_{n+}$, second derivative of position = - twice acceleration

Commentary

(G17) As given by the standard physical interpretation, Appendix (J), out of the set of n eigenvectors, one of them can be physically associated with an acceleration vector, $n-2$ of them as velocity vectors, and one of them with a single position vector, all related by standard calculus relations given in this Appendix. However, there is no calculus in URMT's formulation, only an invariance principle and associated invariance transformations, but absolutely no calculus or difference equations used.

Appendix (H) URM2

The key matrices and eigenvectors in the 2×2 formulation of URMT (URM2) are given following, and quoted without proof as they are very simply derived. URM2 is also a special case of what is termed 'the almost trivial' solution in Appendix C of [1],#3.

The 'unity' root matrix

(H1) $\mathbf{A}_2 = \begin{pmatrix} 0 & P \\ \overline{P} & 0 \end{pmatrix} P, \overline{P} \in \mathbb{Z},\ P \neq 0, \overline{P} \neq 0$.

Two non-zero eigenvalues

(H2) $\lambda = \pm C,\ C \in \mathbb{Z},\ C \geq 1$.

The DCE, $\det(\mathbf{A}_2 - C\mathbf{I}) = 0$:

(H3) $C^2 = P\overline{P}$.

The DCE in kinetic and Potential terms

(H4) $E = C^2 = K + V,\ K = P\overline{P},\ V = 0$.

In URM2, the total energy E is always the kinetic term K, and the Potential V is always zero, with no pre-conditions such as the URM2 Pythagoras conditions, further below.

(H4b) $\mathbf{A}_2 \mathbf{X}_\pm = \pm C \mathbf{X}_\pm$, URM2 dynamical equations

(H4c) $\mathbf{X}_+ = \begin{pmatrix} y \\ z \end{pmatrix},\ \mathbf{X}_- = \begin{pmatrix} y \\ -z \end{pmatrix}$, standard eigenvectors

(H5a) $\mathbf{X}_+ = \begin{pmatrix} C \\ P \end{pmatrix}$, (H5b) $\mathbf{X}_- = \begin{pmatrix} C \\ -P \end{pmatrix}$, $y = C$, $z = \overline{P}$, eigenvector solution

(H6a) $\mathbf{X}^+ = (C \quad P)$, (H6b) $\mathbf{X}^- = (C \quad -P)$, reciprocal eigenvectors.

URM2 **Pythagoras Conditions**

(H7) $\overline{P} = P = C$

(H8) $\mathbf{A}_{20} = \begin{pmatrix} 0 & C \\ C & 0 \end{pmatrix}$,

where the subscript '20' on $\mathbf{A}$ denotes, as usual, URM2 ('2') under Pythagoras conditions ('0').

(H9) $\mathbf{X}_{2+} = \begin{pmatrix} 1 \\ 1 \end{pmatrix}$, $y = 1$, $z = 1$, $\gcd(y,z) = 1$.

(H10) $\mathbf{X}_{2-} = \begin{pmatrix} C^2 \\ -C^2 \end{pmatrix}$.

Note that the gcd condition $\gcd(y,z) = 1$ is why the eigenvalue C in $\mathbf{X}_+$ (H5a) has been moved to $\mathbf{X}_{2-}$ (H10).

(H11a) $\quad \mathbf{X}^{2+} = (C^2 \quad C^2)$, $\quad$ (H11b) $\quad \mathbf{X}^{2-} = (1 \quad -1)$, $\quad$ reciprocal eigenvectors

(H12) $\mathbf{T}_2 = \mathbf{T}^2 = \begin{pmatrix} +1 & 0 \\ 0 & -1 \end{pmatrix}$, the $\mathbf{T}$ operator

(H13) $\overline{\mathbf{X}}_{2+} = \mathbf{X}^{2-} = (\mathbf{T}\mathbf{X}_{2+})^T$, conjugate eigenvector $\overline{\mathbf{X}}_{2+}$

(H14) $\overline{\mathbf{X}}_{2-} = \mathbf{X}^{2+} = (\mathbf{T}\mathbf{X}_{2-})^T$, conjugate eigenvector $\overline{\mathbf{X}}_{2-}$

(H15) $\overline{\mathbf{X}}_{2+}\mathbf{X}_{2+} = \mathbf{X}^{2-}\mathbf{X}_{2+} = (+1 \quad -1)\begin{pmatrix} 1 \\ 1 \end{pmatrix} = 0$, orthogonality.

The only free parameter within URM2 is actually the eigenvalue C, that by definition is unity or greater.

Generally, this URM2 case is considered too simplistic, primarily because the only non-trivial, primitive vector $\mathbf{X}_{2+}$ is the (1,1) pair, which is why URMT generally starts with URM3, since it has the first 'non-trivial' solution $\mathbf{X}_{3+}$ - an arbitrary Pythagorean triple with three non-zero elements and two free parameters.

Neither is there any meaningful variational (or evolutionary) parameter t_2, but this is more of a plus point because it means URM2 cannot shrink further from two to one dimension by setting the parameter t_2 to zero.

Despite URM2 being considered too simplistic, it is not entirely dismissed and, if for no other reason, it is a good illustration of some basic aspects of URMT. Most importantly amongst these is that the above solution (1,1) can be lifted (I6) to a general, 3D Pythagorean solution, see Section (7-2).

Appendix (I) Terminology

This is a subset of the full URMT terminology, covering terms used herein, either uniquely defined within URMT or those in wider use, but with a specific meaning to URMT.

(I1) **Alignment** - see **Flattening** (I5).

(I2) **Divisibility factors**, also known as **scale factors**, are the last three elements of the URMn eigenvector $\mathbf{X}^{n+}$, denoted by α_n, β_n and γ_n, $n \geq 3$, see (A15) URM3, (3.56) URM4, and (C16) UMRn.

(I3) The **Dyadic product** is synonymous with the outer product of two vectors. In the context of URMT, the dyadic product of two vectors $\mathbf{X}_i$ and $\mathbf{X}^j$, $i, j = 1 \ldots n$, gives a square matrix, $\mathbf{M}_{ij} = \mathbf{X}_i \mathbf{X}^j$, of size $n \times n$. The term 'dyadic product' is quite old and is replaced by 'outer product' in modern texts. Nevertheless, the name appears in the earlier URMT literature, e.g. [1], albeit it has been replaced herein with the term 'outer product'.

(I4) An **Excess dimension** is any dimension higher than the third, i.e. the fourth or higher. In an n-dimensional space, an excess dimension r is such that $3 < r < n$; its associated temporal (evolutionary) parameter is denoted by t_r, see Appendix (C).

(I5) **Flattening** is the term used to describe the eigenvector evolution in URM3 whereby the two eigenvectors $\mathbf{X}_{30}$ and $\mathbf{X}_{3-}$ align anti-parallel to $\mathbf{X}_{3+}$ as evolution progresses, i.e. as evolutionary time m (or t_3) increases, see Appendix (B) for the URM3 eigenvector evolution equations. The vector $\mathbf{X}_{3+}$ itself is static and invariant to arbitrary variations in any evolutionary parameter. However, note that $\mathbf{X}_{3+}$ is actually a two-parameter family of integer vectors, parameters k and l (A26), and hence occupies a 2D discrete subspace of 3D; in this sense, the 3D flattens to 2D. Because it is an alignment of vectors, the process is also known as 'alignment' herein, but 'flattening' is used exclusively in earlier, URM3 literature. That they align anti-parallel, and not parallel, is largely a choice of sign convention. See Section (12-8) and [1],#3 for full details of the flattening process.

(I6) **Lifting**, in the context of URMT, is the process of generating eigenvector solutions for an $(n+1) \times (n+1)$ matrix $\mathbf{A}_{n+1}$ using an eigenvector solution to the $n \times n$ matrix $\mathbf{A}_n$, $n \ge 2$. The matrix $\mathbf{A}_n$ is embedded in $\mathbf{A}_{n+1}$ and an eigenvector solution $\mathbf{X}$ to $\mathbf{A}_n$, is also a solution to $\mathbf{A}_{n+1}$, with appropriate zero padding, see Section (7).

(I7) The **magnitude** of a vector is the positive square root of the inner product of a vector with itself, e.g. for URM4's $\mathbf{X}_+$ (1.5), then $|\mathbf{X}_+| = \sqrt{\mathbf{X}_+ \cdot \mathbf{X}_+} = \sqrt{w^2 + x^2 + y^2 + z^2}$, hence $|\mathbf{X}_+| = \sqrt{2}z$. See also the **norm** (I8).

(I8) The **norm** (or **length**) of a vector, using the standard definition of the norm, see [5], is the square root of the inner product of itself with its conjugate (Appendix (E)), e.g. for $\mathbf{X}_+$ the norm, denoted by $\|\mathbf{X}_+\|$, is given by $\|\mathbf{X}_+\| = \sqrt{\mathbf{X}^- \cdot \mathbf{X}_+}$. Normally the positive square root is assumed unless otherwise stated. If $\mathbf{X}_+$ is an n-element Pythagorean n-tuple, the norm is zero since $\mathbf{X}^- \cdot \mathbf{X}_+ = 0$, e.g. (F7). The same remarks also apply to $\mathbf{X}_-$ and its conjugate form $\mathbf{X}^+$, but note that the norm of any one of URMn's, zero eigenvector $\mathbf{X}_{n0j}$, $j = 0 \dots n-3$ is non-zero, as given by $\|\mathbf{X}_{n0j}\| = \sqrt{\mathbf{X}^{n0j} \cdot \mathbf{X}_{n0j}} = +C^2$ (F10). See also the **magnitude** (I7).

(I9) A **Minus eigenvector** $\mathbf{X}_{n-}$, $n \ge 2$, is any vector that satisfies the eigenvector equation $\mathbf{AX}_{n-} = -C\mathbf{X}_{n-}$ for negative eigenvalue $-C$ and unity root matrix $\mathbf{A}$ (1.3a). See also **plus eigenvector** (I11) and **zero eigenvector** (I18). Note that minus eigenvectors are only defined under Pythagoras conditions, e.g. (4.1) for URM4.

(I10) A **Non-trivial eigenvector** is one with all non-zero elements and, in the context of URMT, remains non-trivial for arbitrary variations in all evolutionary parameters. Primarily this applies to the static vector $\mathbf{X}_{n+}$ and its conjugate $\mathbf{X}^{n-}$, since $\mathbf{X}_{n+}$ is the only true, invariant vector.

(I11) A **Plus eigenvector** $\mathbf{X}$, e.g. $\mathbf{X}_{n+}$, $n \geq 2$, is any vector that satisfies the eigenvector equation $\mathbf{AX} = C\mathbf{X}$ for positive eigenvalue C and unity root matrix $\mathbf{A}$ (1.3a). See also **minus eigenvector** (I9) and **zero eigenvector** (I18). Since the eigenvector equation $\mathbf{AX} = C\mathbf{X}$ (1.7) is a fundamental definition in URMT, there is always at least one plus eigenvector $\mathbf{X}$, e.g. (1.5).

(I12) A **primitive Pythagorean n-tuple** is that which has no common factor in its elements, as specified by gcd criterion, e.g. URM4 gcd criterion (1.6). The vector $\mathbf{X}_{n+}$ and $\mathbf{X}^{n-}$ are usually defined to be primitive, with any common factor being absorbed into $\mathbf{X}_{n-}$ and $\mathbf{X}^{n+}$, which are also Pythagorean n-tuples.

(I13) The **Pythagoras conditions** are a set of relations between the standard and conjugate dynamical variables in URM n, which are such that the eigenvectors of the $n \times n$ matrix $\mathbf{A}_n$, for non-zero eigenvalues, are Pythagorean n-tuples. The unity root matrix $\mathbf{A}_n$ is formed exclusively from the dynamical variables, and the conditions make the matrix skew-symmetric in the first $n-1$ rows and columns, and symmetric in the last row and column. All Pythagoras conditions for URM n include URM $(n-1)$ as a subset.

(I14) The **Pythagorean Eigenvalues,** (2.63) et al, comprise the set of two non-zero eigenvalues $\lambda = \pm C$ and all others zero. E.g., for URM4, the four eigenvalues are $\lambda = \pm C$ and $\lambda = 0,0$, i.e. the zero eigenvalue is repeated with a 'multiplicity' [5] of 2. For URM n there the two eigenvalues $\lambda = \pm C$, and the zero eigenvalue is repeated with a multiplicity $n-2$. They are termed Pythagorean because, under URMT Pythagoras conditions, e.g. (4.1) for URM4, the eigenvectors to the non-zero eigenvalues ($\lambda = \pm C$ here), are Pythagorean n-tuples. Strictly speaking, URMT under Pythagoras conditions only actually mandates at least two non-zero eigenvalues and the others may not necessarily be zero. For example, in URM4, if the Potential is non-zero, complex eigenvalues can arise in place of the zero eigenvalues, see (4.13b) for example. In such a case, the complex eigenvectors also satisfy the Pythagoras equation, as do all eigenvectors for non-zero eigenvalues, when under Pythagoras conditions.

(I15) **Scale factors**, see divisibility factors.

(I16) A **Static** quantity in URMT is any quantity (invariably an eigenvector) not dependent on any evolutionary time t_j, $j = 3...n$. The eigenvector $\mathbf{X}_{n+}$, $n \geq 2$ (this includes URM2), is the classic URMT example. A static eigenvector can be a function of none, one or both of the other two URMT parameters, k and l, which are not temporal parameters, e.g. $\mathbf{X}_{n+}$ is a function of both k and l, as for $\mathbf{X}_{3+}$ (A26).

(I17) Vector space, eigenvector space

These two interchangeable terms are used loosely throughout this book since they refer to the eigenvectors of the unity root matrix, but not linear combinations of them, i.e. the whole of URMT currently studies the vector basis comprising the eigenvectors, but not arbitrary vectors generated from them such as linear combinations. Thus, for the purists, the vector space would be better described as an infinite set (space) of n linearly independent eigenvectors, rather than 'vector space' in the true, mathematical sense of the word. See [5] for a strict definition of a vector space. For instance, the URMT discrete, vector space is not generally closed and neither is there a null vector (a zero vector in the conventional sense, not as in (I18)).

The URM3 set of eigenvectors is defined by the infinite set of points termed the 'lattice', symbol '**L**', in [1],#3. The key point is that every vector in the lattice is an eigenvector of the unity root matrix, and arbitrary, linear combinations of eigenvectors do not generally give another eigenvector in the lattice, as is true for any general set of eigenvectors for distinct eigenvalues. For example, the eigenvector sum $\mathbf{X}_{3+} + \mathbf{X}_{3-}$ is not an eigenvector of $\mathbf{A}_{30}$ even though $\mathbf{X}_{3+}$ and $\mathbf{X}_{3-}$ are both eigenvectors, for eigenvalues C and $-C$ respectively, see Appendix (A). That is not to say the n URMT eigenvectors cannot form the basis of an n-dimensional vector space; indeed they can by their linear independence, which is why the term 'vector space' is used loosely. Currently, however, URMT focuses solely on the eigenvectors themselves, but not arbitrary functions of them, such as linear combinations.

That there is no zero vector is quite fundamental to the physical interpretation of URMT. By defining the single eigenvalue C as non-zero (1.7), this constrains all eigenvectors to be non-trivial (I10), and never null or zero, i.e. where all vector elements zero. Put another way, the origin is excluded from URMT on algebraic grounds. Physically speaking, this means that there are no singularities and the zero point (null vector or origin) is simply not required in URMT as currently formulated. This is also discussed in [1],#3.

(I18) A **Zero Eigenvector in** URMT is defined as an eigenvector $\mathbf{X}_{n0}$ satisfying $\mathbf{AX}_{n0} = 0$, i.e. an eigenvector for eigenvalue zero. The eigenvectors $\mathbf{X}_{30}$, $\mathbf{X}_{n0A}$, $\mathbf{X}_{n0B}$, $\mathbf{X}_{n0C}$, $\mathbf{X}_{n0j}$, $j = 0...n-3$ etc., are all zero eigenvectors since they are the eigenvectors for the repeated, zero **Pythagorean eigenvalue** (I14). Generally, zero eigenvectors are only considered under Pythagoras conditions, e.g. (4.1) for URM4.

Appendix (J) The Standard Physical Interpretation

Since the mathematics of URMT is thought to have strong links to the subject of 'Physics in Integers', a standard physical interpretation (or association) is ascribed to all variables. The standard interpretation given here is not the only physical interpretation, but it currently seems the best as regards physical phenomena, and is therefore used exclusively throughout this book.

The physical interpretation, as first given for URM3 in [1], is reproduced immediately below, followed by the extensions for the general, n-dimensional formulation URMn, $n \geq 4$. URM3 and URMn are virtually identical barring the additional variables and extra eigenvectors that come with having more dimensions to work with.

(J1) **URM3**

$\mathbf{X}_{3+}$, x, y, z, acceleration or force per unit mass, LT^{-2}

$\mathbf{A}_{30}$, $\mathbf{X}_{30}$, P, Q, R, C, velocity or momentum per unit mass (J18), LT^{-1}

$\mathbf{X}^{3+}$, $\alpha_3, \beta_3, \gamma_3$, position, L

t_3, m, δ, time, T

C^2, velocity squared or total energy E ($E = C^2$) per unit mass, $L^2 T^{-2}$.

(J2) **URMn**

$\mathbf{X}_{n+}$, v, w, x, y, z (URM5), acceleration, LT^{-2}

$\mathbf{A}_{n0}$, $\mathbf{X}_{30} \sim \mathbf{X}_{n0A}$, $\mathbf{X}_{n0B}$, $\mathbf{X}_{n0C}$, C, M, H, N, J (URM5), S, T, U (URM4), velocity, LT^{-1}

$\mathbf{X}^{n+}$, $\alpha_n, \beta_n, \gamma_n$, position, L

t_n time, T

C^2, as URM3 above.

The physical units for all quantities, as used in the text, are elaborated as follows, with some additional notes at the end. It is emphasized that only those referenced in the book are listed, the units of all other quantities can generally be evaluated using (J1) and (J2).

(J3) $units(v, w, x, y, z) = units(\mathbf{X}, \mathbf{X}_3, \mathbf{X}^{3-}) = $ acceleration $= LT^{-2}$

(J4) $units(C) = $ velocity $= LT^{-1}$ (J17)

(J5) $units(\overline{\mathbf{X}}, s\overline{\mathbf{X}}_3) = $ velocity $= LT^{-1}$

(J6) $units(\mathbf{A}_3) = units(P, Q, R) = $ velocity $= LT^{-1}$

(J7) $units(\mathbf{A}) = units(M, H, N, J, S, T, U) = $ velocity $= LT^{-1}$

(J8) $units(\mathbf{\Delta}, \mathbf{\Delta}_3^P, \mathbf{\Delta}^P) = $ acceleration $= LT^{-2}$

(J9) $units(m) = T$, time,

(J10a) $units(s, \bar{s}) = $ time $= T$ when $\overline{\mathbf{X}}_3 = \mathbf{X}^{3-}$ (3.21a),

or

(J10b) $units(\bar{s}) = $ none,

   $units(s) = $ time $= T$

   when $\overline{\mathbf{X}}_3$ is any function of the dynamical variables, e.g. (3.45), and thus of the same physical units as $\mathbf{A}_3$.

(J10c) $units(t_n) = $ time $= T$

(J11) $units(\mathbf{X}^+, \mathbf{X}^{3+}) = $ length $= L$

(J12) $units(\mathbf{XX}^+) = $ velocity squared $= L^2 T^{-2}$

(J13) $units(\mathbf{E}) = units(\mathbf{XX}^+) = units(\mathbf{X}\alpha) = L^3 T^{-3}$ when using the cubic residual polynomial Section (3-3), otherwise $L^2 T^{-2}$ when using the quadratic form, Section (3-2).

(J14a) $units(K, K_3) = $ energy per unit mass $= L^2 T^{-2}$

(J14b) $units(V, V_3)$ = energy per unit mass = $L^2 T^{-2}$, URM3,4

(J14c) $units(V)$ = energy squared per unit mass = $(L^2 T^{-2})^2$, URM5 (13.25)

(J15) $units(\alpha_3, \beta_3, \gamma_3)$ = $units(\mathbf{X}^3)$ = length = L

(J16) All conjugate quantities have the same physical units as their standard forms, e.g. $units(\mathbf{X}_+) = units(\overline{\mathbf{X}}_+) = units(\mathbf{X}^-)$ = acceleration = LT^{-2}. This is not the case for related reciprocals, e.g. $units(\mathbf{X}_+) \neq units(\mathbf{X}^+)$ and $units(\mathbf{X}_-) \neq units(\mathbf{X}^-)$, except when they are self-reciprocal, i.e. the zero eigenvectors $units(\mathbf{X}_0) = units(\mathbf{X}^0)$. All conjugate and reciprocal units can be determined from the $\mathbf{T}$ operator relations in Appendix (E), using the above units for the standard forms.

(J17) This is big 'C', a fundamental constant in URMT, and not to be confused with little c, the speed of light. Albeit big C also has a physical interpretation as a velocity (speed) constant and, therefore, C^2 has an interpretation as that of energy (per unit mass), see (1.14b). Big C was originally chosen as the first letter of the word 'Constant'. That it appears to be remarkably similar to little c in physical nature is purely coincidental, but there does now seem to be more in this than just coincidence.

(J18) The elements of all unity root matrices, e.g. $\mathbf{A}_{30}$ with elements P, Q, R, are termed 'dynamical variables' as they can be physically associated with velocity (or momentum per unit mass).

Appendix (K) Conventions

Equation Numbering

Unlike most scientific texts, nearly every equation is numbered, whether referenced or not. The reason is primarily for ease of reference when discussing the text.

Equations are given numbers of the form SS.ss where SS always denotes the one or two digit section (chapter) number, but the 'ss' does not necessarily match with the sub-section number.

Many equations are often reproduced (repeated) further in the text to avoid the reader having to flick back and forward. The only downside to this is that the repeat occurrence may appear out of sync with the main section number.

General

All quantities are integers, i.e. $\in \mathbb{Z}$.

All vectors are column vectors unless denoted by a superscript or over-struck bar, which makes them reciprocal or conjugate vectors respectively.

All vectors are eigenvectors, there are no other vectors formed, for instance, from a linear combination of eigenvectors; also see (I17).

All matrices are square $n \times n$ for URM n, e.g. 3x3 = URM3.

A rational expression may appear in intermediate calculations but the results are always in integers. For eigenvectors, any rational elements can cleared by scaling the eigenvector appropriately to clear the denominator, without changing the eigenvectors properties, by their definition.

Usage of raised and lowered indices (superscripts and subscripts) is intentionally analogous to tensor notation for contravariant and covariant quantities. The notational concepts are one and the same, i.e. contravariant quantities are given with respect to a coordinate basis reciprocal (or dual) to the those covariant quantities, and vice versa.

Appendix (L) Notation

Notational equivalence '~'

The oft-used equivalence symbol '~' in this book mostly denotes notational equivalence, and not the more usual mathematical equivalence relation. For example, the URM3 vectors, $\mathbf{X}_+$, $\mathbf{X}_0$ and $\mathbf{X}_-$ in [1], are given the extra, qualifying subscript of '3' to explicitly denote they are URM3 vectors. This is shown by the notational equivalences $\mathbf{X}_{3+} \sim \mathbf{X}_+$, $\mathbf{X}_{30} \sim \mathbf{X}_0$ and $\mathbf{X}_{3-} \sim \mathbf{X}_-$.

Subscripts

A single subscript n denotes the quantity is intrinsic to the URMT incarnation URMn, e.g. $\mathbf{X}_3$ (A3) is the URM3 version of $\mathbf{X}$.

A +, - or 0 in the subscript of a vector $\mathbf{X}$ denotes it is a plus (I11), minus (I9) or zero (I18) eigenvector respectively.

A +, - or 0 in the subscript of a residual matrix $\mathbf{E}$ denotes it is an outer product of a plus (I11), minus (I9) or zero (I18) eigenvector and its reciprocal, Appendix (E), e.g. $\mathbf{E}_+ = \mathbf{X}_+\mathbf{X}^+$ (3.3).

A subscript of zero on the unity root matrix $\mathbf{A}$ denotes its elements, i.e. the dynamical variables, conform to Pythagoras conditions, e.g. $\mathbf{A}_{30}$ is the URM3 matrix (A19) under URM3 Pythagoras conditions (A18).

A capital subscript, A, B or C, on a zero vector e.g. $\mathbf{X}_{40A}$ (11.54a), denotes the first, second or third zero eigenvector (I18) in the solution under Pythagoras conditions, e.g. $\mathbf{X}_{40A}$, $\mathbf{X}_{40B}$ (11.49) (URM4, 5), $\mathbf{X}_{50C}$ (13.52e) (URM5). Note that the notation stops at C because specific incarnations of URMT only go as far as URM5, which has three zero eigenvectors - see (13.52) for the full, URM5 solution. When speaking about the general URMT solution for URMn, Section (14-

9), the subscript i or j is used for a zero vector, e.g. $\mathbf{X}_{n0j}$, $j = 0 \ldots n-3$, albeit $\mathbf{X}_{n00}$, for $j = 0$, is simply denoted by $\mathbf{X}_{n0A}$, see (14.92c) and (14.92d).

Superscripts

A $+$, $-$ or 0 superscript on a vector denotes it is a reciprocal vector, e.g. $\mathbf{X}^{+}$ (3.28). See Appendix (E) on the subject of reciprocal and conjugate vectors.

Superscripts on vectors follow the same conventions as for subscripts.

A superscript prime denotes an initial value, e.g. $\mathbf{X}'_{3+} = \mathbf{X}_{3+}(m = 0)$ (B1a); symbol a'_1 is an exception, see the 'Symbols' sub-section further below.

A superscript on the $\mathbf{T}$ operator, e.g. $\mathbf{T}^n$ (E3), is a raised index, not an exponentiation, i.e. it does not represent $\mathbf{T}$ raised to the power given by the superscript, but rather it denotes the reciprocal matrix operator $\mathbf{T}^n$ (where $\mathbf{T}^n$ is reciprocal to $\mathbf{T}_n$ and $\mathbf{T}_n \sim \mathbf{T}$), and operates on standard vectors, i.e. those with a lowered subscript, such as $\mathbf{X}_{n+}$, to convert them to reciprocal forms. See Appendix (E), equation (E2) onward for more information.

General

An over-struck bar denotes a conjugate quantity, Appendix (E), e.g. $\overline{\mathbf{X}}$.

An emboldened capital denotes a matrix or vector, e.g. $\mathbf{A}$ (1.1) or $\mathbf{X}$ (1.5).

The ith row and jth column of a matrix $\mathbf{A}$ is denoted by $\mathbf{A}(i, j)$.

The kth element of a vector is denoted by a subscript k, e.g. $\mathbf{X}_{n-,k}$ (14.92b) denoting the kth element of the general, n-dimensional minus vector $\mathbf{X}_{n-}$.

The value of a vector $\mathbf{X}$ or matrix $\mathbf{A}$ at evolutionary time m is denoted by $\mathbf{X}(m)$ and $\mathbf{A}(m)$ respectively, e.g. $\mathbf{X}_{30}(m)$ et al in Appendix (B). Given a matrix requires two subscripts to index it, e.g. $\mathbf{A}(i, j)$, and that a vector is subscripted with index k, e.g. $\mathbf{X}_{n-,k}$ (14.92b), then this notation, $\mathbf{X}(m)$ or $\mathbf{A}(m)$, for a vector or matrix at evolutionary time m, should not be ambiguous with $\mathbf{A}(i,j)$. However, note that symbol $\mathbf{A}_m$ is also often used in place of $\mathbf{A}(m)$, e.g. Section (9-4), albeit it will always be the subscript m, unambiguous from the other subscripts, i.e. n and 0, used on $\mathbf{A}$, e.g. $\mathbf{A}_{n0}$.

A solitary hyphen in a matrix or vector is a placeholder, e.g. matrix $\mathbf{A}$ (1.1).

An explicit $+$ sign prefixing a quantity, e.g. $+C$ is added for emphasis, or even just layout reasons, i.e. equal spacing of elements in a vector but, otherwise, the sign prefix has no algebraic significance.

Mathematical

$Tr(\mathbf{A})$ trace of matrix $\mathbf{A}$

$units(x)$ physical units of x, see Appendix (J)

$|x|$ absolute, positive value of scalar x

$|\mathbf{X}|$ magnitude (I7) of vector $\mathbf{X}$

$\|\mathbf{X}\|$ norm (I8) of vector $\mathbf{X}$

[1],#N denotes paper number N, N$=$1..6, in Reference [1].

Dot product versus matrix notation

The inner (or dot) product operation between two vectors, which gives a scalar result, is often denoted in mathematical texts by a centred dot, e.g. between $\overline{\mathbf{X}}$ and $\mathbf{X}$ it is denoted by $\overline{\mathbf{X}} \cdot \mathbf{X}$ (2.9). However, in URMT, this dot is often omitted because $\overline{\mathbf{X}}$, e.g. (2.8), is a conjugate, $1 \times n$ row vector and $\mathbf{X}$ is an $n \times 1$ column vector, so the matrix product $\overline{\mathbf{X}}\mathbf{X}$ also gives a scalar result, identical to the inner product by the rules of matrix multiplication. The only downside to omitting the dot is that the order of multiplication of the two vectors becomes important. Secondly, the multiplication must be between a conjugate (or reciprocal) row vector and a standard column vector, in that order; see Appendix (E) for details on conjugate and reciprocal vector forms. When this isn't the case, or for some reason the inner product operation must be explicitly shown, then a dot is used, i.e. $\overline{\mathbf{X}} \cdot \mathbf{X}$. The classic example here is the magnitude of a vector, which is the dot product with itself; see (I7).

Symbols

$\mathbf{0}_3$ 3x1 column vector with all elements zero

$\mathbf{0}_{33}$ 3x3 matrix vector with all elements zero

a arbitrary parameter (6.5b) in the 2a2p1 solution, Section (6)

a_0 coefficient (1.10c) of the URM4 characteristic equation (1.9)

a_1 coefficient (1.10b) of the URM4 characteristic equation (1.9)

a_1' partial term (1.19b) of a_1

a_2 coefficient (1.10a) of the URM4 characteristic equation (1.9)

A subscript on a zero vector, e.g. $\mathbf{X}_{40A}$ (5.54a)

$\mathbf{A}$ unity root matrix, e.g. (1.3a) URM4

$\mathbf{A}_n$ unity root matrix URMn, $\sim \mathbf{A}$

$\mathbf{A}_m$ unity root matrix at evolutionary time m, e.g. (9.42), $\sim \mathbf{A}(m)$

$\mathbf{A}_{30}$ URM3 unity root matrix (A19) under Pythagoras conditions (4.1)

$\mathbf{A}_{n0}$ URMn unity root matrix under Pythagoras conditions, e.g. (4.1) UMR4

$\mathbf{A}_2$ URM2 unity root matrix (H1)

$\mathbf{A}_3$ URM3 unity root matrix (A1a)

$\mathbf{A}_4$ URM4 unity root matrix $\sim \mathbf{A}$ (1.3a)

$\mathbf{A}_5$ URM5 unity root matrix (13.1)

b arbitrary parameter (6.5b) in the 2a2p1 solution, Section (6)

B subscript on a zero vector, e.g. $\mathbf{X}_{40B}$ (5.54b) URM4

c arbitrary parameter (6.5b) in the 2a2p1 solution, Section (6)

C 1. eigenvalue (1.7)

C 2. subscript on a zero vector, e.g. $\mathbf{X}_{50C}$ (13.52e) (URM5)

d arbitrary parameter (6.5b) in the 2a2p1 solution, Section (6)

E total energy (per unit mass) (1.14b) of the DCE (1.14)

$\mathbf{E}$ residual matrix, outer product $\mathbf{XX}^+$ (3.3)

$\mathbf{E}_+ \sim \mathbf{E}$

f parameter (5.20a) in the PS+RU solution, Section (5)

g parameter (5.20b) in the PS+RU solution, Section (5)

H URM5 dynamical variable (13.2)

i 1. arbitrary parameter, e.g. URM4 solution (5.23b)

i 2. integer factor, (8.76)

$\mathbf{I}$ identity matrix

j 1. arbitrary parameter, e.g. URM4 solution (5.23b)

j 2. integer factor, (8.59)

J URM5 dynamical variable (13.2)

K Kinetic term in the DCE, (1.8) URM4, (13.24) URM5

K_3 URM3 Kinetic term (A8) in the DCE (A7)

k 1. arbitrary parameter in URM3 solution (A26)

k 2. polynomial $k(x, y, z)$ in the URM3 coordinate equation (2.21)

k 3. integer factor, (8.66)

l arbitrary parameter in URM3 solution (A26)

L dimensional unit of length, Appendix (J)

m URM3 evolution parameter, Appendix (B), $m = 0$ denotes the initial time

M URM5 dynamical variable (13.2)

$\mathbf{M}_n$ $n \times n$ constant matrix (14.91) in the URMn solution

n URMn, $n \geq 2$, URM2 Appendix (H), URM3 Appendix (A), URM4 Part I, URM5 Part II.

N URM5 dynamical variable (13.2)

p arbitrary parameter (2.69)

P URM3 dynamical variable (A1b)

q arbitrary parameter (2.72)

Q URM3 dynamical variable (A1b)

R URM3 dynamical variable (A1b)

s 1. arbitrary parameter in URM3 solution (A28a), URM4 solution (5.23b)

s 2. variational parameter in URM4 matrix **A** (2.25)

S 1. URM4 dynamical variable (1.3b)

S 2. arbitrary parameter (2.65), not in common use.

t arbitrary parameter in URM3 solution (A28b) , URM4 solution (5.23b)

t_n evolutionary parameter URM n, e.g. t_4, t_5 (13.31d)

T 1. URM4 dynamical variable (1.3b)

T 2. dimensional unit of time, Appendix (J)

T 3. root sum of squares of evolutionary times t_n (14.69a)

T T operator matrix (E3), (E4)

U URM4 dynamical variable (1.3b)

v coordinate of URM5 eigenvector **X** (13.4a)

w coordinate of URM4,5 eigenvector **X** URM4 (1.5), URM5 (13.4a)

V_3 URM3 Potential term (A9) in the DCE (A7)

y coordinate of URM n eigenvector **X**, see $\mathbf{X}_{n+}$ (14.92a) and $\mathbf{X}_{3+} \sim \mathbf{X}_3$ (A3)

x coordinate of URM n eigenvector **X**, see $\mathbf{X}_{n+}$ (14.92a) and $\mathbf{X}_{3+} \sim \mathbf{X}_3$ (A3)

X 'Plus' eigenvector (I11) for eigenvalue C, e.g. (1.7)

$\mathbf{X}_+ \sim \mathbf{X}$

$\mathbf{X}_n \sim \mathbf{X}$

z coordinate of URMn eigenvector $\mathbf{X}$, see $\mathbf{X}_{n+}$ (14.92a) and $\mathbf{X}_{3+} \sim \mathbf{X}_3$ (A3)

α scale factor (3.56a), element of $\mathbf{X}^+$, and also $\mathbf{X}_-$ under Pythagoras conditions

β scale factor (3.56b), element of $\mathbf{X}^+$, and also $-\mathbf{X}_-$ under Pythagoras conditions

γ scale factor (3.56c), element of $\mathbf{X}^+$, and also $\mathbf{X}_-$ under Pythagoras conditions

δ, variational parameter of $\mathbf{\Delta}$, $\mathbf{\Delta}_3$ (9.21)

δ_a δ_b, δ_c, δ_d URM4 variational parameters (8.119) under URM4 Pythagoras conditions (4.1)

δ_{ij}, variational parameter, element of $\mathbf{\Delta}^P(i, j)$, $i, j = 1 \ldots n$, (8.16)

$\mathbf{\Delta}$ variational matrix (8.16)

$\mathbf{\Delta}^P$ variational matrix under Pythagoras conditions, URMn, e.g. URM4 (8.121)

$\mathbf{\Delta}_P$ finite difference (9.14b) in dynamical variable P, not in common use.

$\mathbf{\Delta}_R$ finite difference (9.14c) in dynamical variable R, not in common use.

$\mathbf{\Delta}_S$ finite difference (9.14a) in dynamical variable S, not in common use.

$\mathbf{\Delta}_U$ finite difference (9.14d) in dynamical variable U, not in common use.

Δt_i finite difference, temporal parameter t_i (G6)

$\mathbf{\Delta}_3^P$ URM3 variational matrix under Pythagoras conditions (8.133)

$\mathbf{\Delta}_3$ URM3 variational matrix (8.4)

ε estimate of relative error in approximation (14.68)

ε URM3 variational parameter of $\mathbf{\Delta}_3$ (8.4)

ε_- absolute error in approximation (14.72)

η URM3 variational parameter of $\mathbf{\Delta}_3$

κ curvature, (12.90a) to (12.90d)

λ general eigenvalue, Section (1-4)

χ compactification ratio, e.g. χ_j (14.23)